Neglected and Underutilized Crops - Towards Nutritional Security and Sustainability

Sajad Majeed Zargar • Antonio Masi •
Romesh Kumar Salgotra
Editors

Neglected and Underutilized Crops - Towards Nutritional Security and Sustainability

Springer

Editors
Sajad Majeed Zargar
Proteomics Laboratory, Division of Plant Biotechnology
Sher-e-Kashmir University of Agricultural Sciences and Technology of Kashmir
Srinagar, Jammu and Kashmir, India

Antonio Masi
Department of Agronomy, Food, Natural Resources, Animals and the Environment (DAFNAE)
Università degli studi di Padova
Padova, Italy

Romesh Kumar Salgotra
School of Biotechnology
Sher-e-Kashmir University of Agricultural Sciences and Technology of Jammu
Srinagar, Jammu and Kashmir, India

ISBN 978-981-16-3875-6 ISBN 978-981-16-3876-3 (eBook)
https://doi.org/10.1007/978-981-16-3876-3

This Springer imprint is published by the registered company Springer Nature Singapore Pte Ltd.
The registered company address is: 152 Beach Road, #21-01/04 Gateway East, Singapore 189721, Singapore

Foreword

There are more than thousand edible plant species in the world, yet global agriculture has been dominated by only four major crops including rice, maize, wheat and potato to meet the calorific needs of human population worldwide. The predominance of only four food crops has led the agriculture sector extremely vulnerable to environmental vagaries arising due to impending climate change. Additionally, these food crops though major contributors of carbs in our diet are poor in good quality protein, minerals and other nutrients. Neglected and underutilized crops could overcome both these challenges by diversifying cropping regimes and supplementing the diet owing to their nutrient-rich seeds and plant parts. Most of the underutilized crops, e.g. buckwheat, a versatile pseudo-cereal cultivated for its seed as well as greens (food and feed), have the potential to be the golden crop in coming years owing to their nutritional superiority in terms of balanced amino acid profile, nutraceuticals such as bioactive flavonoids and the ability to thrive well under marginal and low input areas. These neglected and underutilized crops can bridge the nutritional gap to produce foods that are needed for a healthy diet. As per a recently published paper, the available information about the "neglected and underutilized crops" is very limited that is attributed to the lack of clear research goals, limited funding directed for these crops and journal apathy towards publishing work on neglected and underutilized crops. Taking into consideration these crops as a source of "nutritional powerhouse" and their ability towards attaining nutritional security and sustainability, the publication of this book entitled *Neglected and Underutilized Crops: Towards Nutritional Security and Sustainability* is the need of the hour. This book will help not only students but also professionals and crop researchers.

This book consists of 11 different chapters covering the status and future prospects of neglected and underutilized crops and the genetic and genomic resources of various neglected and underutilized crops (vegetables, cereals, fruit crops). There are also chapters on the importance of underutilized crops in the scenario of sustainable agriculture and how recent technologies can be utilized for improvement of these crops.

I applaud Dr. Sajad M. Zargar (SKUAST-K, India), Prof. Antonio Masi (University of Padova, Italy) and Dr. R K Salgotra (SKUAST-J, India) for the hard work in compiling this useful book with relevant topics.

Division of Plant Biotechnology, SKUAST-Kashmir
Srinagar, Jammu and Kashmir, India

F. A. Nehvi

Preface

A major milestone in the history of mankind was represented by the establishment of agriculture and domestication of plants, which started several thousand years ago and resulted in the cultivation of many different species from diverse ranges of taxa and in different parts of the globe. Despite such a huge diversity of resources, cultivation of different crops provided food and fodder for centuries. Different crop plants were locally grown, which implied the cultural transition from a nomad to a residential lifestyle, the development of agricultural practices and technology, the evolution of dietary styles and culinary traditions. Until a few centuries ago, the agricultural landscape was characterized by thousands of cultivated crops around the world, which was significantly different from the current scenario where only three major crops are widely and intensely cultivated and dominate 50% of total foodstuff, and 30 species represent 95% of total raw food supply. The reason for this drastic shift may ultimately be linked to the market: as a major effect of globalization, farmers were more inclined to cultivate only a limited number of crop species that were more productive and could generate more profit at a lower cost.

As a consequence, a huge number of crop species were cultivated less and less, their importance was reduced globally in terms of production and market value, and they were finally abandoned because they were not enough profitable; they became "neglected" (from the Latin: *neglectus*, forgotten). But forgetting also means losing something. Not only some important traditional carbohydrate and protein sources were lost but also agrobiodiversity was strongly reduced. And from a broader perspective, also some knowledge, traditions, and linkage with the cultural heritage were lost.

It is generally accepted that the risks of globalization represent a major challenge for future generations. Climate change and loss of diversity call upon a shift in the human relation with the natural and agricultural environment. In this context, when referred to crops, the word "**underutilized**" also contains the significance of a potential for exploitation. The possibility to revert the negative trend of reduction in cultivation is documented, and several "minor" crops that were underutilized in some regions and in some periods of the history (such as kiwi, oil palm and soybean) were later reconsidered and they are now major staple foods again.

Despite their economic potential and importance for dietary diversification, still they attract little attention from national policy and decision makers, stakeholders and consumers; they are ignored by the private sector and research suffers from the lack of adequate funding sources. In many cases, these crops are maintained by socio-cultural preferences and traditional uses and contribute to food identity of many populations. The problem with these neglected crops is that they are often low yielding when compared with major crops, but in many cases they may become economically viable. To achieve that, however, efforts are required to increase the available germplasm and to promote breeding programmes. In fact, the lack of genetic improvement for important agronomic traits represents a strong limitation to the development of these crops.

Thus, this book is aimed to provide comprehensive state-of-the-art information and highlight the vast untapped potential on these neglected and underutilized crops, providing directions for future studies. It will be a valuable resource to encourage research on these crops, which is necessary to improve their traits and to assess their nutritional value; in turn, this may increase consumer awareness on the importance of nutrition and dietary diversification on health and may contribute to food security and income generation especially in marginalized areas, where these crops already contribute to an extent to sustainable agricultural production, ecosystem conservation and poverty alleviation.

The impact on society is not secondary; a major reason to engage in this task, in fact, also stems from the concept that underutilized crops may play an important role in the subsistence of local communities and economy of poor countries, which were in many cases their centres of origin. The advantage of these crops is that they often occupy particular agroecological niches where they require very limited or no external inputs and have multiple uses, being not only nutritious but also having medicinal properties.

This book contains 11 chapters that cover different underutilized crops with detailed information about their available genetic and genomic resources. The book also covers few chapters related to advanced technologies that can possibly be utilized for improving the yield potential or other traits of economic interest.

Srinagar, Jammu and Kashmir, India — Sajad Majeed Zargar
Padova, Italy — Antonio Masi
Srinagar, Jammu and Kashmir, India — Romesh Kumar Salgotra

Acknowledgements

Sajad Majeed Zargar acknowledges *National Mission on Himalayan Studies* [NMHS] implemented by the Ministry of Environment, Forest & Climate Change [MoEF&CC] vide project sanction no.: GBPNI/NMHS-2017-18/SG24/622 for financial support to undertake research on underutilized crops of Northwestern Himalayas.

Sajad Majeed Zargar and Antonio Masi acknowledge INPPO (International Plant Proteomics Organisation) for establishing scientific collaboration among editors. They also acknowledge the European Union for the award of Erasmus + Programme that involves staff and student mobility between University of Padova, Italy, and SKUAST-Kashmir, India. We are also grateful to Dr. Rakeeb Ahmad Mir for his valuable inputs in framing this book.

Contents

About the Editors

Sajad Majeed Zargar, Ph.D., is an Assistant Professor at Sher-e-Kashmir University of Agricultural Sciences & Technology of Kashmir (SKUAST-Kashmir) in India and visiting Professor at the University of Padova, Italy. He was previously a Visiting Professor at the Nara Institute of Science & Technology, Japan. He has worked as an Assistant Professor at SKUAST-Jammu and Baba Ghulam Shah Badshan University, Rajouri (BGSB), in India. He has also worked as scientist at Advanta India Limited, Hyderabad, India, and TERI (The Energy & Resources Institute), New Delhi, India. Dr. Zargar is recipient of the CREST overseas fellowship from DBT, India; Goho grant from Govt. of Japan; and Erasmus Fellowship from the European Union. He has received several awards for his work and research and is the recipient of a young research grant from SERB-DST, India. He is also the member and representative of INPPO (International Plant Proteomics Organisation). Dr. Zargar has established well-equipped Proteomics Laboratory at the Division of Plant Biotechnology, SKAUST-Kashmir, to undertake basic plant proteomics research. He has chaired a session in third INPPO World Congress held at the University of Padova, Italy, in 2018. His editorial activities and scientific memberships include publishing research and review articles in international journals and also as a reviewer of reputed journals.

Antonio Masi, Ph.D., is an Associate Professor at DAFNAE—Department of Agronomy, Food, Natural Resources, Animals and Environment, University of Padova, Italy. Following studies in Agricultural Sciences and a PhD in Photobiology at the University of Padova, he carried out studies at the Pflanzen Physiologisches Institut in Bern, Switzerland (1990), and in the Department of Plant Biology at the University of California at Berkeley (1994–95) in the field of plant molecular biology, plant physiology and biochemistry. He was a recipient of a Fulbright grant for research in plant proteomics (2004) at Cornell University, Ithaca, NY. He was visiting professor at Tribhuvan University (2010) in Kathmandu and Pokhara, Nepal, within the European programme "Sutrofor"—Sustainable Tropical Forestry. His research activity has mainly focussed on oxidative stress and antioxidants in plants; plant abiotic stress in crops, plant sulphur metabolism and plant thiols; functions of gamma-glutamyl transferase in glutathione transport and degradation; biological effects of humic substances and biostimulants; comparative plant proteomics and application of proteomics in agricultural and food science; and factors affecting the uptake and distribution of persistent organic pollutants in plants.

Romesh Kumar Salgotra, Ph.D., is a Professor and Coordinator at the School of Biotechnology at Sher-e-Kashmir University of Agricultural Sciences & Technology of Jammu (SKUAST-Jammu) in India. Dr. Salgotra received his Ph.D. in Plant Breeding and Genetics in 1999 from CSK Himachal Pradesh Krishi Vishwavidyalaya, Palampur, India. He was previously an Endeavour Executive Fellow of Queensland Alliance for Agriculture and Food Innovation (QAAFI) at the University of Queensland, Australia. He has worked in different capacities as Junior Scientist and Senior Scientist at SKUAST-Jammu under All India Co-ordinated Rice Improvement Project, National Rice Research Institute (NRRI), Hyderabad, and *Central Research Institute for Dryland Agriculture* (CRIDA), Hyderabad, India. He has also worked as Agriculture Extension Officer in Govt. of Jammu & Kashmir, India. Dr. Salgotra is recipient of BOYSCAST Fellowship, Tennessee University, USA. He was awarded with prestigious Norman E. Borlaug Fellowship by the United States Agriculture Development (USDA) during 2015. He has participated in Genetic Resources and Intellectual Property Rights Programme (GRIP) under Swedish International Development Agency (SIDA), Sweden, and a Follow-up programme at Pretoria, South Africa. He has received several awards for his work and research. He was the member and representative of the Global Forum for Innovations in Agriculture (GFIA), Abu Dhabi, UAE. Dr. Salgotra was also one of the members of delegation in the Joint Committee on Science and Technology, Govt. of India, held at Belgrade, Serbia, in 2017. He has chaired a session in International Association of Science & Technology Development for Computational Bioscience held at Cambridge, Massachusetts, USA. He has more than 50 peer-reviewed publications and has delivered numerous oral and poster presentations in national and international meetings/conferences.

Neglected and Underutilized Crop Species: Are They Future Smart Crops in Fighting Poverty, Hunger and Malnutrition Under Changing Climate?

1

Akbar Hossain, Mst. Tanjina Islam, Sagar Maitra, Debjyoti Majumder, Sourav Garai, Mousumi Mondal, Asgar Ahmed, Anirban Roy, Milan Skalicky, Marian Brestic, and Tofazzal Islam

Abstract

Throughout history, more than 5538 crop species have been used as food for human beings, whereas only 12 crop species share the major percentage of food security globally. Among these crop species, three kinds of cereal, such as rice,

A. Hossain (✉)
Bangladesh Wheat and Maize Research Institute, Dinajpur, Bangladesh

M. T. Islam
Department of Agronomy, Hajee Mohammad Danesh Science and Technology University, Dinajpur, Bangladesh

S. Maitra
Department of Agronomy, Centurion University of Technology and Management, Khurda, India
e-mail: sagar.maitra@cutm.ac.in

D. Majumder
Uttar Banga Krishi Viswavidyalaya, Cooch Behar, West Bengal, India

S. Garai · M. Mondal
Department of Agronomy, Bidhan Chandra Krishi Viswavidyalaya (BCKV), Nadia, West Bengal, India

A. Ahmed
Division of Maize Breeding, Bangladesh Wheat and Maize Research Institute, Dinajpur, Bangladesh

A. Roy
Department of Genetics and Plant Breeding, Bidhan Chandra Krishi Viswavidyalaya (BCKV), Nadia, West Bengal, India

M. Skalicky
Department of Botany and Plant Physiology, Faculty of Agrobiology, Food and Natural Resources, Czech University of Life Sciences Prague, Prague, Czechia
e-mail: skalicky@af.czu.cz

S. M. Zargar et al. (eds.), *Neglected and Underutilized Crops - Towards Nutritional Security and Sustainability*, https://doi.org/10.1007/978-981-16-3876-3_1

wheat and maize share >50% of calories across the globe. Besides these, more than 1000 neglected and underutilized crop species (NUFCs) across the world have been estimated as upcoming survival crop species as these crop species are enriched in nutrients and have wider adaptability to various stresses in the modern era of climate change. Although their importance and potentiality have been still unknown to various stakeholders. Recently, these NUFCs crops have received appreciation due to their prospective role for the sustainability of crop production through alleviating agricultural risk. Researchers also recognized that the extensive utilization of underutilized minor crops to reform time-based and dimensional heterogeneity into uniform farming systems will expand the pliability to both biotic and environmental stresses. The utilization of biotechnological approaches such as genotyping by sequencing and arrays and pan-genomics tools are important tactics for evaluating the secreted prospects of the NUFCs. The application of these genomic approaches has been already proved to improve the productivity of numerous climate-resilient NUFCs including sweet potato, cassava, yam, coconut, sorghum, groundnut, cowpea, common bean, chickpea, cacao, etc. Unlocking the real potentials of the NUFCs by employing advanced breeding technologies is imperative for global food and nutritional security and also for attaining Sustainable Development Goals (SDG). This chapter highlights the prospects of NUFCs as genetic resources for fighting poverty, hunger and malnutrition for the food security of the increasing population. This chapter also discusses the important roles of NUFCs in advancing defensible agricultural improvement beyond the Green Revolution in the changing environment.

Keywords

Neglected · Underutilized crop species · Food security · Biotic and abiotic stresses

1.1 Introduction

To meet the food demand of the increasing population under changing climate, the agricultural system is under stress for producing more food, feed and biofuel under limited land resources (Godfray et al. 2010). It is estimated that food production has to be enlarged by 70% to meet the food and nutritional security of nine billion people by the year 2050 (Bruinsma 2009). While 90% of the increased demand can be

M. Brestic
Department of Botany and Plant Physiology, Faculty of Agrobiology, Food and Natural Resources, Czech University of Life Sciences Prague, Prague, Czechia

Department of Plant Physiology, Slovak University of Agriculture, Nitra, Nitra, Slovak Republic
e-mail: marian.brestic@uniag.sk

T. Islam
Institute of Biotechnology and Genetic Engineering (IBGE), Bangabandhu Sheikh Mujibur Rahman Agricultural University, Gazipur, Bangladesh

mitigated through enhancing cropping intensity and crop productivity; the remaining demand has to be achieved by the utilization of land, presently not used for crop production (Wu et al. 2018).

Recently, more than 1000 neglected and underutilized crop species (NUFCs) across the world have been recognized as forthcoming survival crops (Ravi et al. 2010; Mayes et al. 2012), since these crop species are enriched in nutrients and also have wider adaptability against numerous stresses (IPGRI 2002; Chivenge et al. 2015). These NUFCs crops have received appreciation by researchers due to their prospective role for the sustainability of crop production through alleviating agricultural risk (Pudasaini et al. 2013; Muthamilarasan et al. 2019; Zhang and Batley 2020). Among these NUFCs, numerous underutilized vegetables and legumes are the important sources of minerals, vitamins, micronutrients and proteins, thus, appreciated crops to achieve dietary safety (Ebert 2014; Ruiz et al. 2014). Therefore, scientists across the globe have been already recognized for widespread exploitation of NUFCs into uniform farming systems, which increase the flexibility against both biotic and abiotic stresses (Mal 2007; Esfeld et al. 2013). However, it is important for unlocking the real potentials of the NUFCs by employing advanced breeding technologies that is imperative for global food and nutritional security. The use of biotechnological approaches such as genotyping by sequencing and arrays and pan-genomics tools have great potential for evaluating the secreted prospects of the NUFCs to find out the potential genetic resources for the development of high-yielding and stress-tolerant crop cultivars (Jain and Gupta 2013; Tiwari and Kumar 2013). This chapter highlights the potentials of NUFCs as genetic resources for alleviating hunger and malnutrition. This chapter also discusses the potential role of NUFCs for the sustainability of agricultural improvement beyond the Green Revolution in the changing environment.

1.2 Concept of Neglected and Underutilized Crop Species

The group of crops has been described as underexploited potential crops having nutritional and medicinal values including income sustainability. These crops are environment suitable with major food security and have sporadic existence in the food chain. Even if having a food value throughout the world in diverse crops are being neglected unnecessarily (Magbagbeola et al. 2010) categorized as underutilized (Kour et al. 2018) and neglected since long and in long-distance with the national policies. Some crops are important in case of local requirement in their food system, but minimum characterization (Thies 2000) and only centre of origin and diversity-based distribution led these crops to be categorized as neglected. Similarly uncompetitive assumption of these crops by the farmers in the competitive large scale farming made them underutilized (IPGRI 1998). FAO emphasized the importance of these crops in the subsistence of the economy, even once these crops were involved in the conventional system. Similarly Bioversity International (Previously IPGRI) mentioned their restricted niche distribution based on local need. Past involvement of the crops in marginal ecology remained as an income-generating source, especially in developing countries where these crops are associated with the people culturally in terms of history and religious belief. The status of nourishment throughout Sub-Saharan African countries is at a vulnerable stage consisting of 23.8% undernourishment children <5 age children in East Africa (FAO 2014). The green revolution has set the food security at a level which only prioritized the major cereals and diverted the research direction towards these crops instead of diversifying the sustainability issue of food bowl (Hazell 2003) and dependency needs to be shifted for sustainability and resilience issue from monoculture related crops to diversified one. Underutilized crops also termed as development opportunity crops possess tremendous potentiality from the undervalued status (Kahane et al. 2013). To meet the unfulfilled demand of Millennium Development Goals (MDGs) at the end of 2015 and realizing the next phase as Sustainable Development Goals (SDGs) for future sustainability through increasing variability in the system for the long term has been initiated and biodiversity-based nutrition has been emphasized. Underutilized crops have been a part of various SDGs, Biodiversity projects, Programmes covered under the International Treaty on Plant Genetic Resources for Food and Agriculture (ITPGRFA) and UN for sustainability issue. To enrich the food security, emphasis has been given to the crops which are unexplored and have huge potential with series of nutrient status to feed the undernourished human populations and needs very fewer efforts to incorporate in crop growing scenery. These crops are not in mainstream production along with conventional crop species and even less emphasize in terms of research involvement and occupy marginal farming community. In most cases, minimum awareness is the major factor in controlling the status of the mentioned crop as underutilized (Kaul et al. 2019). Underutilized crops can be the best candidate groups of crops under contingency crop plans. The recent emergence of Buckwheat, yam, colocasia, lemon, pumpkin, amaranthus, cassava, faba beans, millets, minor legumes and pulses, traditional vegetables, etc., have diversified the food consumption profile as an underutilized and marginalized neglected crop species. Though these crops are fit in the marginal area but have a strong place in local food basket value with sufficient nutrient component (Jain and Gupta 2013). Maintaining genetic diversity by restoring a wide set of variability through domestication holds an important phenomenon that we can exploit to reduce mono-cropping and homogeneous dietary plants (Padulosi et al. 2019a, b). These crops possess various sources of nutraceutical shaving immunity booster properties, which is in tremendous need under changing dietary plans of the modern era to diversify lifestyle, reducing the prevalence of the diseases (Jahan et al. 2020). Under SDG goal of 3, i.e. good health and well-being, using sources of various phytochemicals shown by the indigenous people can be implemented for developing next-generation food to boost the traditional health care system. Mabhaudhi et al. (2016) defined underutilized crops as crops that cover a wide range of species but have a high potential for improvement. These crops are best fit in the marginal area with suitable adaptation as evolved for a long time in the specific ecology (Padulosi et al. 1999). The lack of intensive practices of agriculture, which maintained ecology driven practices managed these sites to sustain as a hotspot for those crops. The disappearance of traditional community and cultural erosion has led to downwards growth in terms of exploitation of such crops and genetic diversity at species and cultivar level (Padulosi et al. 2019a, b) even reduced representation of germplasms related to underutilize and neglected crops are the matter for concern. Although a limited number of major crop species, i.e. mono-genotyping are fully exploited, minimum research efforts are paid to the neglected crops. Genomic analysis is needed to dissect the valuable traits in the neglected crops for their utilization in biotechnology(Kaul et al. 2019). These traditional crops remained challenged due to avoidance in terms of international market value, as lack of uniformity and unnecessary negligence kept them localized (Stamp et al. 2012). As of recent changes in the scenario in understanding the importance of underutilized crops in America is gaining importance at the global level (Galluzzi and Noriega 2014). Different regions in the world have set to prioritize the underutilized crops in response to sustainability issues like in Southern Brazil (Barbieri et al. 2014), Brazil (Siqueira 2011), Latin America (Galluzzi and Noriega 2014), Andean region and Bolivia (Silva Ramos 2002), Ecuador (Espinosa 1996), Spain (Hernández-Bermejo 2013) and even IFAD based NUS project for north-west Africa, West and South Asia, Latin America, etc. has been created. These underutilized crops are now increasing the sustainability of the marginal local community as observed in India through the millet programme (Padulosi et al. 2015) and the Andes through grain centric programme (Padulosi et al. 2014) of IFAD. The neglected crop has the climate resilience potentiality because these crops have been evolved in the marginal and harsh condition with stress environments, which is the key target for the agricultural policy planner at present (Thompson et al. 2007). An organization dealing with NUFCs, i.e. crops for the future (previously International Centre for NUFCs) and Global facilitation unit for NUFCs mentioned that NUFCs increase substantial farming income from low to high, increase harmony in family farming, food sovereignty for equal distribution, biodiversity-based collaboration having semi-domestication status, which increases new market opening ability with minimum external input. Among the NUFCs, those having higher adaptability having higher potentiality for providing future nutrition can be classified as a future smart food crop (Li and Siddique 2018). Changing climate scenario demands the development of region bound staple food crop in addition to existing food crop for sustainability, this enrichment can be done from the underutilized and future smart crop species (Joshi and Shrestha 2018).

1.3 Production Trends and Consumption

An estimation of nine billion populations by 2050 and no scope in horizontal land intensification accelerate agriculture towards the investigation of new cropping systems that would able to secure the food and nutrition demand (Garai et al. 2020; Mondal et al. 2020). Nutritional security contributes to the goal of sustainable agriculture, health, poverty alleviation and gender development (Adhikari et al. 2017). Only 12% of plant species are cultivated for human consumption and 9% contributes to supply 75% of global plant-derived energy (FAO 2009). Adhikari et al. (2017) reported that more than 100 edible crop plant species are deserted with respect to their nutritional values. However, these NUFCs have more disease-pest and stress tolerance ability as well as they are a very nutritionally valuable food source and could be grown in diverse agroclimatic situations (Schmidt et al. 2010). The underutilized food sources include minor pulses, cereals, tuber and root crops and fruits and vegetables (Keatinge et al. 2010). Rather than the commercial, medicinal and nutritional values, the NUFCs reduce the negative impact of the production system on the environment as they are adapted to the marginal environment and can be sustained with low external input (De la Peña et al. 2011).

1.3.1 Underutilized Cereals

Millets are the eldest cereal, which is very well-known by humans and perhaps it is the first cereal grain to be utilized for staple food purposes. However, at the time of the green revolution, the use of high-yielding crop cultivars, heavy external inputs and higher food demand pushed the millet production to back foot. Millets are highly nutritious, easy to digest as compared with traditional major cereals. As an example, finger millets contain ten times more calcium than polished rice (Michaelraj and Shanmugam 2013). Millets are popular foods employed in developing countries due to their hardy nature and adaptation to diverse climatic conditions. Trends of millet consumption in India are demonstrated in Fig. 1.1a. The average production of minor millets in Baluchistan has been steadily decreased from 1995 to 2005 and after that increased steadily as illustrated in Fig. 1.1b (Adhikari et al. 2017).

The trends of underutilized crop production in the Himalayan region of Nepal are shown in Fig. 1.2a. Quinoa (*Chenopodium quinoa*), which belongs to the Amaranthaceae family, is categorized as an underutilized or neglected cereal, rich in protein and fibre (Murphy and Matanguihan 2015). It contains 14% protein, 9 essential amino acids, a decent source of several vitamins, vital fatty acids. Besides these, it is gluten and cholesterol-free (Murphy and Matanguihan 2015). Recently, the status of quinoa is going to be upgraded from underutilized to commercial crops concerning its trends of production. The world quinoa production has been reached 192.5 metric tons in 2014 from 32.4 metric ton in 1961 (Solh 2016).

1.3.2 Underutilized Pulses

The cultivation of underutilized pulse crops, viz. mung bean, lentil and pigeon pea are mainly confined into the Southeast Asian countries. Recently, the mung bean is cultivated in more than six Mha globally. Though, its production reaches a significant level as an underutilized crop but considered insignificant when compared with major cereals. India contributes more than 50% of total mung bean production from the area of 3 m ha globally (Nair et al. 2012). The excessive benefit of this pulse crop is it can be fitted well into the fallow period between two major crops as it has short duration in growth habit. Mung bean production enhanced from 2.3 MT to more than

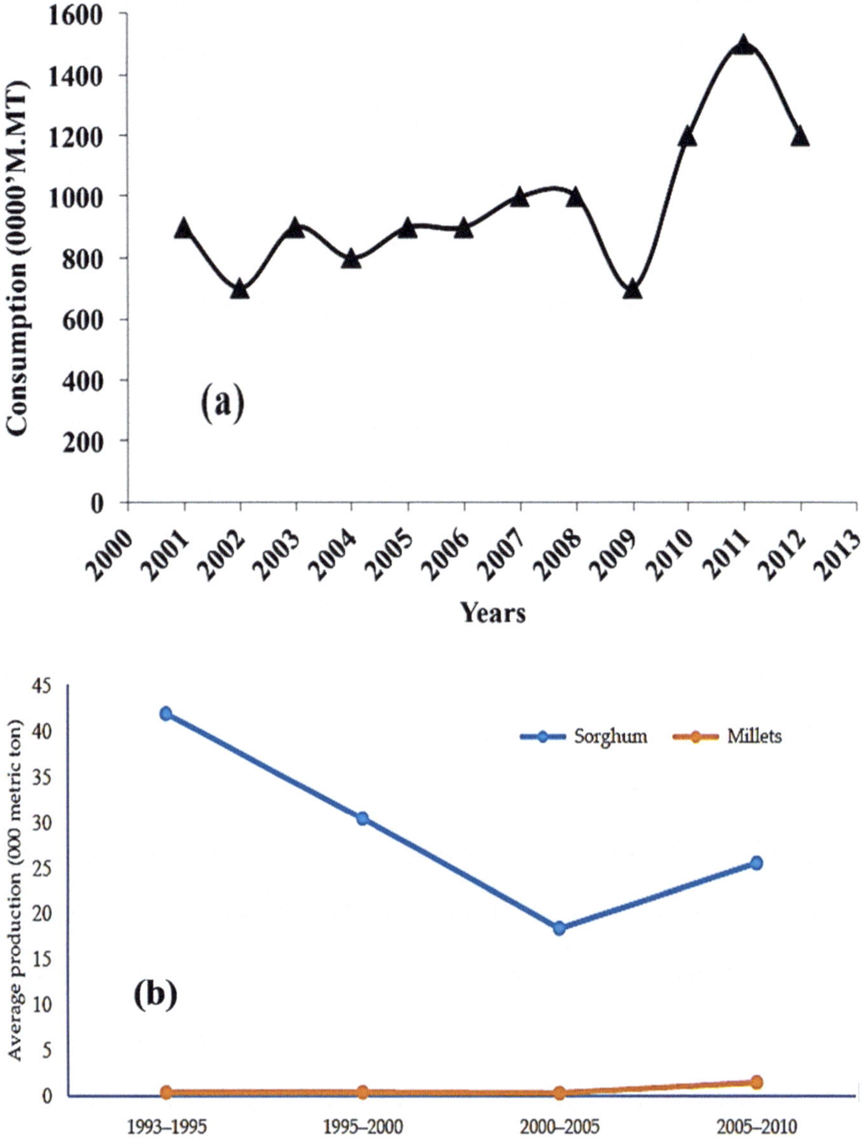

Fig. 1.1 Trends of millet consumption in (**a**) India and (**b**) Baluchistan. (Source: Adhikari et al. (2017))

3 MT from1985 to 2010 in Asia, as a result of improved breeding lines (Weinberger 2003). In between this period, it has been realized the 28–55% yield increment of mung bean in Asian countries. Additionally, the trend of consumption has also increased from 22 to 66%, benefitted 1.5 million anaemic children and accounted for four million USDs additional income generations during the same period of

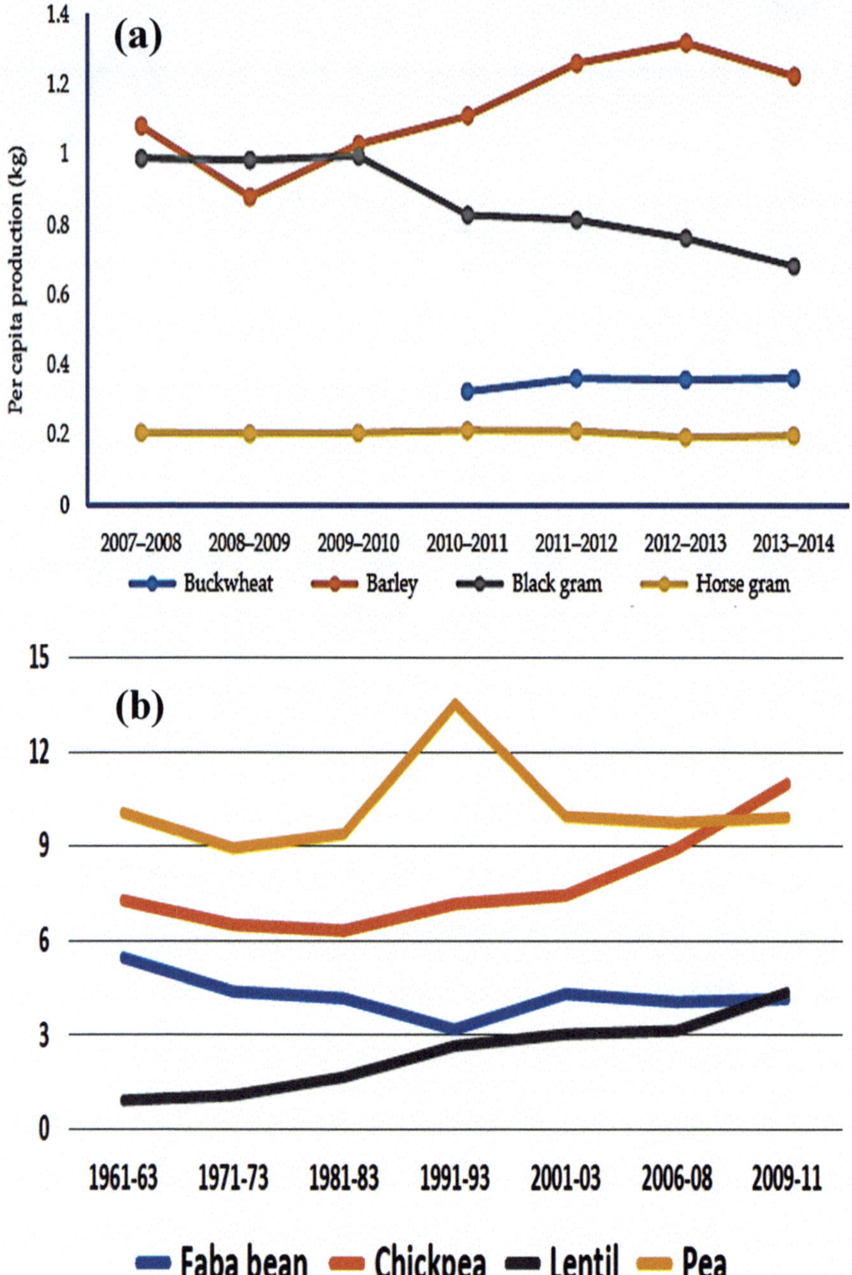

Fig. 1.2 Trends of (**a**) millet production in the Himalayan region of Nepal (Source: Adhikari et al. 2017) and (**b**) pulse production (million tonnes) in Asia. (Source: Solh 2016)

observation (Shanmugasundaram et al. 2009). The trends of global underutilized pulses production are depicted in Fig. 1.2b.

Global lentil production along with total area and productivity is illustrated in Fig. 1.3a.In India, lentil production has been increased from 0.35 mt in 1961 to more than 1 mt in 2014. The trends of productivity fluctuated in different eras and accounted maximum from 1991 to 1993. Additionally, the total area of lentil cultivation in India continuously increases from the pre-green revolution era to 2014 (Fig. 1.3b). Another underutilized neglected pulse crop, grass pea is now catching up concern due to its stress-tolerant nature and well adaption capacity in marginal resources. Additionally, it meets the demand for protein in human and animal feed (Solh 2016). India, Bangladesh and Ethiopia contribute to global grass pea production (Solh 2016). The increasing trend of area and production of grass pea has been reported in Ethiopia (Fig. 1.4a), while the opposite trend has been observed in India (Fig. 1.4b).

1.3.3 Underutilized Vegetables

Among the underutilized leafy vegetables, Amaranth is widely cultivated in tropical countries like Southern parts of Asia, Africa, South America and Mexico (Dehmer 2003). Its nutritional value is very much higher than cabbage (Ebert et al. 2011). Amaranth has attained the importance increasingly due to the increasing trend of domestic consumption as well as the marketable production in several Asian countries. It has been reported that a 500 m^2 small plot of Amaranth supplemented 250 USD in a year (Hughes and Ebert 2013). Another reason for its popularity is Amaranth can be cultivated with comparatively low external inputs and also has a less threat of crop failure; thus, it is very much familiar to small scale farmers. Drumstick is another nutrient and minerals rich neglected plant, widely grown in tropical Asia, Sub-Saharan Africa, Latin America and the Caribbean. It has already gained the status of a profitable crop of India, Philippines, Ethiopia and Sudan for a resource-poor farmer (Fahey 2005). Most of the parts are used for consumption, such as leaves and flowers are used as salad, vegetables, making soups and sauces (Duke 2013). The young pods are marketed as high-value vegetables and also used for making pickles (Ramachandran et al. 1980). The root bark consumed as a condiment, having a pungent taste. The priorities on underutilized crop production depend upon the food habit, agroclimatic conditions and market demand of the particular area. Some examples of location-specific underutilized crops are enlisted in Table 1.1.

1.3.4 Underutilized Root and Tuber Crops

The root and tuber crops are adorned as the second-highest source of carbohydrate after cereals (Chandrasekara and Kumar 2016). Globally, approximately 836 million tonnes of root and tuber crops are produced that indicates their importance as a

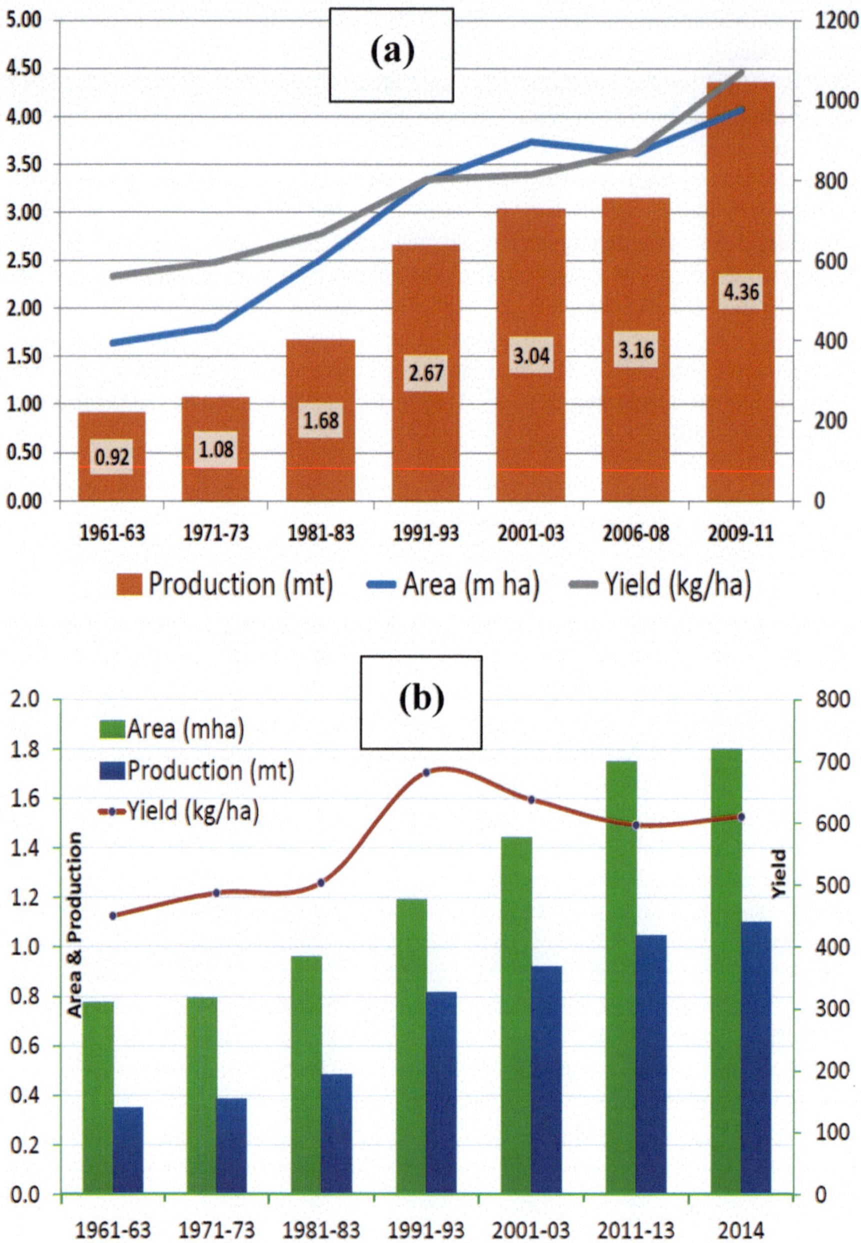

Fig. 1.3 Area, production and productivity of lentil globally (**a**) and in India (**b**). (Source: Solh 2016)

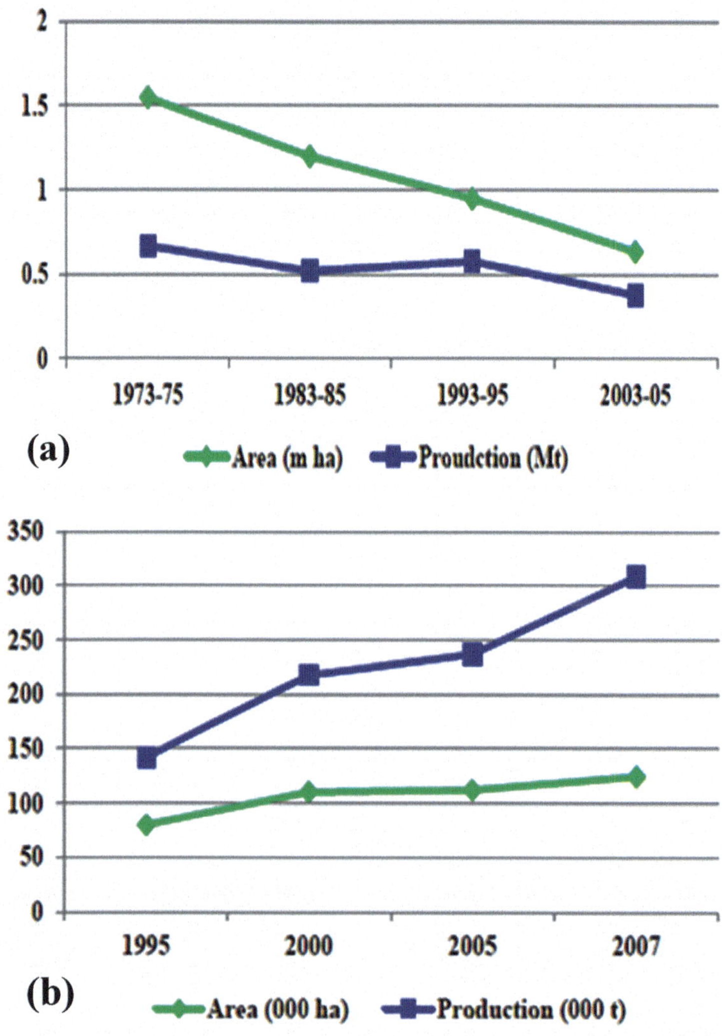

Fig. 1.4 Area and production of grass pea in India (**a**) and Ethiopia (**b**). (Source: Solh 2016)

considerable portion of the food security in the world (FAOSTAT 2013). Among different underutilized roots and tuber crops, sweet potato, cassava, Yams and aroids constitute 90% of the production and consumption rather than potato. Asian countries are the major producers of these crops, followed by America and Europe (FAOSTAT 2013). More than 40% of production is exclusively contributed by Asian regions (Chandrasekara and Kumar 2016). National Aeronautics and Space Administration (NASA) includes sweet potatoes into the food habit of astronauts during space missions because of the higher nutritional values (Bovell-Benjamin 2007). The crop cassava is originated in South America, but it is recently extensively

Table 1.1 List of cultivated underutilized crops in a specific location

Underutilized crop categories	Crops name (Indigenous)	Crops name (Exotic)
Cereal and pseudocereal	*Fagopyrum esculentum*	*Amaranthus* sp.
	Panicum miliaceum	*Eleusine coracana*
	Echinochloa frumentacea	*Triticale*
	Paspalum scrobiculatum	*Chenopodium quinoa*
	Dimocarpus longan	
Fruits and nuts	*Dimocarpus longan*	*Passiflora edulis*
	Carya cathayensis	*Carica papaya*
	Hovenia dulcis	
Vegetable and pulse crops	*Ipomoea aquatica*	*Vigna subterranea*
	Amaranthus spp.	*Crambe cordifolia*
	Corchorus sp.	*Lagenaria* sp.
	Lablab purpureus	
	Chenopodium album	
	Emilia spp.	
	Macrotyloma uniflorum	
	Curcuma spp.	
	Momordica spp.	
	Vigna umbellata	
	V. umbellata	
	Moringa oleifera	
	Lathyrus spp.	
	Vegetable *Glycine*	
Root and tuber crops	*Colocasia esculenta*	*Xanthosoma sagittifolium*
	Alocasia spp.	*Pachyrhizus erosus*
	Dioscorea spp.	

Source: Williams and Haq (2002)

civilized in tropical and subtropical countries of Asia and Africa (Blagbrough et al. 2010). More than 500 million peoples consume cassava as a staple food for its unique nutritional constituents and high carbohydrate content (Blagbrough et al. 2010). The Yam production and consumption are mainly confined to Southeast Asia, West Africa and Caribbean regions (Liu et al. 2007). Among Aroids, Colocasia is most popular in food habits in India and considered as a staple food in most of the islands of the South Pacific namely Tonga and Western Samoa and in New Guinea. Additionally, Taro is mainly grown in Asia, Africa and Pacific regions (Chandrasekara and Kumar 2016).

1.4 Importance of Neglected and Underutilized Crop Species

As per the report of Cheng et al. (2017), more than 50,000 identified edible crop species exists globally. Despite such wide variability mankind is confined mainly to wheat, Maize and Paddy in cereal crops, which covers more than 60% of global intake (FAO 1999). These limitations have resulted in global food insecurity accompanied by nutritional insecurity (Khoury et al. 2014). Mabhaudhi et al. (2017a, b) suggested that wider spread diversification and promotion of rural underutilized, potentially vigour crop species is an urgent need of the time. Aboagye et al. (2007) described the crops/vegetables/fruits, etc., in spite of having tremendous ability or potentiality and other benefits have failed to contribute to national income generation. The underlying cause may be due to lesser exploration, lesser attention, small production and also very less consumption. Even in the twentieth century food insecurity and malnutrition continue to be the most pronounced and sound hindrance in developing as well as underdeveloped nations (Godfray et al. 2010; Da Silva 2014). It is evident from many case studies that in developed and developing nations like the USA, UK, India, Bangladesh there has been a theatrical upsurge in critical illnesses, such as cancer, hypertension, diabetes as a result of unhealthy food habits and improper dietary habits. It is well established that malnutrition often retards the mental and physical development of a child which often leads to unhealthy body development (Triantaphylides and Havaux 2009). Ransom and Elder (2003) recommended that poor nourishment at the time of childhood can lead to diseases like iodine, iron, vitamin deficiency, which often leads to anaemia, blindness, cardiovascular ailments and life-threatening diseases like cancer, etc. It is noteworthy to mention that it becomes very difficult for middle or lower-income people to access to animal source of nutrients through meat, milk, or other sources particularly in underdeveloped and developing nations (Yang and Keding 2009). So, here comes the importance of developing or either nutrition-rich crops or vegetables, fruits, medicinal and aromatic plants, etc. (Baldermann et al. 2016). The focus should be not only concentric on producing more crops, but on accessibility and availability of nutritionally rich and cheap crop plants. It is also significant to mention that food habits differ from people to people and place to place, hence food sources should be such that it matches the taste and acceptability of the local people. Globally there are several underutilized or less popular agricultural and horticultural crops, which are nutritionally very rich and potent. Popularizing and conserving such type of foods and slowly incorporating them into daily diets may serve as very good alternatives. Moreover, under global climatic changing scenarios there lies additional advantage for promoting orphan, underutilized crops. The underutilized crops are always grown naturally and aborigine to particular geographical areas which are potent towards tolerating various stresses and require little agronomic management practices (Thompson et al. 2007). The top leading multinational seed companies holding two-third of the entire monopoly in seed production promote production and multiplications of principal cereal crops to ensure maximum possible returns from the invested amount (Stamp et al. 2012).

1.4.1 Potentiality and Nutritional Importance of Underutilized Crops

As it is well accepted that climate change is not a hoax but reality hence, there lies the importance of diversification (Keatinge et al. 2012). Wider crop diversification both temporally and spatially will lead to better adaptability towards various stresses (Newton et al. 2011). Several pieces of research have been shown that wider diversification and mixed agro-ecosystem have shown better resistance and suppression ability against pests and diseases (Lin 2011). Several strategic measures may lead to the adaptation of a diversified cropping system. These include: (1) promoting intra-specific genetic monoculture, (2) alternating individual field with the cultivation of legumes and broadleaf crops, (3) promoting mix farming, (4) integrating agriculture, horticulture, animal husbandry, fisheries, forestry, etc. Studies conducted by Bobojonov et al. (2013) showed that nearly 2/3rd of Uzbekistan is dominated by lowland irrigated cotton followed by wheat cultivation as part of the government-run schemes. Other nutritionally rich crops like potato, sorghum, mung bean, maize, etc., are left behind. These crops have tremendous potential for economic, social and ecological prospects, which can lead to promoting sustainable agriculture. Arora (2014) listed around 778 NUFCs around the Asian-Pacific region which include fruits (261), vegetables (213), root/tuber types (55), grain legumes/pulses (14), industrial crops (25) and several other crops. A research trial conducted by Liebman et al. (2013) in the corn belt of the USA showed that the cultivation of Maize-Soybean cropping pattern significantly reduced dependency on agrochemical and fossil fuels without hampering the yield and productivity. Similarly, incorporating inherent perennial trees, grasses, fodder crops in the cropping pattern has several benefits like: (1) acting a source of biomass, (2) enriching the carbon sink of the soil and (3) reduce nitrogen emission from the irrigated system into the atmosphere. Gene modification and biotechnological interventions of the crop wild relatives have resulted in developing modern high-yielding crop cultivars, developing tolerance to various stresses, etc. (Jackson et al. 2007; Frison et al. 2011; McCouch 2013). It is well-known to all that presently various minor cereal grains, tuber and root crops, vegetables, non-timber trees, etc. have huge potentials to provide food and nutritional security. They provide a better contribution to climate abnormalities, better market price, enhancing sustainable production (Keatinge et al. 2010).

The economically strong segment of people of some parts of Africa and Southeast Asia are now opting out more for vegetables and shifting it from traditional under nutritional food crops to commercially and nutritionally rich vegetables, food crops (Weinberger 2007). The growing market demand has resulted in multinational companies exploring and developing the possibilities for producing seeds of popular crops (AVRDC 2008). It is also noteworthy to mention that all the underutilized crops cannot be a success at a glance commercially. Research, commercial breeding programmes, selections of genomes, seed multiplications and adaptation of the underutilized crop plants are of utmost importance (Stamp et al. 2012; Ochatt and Jain 2007). Malkanthi (2017) mentioned that in Sri Lanka several NUFCs like finger millets (*Eleusine coracana*), amaranthus (*Amaranthus bicolorata*), Alupuhul (*Benincasa hispida*), Kekiri (*Cucumis sativus*), Thibbatu (*Solanum torvum*), Bitter gourd (*Momordica charantia L.*), Diyalabu (*Lagenaria siceraria*), leafy vegetables, Sarana (*Boerhavia diffusa*) Thebu (*Costus speciosus*) and Kirihenda (*Celosia argentea*); Legumes: Black gram (*Vigna mungo (L.)*; Oil crops: Gingili (*Sesamum indicum*), Mustard (*Brassica juncea*); Root and Tuber crops: Welala (*Colocasia nymplimfolia*), Kiri ala (*Dioscorea alata*), Gahala (*Colocasia esculenta)*, Buthsarana *(Canna indica);* Fruit crops: Sidaran (*Citrus medica*), Anoda (*Annona muricata*), Jambu (*Syzygium jambos*), Lawulu (*Pouteria campechiana*), Bilin (*Averrhoa bilimbi*), Beli (*Aegle marmelos*), Kamaranga (*Averrhoa carambola*), Nelli (*Phyllanthus emblica*), etc. Herbal crops: Ginger (*Zingiber officinale*), Endaru (*Ricinus communis*), Welpenela (*Cardiospermum halicacabum*), etc. These underutilized crops are generally cultivated in a small scale as backyard or homestead farming but not as commercial crops. Underutilized crops like bitter gourd (*Momordica charantia* L.), Weralu (*Elaeocarpus serratus)*, Ginger (*Zingiber officinale)*, Tippili (*Piper longum*), Wadakaha (*Acorus calamus*), Kohila (*Lasia spinosa*) are traditionally used as medicines. These crops are highly potent of rich medicinal values such that it saves the expenditure for medical consultation in the poor households. These crops apart from having medicinal values also consist of several micronutrients which have proved to be a weapon against malnutrition.

1.4.2 Nutritional Importance

Amaranthus (*Amaranthus bicolorata*), for example, is a widely grown leafy vegetable which are grown in tropical regions, Central and South American countries, Mexico and some parts of Southeastern Asia (Weinberger 2007). *Amaranthus cruentus* is cultivated as a cereal in parts of Africa and Asia. It is a short and quick maturing crop, which almost get ready on an average within 30–45 days after transplanting (Ebert et al. 2011). The good economic return has also encouraged the farmers of Southeast Asia to cultivate the crop under a play-house or shade net to prevent the losses due to the occurrence of Typhoon in east Pacific ocean regions during the month of March–April. The higher temperature inside the greenhouse, i.e. up to 40 °C does not result in significant yield reduction. Amaranthus, water spinach (*Ipomoea aquatica*), cucurbits are thus an exclusive example of heat-tolerant crops of which water spinach is highly heated tolerant, amaranthus on the other side is moderately heat-tolerant while most of the green vegetables have shown lower heat tolerance (Kuo et al. 1992). Amaranthus being a C_4 plant can sustain high photosynthetic activities and are very high in water-use efficiency. Amaranthus grains contain about 4% protein, 10% fat, 2.5% ash, 64% starch and 8% of dietary fibre. The black seeds, however, have higher fibre content (16%). It also contains certain essential amino acids like lysine (Pedersen et al. 1987). Thus, taking into consideration the nutritional facts several multinational seed companies have shown keen interest in the inclusion of amaranthus seed under their business and marketing strategy.

Drumstick plant belonging to the Moringaceae family which in all total comprises around 13 different species with clear marked geographical origins (Olson 2013). Four species can store water in their trunks, such as *Moringa drouhardii* (Madagascar), *M. hildebrandtii* (Madagascar), *M. ovalifolia* (Namibia and southwest Angola) and *M. stenopetala* (Kenya and Ethiopia). Another three species which have narrow trunks and are tuberous during the juvenile stage are *M. concanensis* (India), *M. oleifera* (India) and *M. peregrina* (Red Sea, Arabia, Horn of Africa). The rest of the *Moringa* species are found in northeast African countries.

Moringa oleifera is one of the most widely cultivated species around the globe with its centre of origin in the sub-Himalayan ranges of India, Pakistan, Bangladesh and Afghanistan. It has already been identified as a high revenue-generating crop in Southeast Asian countries like India, the Philippines and African countries like Sudan, Ethiopia, etc. (Fahey 2005). Different parts of the drumstick plant have varied levels of human usage. Starting from leaves to flowers and pods are used in making various food items in different regions of the Asian subcontinent (Ramachandran et al. 1980; Bosch 2004). The properly dried leaves of *Moringa* are also often used as a supplement for children, and women after giving birth (Duke 2013). The *Moringa* is a highly versatile and fast-growing plant that can adapt itself from plains to hills, dry to occasional wet conditions (Polprasid 1996; Patricio et al. 2012).

Some underlying benefits of *Moringa* still are unknown which are tremendous potential for fighting against hunger and poverty. It can play a vital role in minimizing malnutrition. Moreover, it has a good resource generation ability of the farmers to uplift the income and social status. Diabetes, hypertension and obesity-related health issues are one of the most pronounced effects of the modern world and are increasing at an alarming rate throughout the globe. Thus, to combat these, the focus has been always on biofortification of the whole grain cereals, enhancing the availability of dietary fibres, etc. (Shobana et al. 2007). Under these context Finger millets (*Eleusine coracana)* and pearl millets having a low glycaemic index, anti-mutagenic, anti-oxidative properties with high dietary fibres can be utilized for producing various health foods (Friedman 1997; Jones and Engleson 2010; Tadele 2018). It is an important staple crop in the Northern and central parts of India mainly for poor section of the people. It is high in various minerals and nutritionally it contains the highest calcium concentration among the cereal around 0.38%, dietary fibres (18%) and phenolic compounds (0.3–3%). As per Chethan and Malleshi (2007), finger millet additionally contains 5–8% protein, 1–2% ether extractives, 65–75% carbohydrates, 15–20% dietary fibres and 2.5–3.5% minerals.

Around the globe pearl millet is utilized in diverse recipes: such as roti or chapatti, porridge, gruel, dessert, etc., and it is often referred to as a poor man's bread (Burton et al. 1972). It has several multipurpose usages, which have the potential to replace refined flours by almost 10–20% (Dahlberg et al. 2003). The extracts from pearl millets which are rich in phenolic compounds showed promising result against carcinoma (Chandrasekara and Shahidi 2011). It is proven that millets are either nutritionally at par or rather superior with respect to principal cereal grains. Millets are not only high in dietary fibre content rather they have a very low glycemic index,

Table 1.2 Water-use efficiency (WUE) of selected underutilized cereal and legume food crops

Sl No.	Underutilized cereals and legumes	Water use (mm)	WUE (kg ha^{-1} mm^{-1})
1	*Arachis hypogaea* L.	697–809	3.96–5.25
2	*Vigna faba* L.	101–261	1.7–12.5
3	*Vigna subterranean* L.	582–856	0.90–3.60
4	*Vigna unguiculata* L.	78–258	0.11–0.20
5	*Lupines albus* L.	178–272	21–8.5
6	*Cicer arietinum* L.	150–340	1.90–3.60
7	*Eragrostis tef* L.	450–550	4.2–11.2
8	*Zea mays* L.	500–800	3.1–5.7
9	*Eleusine coracana* L.	450–650	5.1–10.4
10	*Sorghum bicolor* L.	450–650	12.4–13.4

Data adapted from Chibarabada et al. (2017) and Hadebe et al. (2017)

rich in bioactive components and also gluten-free which is beneficial for reducing obesity and other health issues (Kannan et al. 2013). It is to be specially mentioned that of all barnyard millets have the lowest carbohydrate contents (Leder 2004; Saleh et al. 2013). The importance and health benefits of the millets in our regular diet need to be hence promoted by enhancing biofortification with other flours for better acceptability and producing value-added products. The bio-fortified value-added products will encourage more farmers to grow millets and boost up the performance. The inclusion of millet-based products in our daily dietary habit may thus help to overcome nutritional deficiencies of proteins, calcium, iron, zinc and other micronutrients in poor and developing countries like India, Bangladesh, Nepal, Sri Lanka, Pakistan, etc. (Kumar et al. 2018).

In India, several agroclimatic zones such as hot deserts, coastal saline zones, hilly mountainous region and vast stretch of fertile alluvium are present. Varied agronomic practices and cropping pattern is followed across the country. Some legumes, such as *Canavalia ensiformis* and *C. gladiata* have been identified as good nutritional food (Rai et al. 2007). Similarly, a very uncommon sword bean has also proved to have several nutritional benefits which contain certain flavonoids and anti-carcinogenic properties (Morris 2007). The vegetables have also shown drought-tolerant ability and several underutilized less popular crops need less water and have comparatively high water-use efficacies (Table 1.2) as per studies conducted by Chibarabada et al. (2017) and Hadebe et al. (2017).

It is to be mentioned that land suitable for the cultivation of some well-known crops may be suitable for introducing adjustable NUFCs (Mabhaudhi et al. 2017a, b). Sweet potato (*Ipomoea batatas*) is one of the oldest crops which have been domesticated, but it still has been considered as an underutilized orphan crop. Sweet potato, elephant foot yams, cassava are some of the most popular and important crops utilized in developing countries (Lebot 2009). The crop has been very popular and still is one of the greatest chief food crops in several African countries (Mukhopadhyay et al. 1990). Sweet potato has proved to be very efficient in terms of lower input use production system, drought tolerance ability and wider adaptability regardless of prevailing weather conditions (Motsa 2015). The plants as a whole comprising of the root, leaves are often eaten as a vegetable. It is often used by the poor house holdings as nutritional food, highly rich in magnesium, calcium providing food security (Mukhopadhyay et al. 2011; Low et al. 2007). Sweet potato, elephant foot yams, cassava are also rich in β-carotene, which acts as a precursor of Vitamin-A. Several studies revealed that the inclusion of sweet potato in regular diets of children in African countries significantly reduced vitamin A deficiency (Amagloh et al. 2012; Laurie and Van Heerden 2012). Detailed studies need to be conducted regarding the beneficial and nutritional aspects of the crop.

Vigna unguiculata commonly referred to as cowpea is one of the oldest legume crops belonging to the Fabaceae family and has been domesticated in the African regions (Verdcourt 1970). It has served as an important source of protein in the surrounding regions of African countries (El-Jasser 2011). The legumes can be eaten as vegetables, grain legume, etc. It also has a high potentiality in terms of providing vitamins and minerals (Bressani 1985). It has also some ability to provide vitamins, minerals and proteins and also often used as the source of green fodder, pastures (Sebetha et al. 2010; Modi and Mabhaudhi 2013). This crop is having a long tap-root, which enables it to draw moisture from deeper soil depth and a greater probability for surviving during extreme drought conditions. However, detailed research needs to be conducted regarding the drought-tolerant ability of the crop, such that it may be included in the diet of human beings and also an alternative crop for climate change adaptation programme.

Mung bean (*Vigna radiata*) is a vital pulse in the Indian subcontinent as well as grown on a small scale in Africa and in the American subcontinent. Half of the world's mung bean is produced in India next to China and Myanmar (Nair et al. 2012). Though its production is quite apt but the area under the crop is still insignificant. The crop is very much rich in dietary protein and also serves rich in iron (Keatinge et al. 2011). It has also a tremendous potential to fix atmospheric N into the soil, thereby increasing the potential yield of the succeeding crops. Due to its proliferated growth rate it also checks the infestation of pest and diseases (Weinberger 2003). During the late ninetieth century, mung bean was cultivated with less effort and little agronomic management practices in marginal or low productive land. Being indeterminate type, several harvesting were required which made them economically unsound (Shanmugasundaram et al. 2009). They were also highly susceptible to pest and disease incidence. However, with modern breeding technologies, lower photosensitive cultivars, pest and disease-resistant varieties have been developed. Mung bean being fast-maturing crops provides an easy opportunity for the farmers to cultivate it between two crops. It has shown promising result for increased production through small or marginal farmers, who are often unskilled and always devoid of input materials, thereby generating or increasing remunerative margins.

The wild mustard species, such as *Brassica nigra* and *Brassica juncea* are indigenous crops of South Africa, cultivated under widely diverse conditions. It is a rich source of several vitamins and other essential nutrients. The crop is also cultivated for its oil content and protein (Chweya and Eyzaguirre 1999; Burton et al.

1999). As early as in the nineties scientist like Woods et al. (1991) showed that the crop has a drought tolerance ability underwater scarce situations and hence, enough further research needs to be conducted regarding its potentiality under climate change conditions.

In the United States of America, *Arracacha esculenta* is an underutilized crop, which most of people have not heard off. It was originally cultivated around the region of Andes. Being a tuber crop the root is often used as an alternative to the potato being used in several dishes. It is also used as salads and boiled vegetables (Lim 2015). Similarly, Cashew apple has also shown promising alternatives for producing beverages. It is also nowadays processed for the preparation of jam, syrup, chutney, juices, etc. (De Abreu et al. 2013). It has also shown promising results in curing sore throats, chronic diarrhoea, along with several benefits such as for producing pain-relieving ointments, along with some anti-bacterial, anti-inflammatory properties (Morton 1987; Kubo et al. 1999; Da Silveira Vasconcelos et al. 2015). Therefore, it is established that NUFCs can contribute significantly to the people by enhancing the income particularly for the poor women who rely upon agricultural activities (Chivenge et al. 2015; Chibarabada et al. 2017; Kour et al. 2018).

1.4.3 Diversification for Food Security: The Role of Underutilized Crops

Human livelihood like any other living beings requires three basic fundamental compositions for sustainable development, i.e. carbohydrate, protein and fat. The majority of the population highly relies on three principal cereal crops, namely rice, wheat and maize for fulfilling the daily carbohydrate requirements. For protein source, we mainly rely on animal components and by-products, such as fish, meat, milk, etc. The ever-increasing population globally has compelled researchers to modify the existing cropping system by introducing transgenic and conventional breeding techniques. Interestingly, despite focusing merely on increasing global productivity by technological interventions to meet the food requirement, foods insecurity and nutritional requirement cannot be fulfilled. These interventions also make the cereals crop more prone to several biotic and abiotic stresses.

As per studies conducted by Padulosi (2006), underutilized crops are labelled to the crop species which are indigenous rather than which have been adapted or non-native to the region. The underutilized crops also contribute to a large part of the dietary and cultural aspects of the people. As per FAO (1998), there are approximately 7000 species that once or twice have been part of the people cultivation practices as cultivated crops.

In the modern decades, climate change-related incidences, such as flood, hailstorm, droughts, extreme heat and cold waves, cyclones, etc. have drastically caused an impact on the livelihood of the farming community. The small and marginal farmers are prone to more vulnerability due to lesser risk and adaptability (Lin 2011; Makate et al. 2017). Under such circumstances, the small and marginal farmers are compelled to undergo or rather diversify cropping practices. It is well established that crop diversification often leads to climate-resilient agriculture for the agrarian community. It is also one of the most economically feasible, rational way out for mitigating climate risks (Njeru 2013). According to Falco and Chavas (2009) diversifying crop species also helps in achieving higher price by reducing marketing risks.

1.4.3.1 Medicinal Values

Since ancient times, the wild species of edible fruits often play an important role in our regular dietary habits, which acts as a supplement around the globe. Apart from food intake the wild cultivars have varied levels of health benefits as it is one of the rich ingredients for enhancing immunity. The ancient Indian Ayurveda treatment is mostly developed from wild fruits and plants species. The wild edible fruit crops refer to those that are neither cultivated nor domesticated, but originated from wild natural habitat and often used as food (Beluhan and Ranogajec 2011). Most of the underutilized orphan fruit crops are often used in Southeast Asian countries including India and China for producing many herbal medicines often known as 'Unani', Homoeopathy and Ayurveda. As per researches (Ganry 2006) intake of these underutilized fruits and vegetable crops often helps to prevent non-communicable diseases like diabetes type II, hypertension, cancer, rheumatic arthritis, etc.

Aonla is one of the underutilized fruit with the highest level of vitamin C content. It is richer than vitamin C contents of other popular mainstream fruits, such as guava, citrus and tomato. In fact, the juice extracted from the fruit is nearly 20 times richer in vitamin C than orange juice. A small amount of aonla juice daily could prevent us from several critical ailments like asthma, bronchitis, tuberculosis, anaemia, hypertension and many more. High acidity often disengages them from consuming it in raw form. They are rather widely used in the preparation of many value-added products, such as Chawanprash, Triphala, etc. The fruit is useful for the treatment of ailments like Leucorrhoea, blood discharge from uterus and haemorrhagic symptoms (Hasan et al. 2010).

Indian jujube (*Zyziphus mauritiana*) locally called as 'Ber' from the Rhamnaceae family is a native fruit crop of the Indian subcontinent. The fruit is very common in India but rarely cultivated commercially. As per the early analytical studies conducted by Tiwari and Banafar (1995), the fruit has a very high content of mostly all the essential vitamins, elemental nutrients like Ca, K, Br, Rb, etc. due to its universal nutritious role it is often termed as 'Poor Man's fruit'. The extraction of the bark and root is used as a good alternative for medicines against dysentery, whereas extraction from the leaves is used as remedies against treating sore throats, curing ulcers, fevers, wounds, etc. It is also used for blood cleansing and appetizers (Baloda et al. 2012; Rathore et al. 2012).

Karonda (*Carissa carandas* L.) of Apocynaceae is primarily another native crop from India which contains a good amount of vitamin C, carbohydrate, protein, calcium, fibres, etc. However, it is mainly recognized for its high richness in Iron. The fruits after drying are widely used as a substitute of raisins (Nalawadi and Jayasheela 1975), treatments of malaria, nerve disorders, pain reliever, leprosy, etc.

(Rahmatullah et al. 2009). The high contents of alkaloids, flavonoids, Saponins, such as Lupeol, carionol, ascorbic acid, Triterpenoids, etc. have proved to be very effective against treating scabies, deworming, anthelmintics (Virmani et al. 2017).

M. tuberosa (kasarkaya) *M. tuberosa* (Kaasara kaya) is underutilized in southern parts of India comprising of Tamil Nadu, parts of Andhra Pradesh (A.P) and Karnataka. The crop is well suited in black soil regions of A.P mainly comprising of Rayalaseema regions is cultivated mainly during the monsoons. The crop is mainly propagated through root. The crop is often valued for medicinal purpose. As per Reddy et al. (2007), the crop has been used extensively used for manufacturing and producing indigenous medicine as a remedial measure for curing diabetes (Reddy et al. 2007). Similarly, the fruits are also considered as emetic, laxative, antibilious, tonic, stomachic, stimulant and alternative medicine. The fruit also has been found useful for its unprecedented role in gout healing, rheumatism, moderate cases of the spleen and liver ailments and others (Koneri et al. 2006). Chironji Chironji (*Buchanania lanzan*) belonging to the family Anacardiaceae is a potential source of proteins, fats, dietary fibres and serves as a high energy food. It is a very good source of phosphorus, calcium, magnesium and iron. The fruit is generally juicy and sweet and is also processed for the manufacturing of various value-added products, such as squash, health drinks. It is very much enriched in polyphenols, which have proved its efficacy in enhancing immunity along with a rich source of antioxidants.

Another very common yet underutilized multipurpose tree locally known as Tree Bean (*Parkia roxburghii*) belonging to the Mimosaceae family having its origin in the northern states of India mainly Mizoram and Manipur has varied role and medicinal values. Tree bean is a multiple branched medium height legume which rarely exceeds a height of 10–12 m. The fruits are generally soft, tender and bright green in colour during the early developmental stage and gradually turn black at the mature stage. Apart from having culinary uses, the tree is often used in local medicines. The bark, fruit skin and leaves after decoction are used for curing diarrhoea and dysentery.

Rice bean/climbing bean/haricot bean/red bean/Jerusalem Pea (*Vigna umbelata*) whatever it may be called is a highly branched semi-erect vine. It is considered as an underutilized and neglected legume crop in spite of having several multipurpose roles (Joshi et al. 2008). Arora et al. (1980) reported it as the superior most among all the legumes. The legumes are very much rich in protein content with two limiting amino acids such as methionine and riboflavin (Singh et al. 1980). Additionally, the protein digestibility inhibitory compound such as phytinis low in legumes makes them more suitable for human consumption.

Dioscorea bulbiferous locally known as Aerial yam or Air potato is commonly used as vegetables in southern states of India and propagated by bulbil. Traditionally, after removing off the hardback during cooking generally, the bulb is eaten. The crop has been traditionally/locally used as an indigenous rather conventional way for treating conjunctivitis, diarrhoea and dysentery. The crop has shown promising results in treating diseases like syphilis, gonorrhoea, hydrocele, goitre, piles, etc.

Similarly, *Ipomoea aquatica* commonly known as 'water spinach,' water convolvulus or kangkong is a herbaceous aquatic or semi-aquatic trailing type crop, of which the tender fleshy leaves are consumed as vegetables in different parts of Kerala and Tamil Nadu in India. Generally, propagated by utilizing seed and herbaceous cutting, it is used for treatment against diabetes as an emetic, diuretic, purgative, liver function abnormalities, etc. (Uawonggul et al. 2006). The Indonesian people intake it as a sleeping pill. Doctors often prescribed it as a supplement for enriching the Fe contents within the body to anaemic patients (Jayaweera 1982; Malalavidhane et al. 2001).

Underutilized vegetables, fruits, trees, etc. which are rich in nutrient and have the potential to stand adverse weather conditions not only will enrich the economies of the country, but also help in employment generation and bring out the traditional methodologies for treating critical ailments taking into considerations the varied medicinal role of these crops.

1.4.3.2 Genetic Resources for the Development of Stress-Tolerant Crops

It has been identified that one of the most feasible and viable strategies to identify the role of underutilized crops is trait identification, finding and characterizing the stress-tolerant traits which are superior to the existing traits in major crops are very challenging. It is well-established fact that we need to cultivate crops under the challenges of changing climate as well as increasing productivity at the same time. At the same time, we also need to focus on the acceptability of the crop and market demands for the cultivation of any crop. Mere focusing just on the environmental adaptability of the crop would not serve the purpose. Several crops often face hurdles regarding their increased uptakes of nutrients from the soil solution. The underlying causes are reduced shelf life, anti-nutritional factors, unattractive look, etc. There are also incidences where lack of value-added products in the local market also has caused restrictions for the poor to obtain their staple foods. Some of these issues may be addressed by laboratory-based researches, while others may be simply done in fields. But overall, market-based advocacy and developmental measures are the need of the hour for addressing such issues.

In breeding the genetic potential of a crop is determined by combining several genes which comprise the crop, mutual interactive genes and interactions with the ecosystem to produce the phenotypic expressions to assess its ability against several environmental stresses, such as, heat, cold, flood and drought. It also helps in developing resistance against insect and pest diseases. Agronomic management practices often play a pivotal role in creating ambient crop growing conditions. The success of the agronomic management, however, will depend upon the resource intensiveness and crops selected. Under these contexts, the farmers are encouraged to reduce non-recycled fossil generated input products especially for farm machinery and enhance natural farming. However, in reality, the goal is way distant. Precision agriculture can somewhat act as an alternative strategy for the development of versatile crop plants. For most of the underutilized crops, the maximum benefit through agronomic management practices could not be harvested. The most cost cutting edge for the improvement of a crop lies in its genetic material and compositions. Varshney et al. (2009) showed that during the last few years the crop genomics and systems biology have significantly altered through technological interventions, such as next-generation gene sequencing (NGS), etc. The technology has helped in achieving gene sequencing data by at least 100 folds. These data help us to generate significant molecular data, such as DNA markers based information at a lesser cost for crops having very little or no data. It helps in comparing the underutilized crops to that of its predecessors. Several types of research are mostly done on staple or principal crops. The researchers have generated a huge database on the genetic control of traits and environmental interactions. NGS and similar technologies help us to have a vivid and detailed depth in studies of underutilized crops and also finding its potential. Thus, gradually a paradigm shift from individual crop study to a group study of several crop species is taking place. However, several kinds of findings have raised concerns about whether genetic modification of characters of underutilized orphan crops will result in acceptability and domestication by the farmers. The answer is clear that we must focus on trait-specific studies which will enable to match them in different environments.

In recent decades, molecular technologies for the domestication of crop species have started to study the common changes in domesticated species with their wild ancestors (Doebley et al. 2006; Vaughan et al. 2007). Several studies showed that very less genetic changes are required for crop species to be domesticated. Most of the principal crops, viz. wheat, maize, rice species are ancient and hence are subjected to continuous adaptation and optimization. Thus, the crop species have shifted gradually from the centre of origin to new areas of domestication. For example, the domestication of wheat has resulted in the adaptation of the crop to certain environmental conditions through changes in photoperiodic traits and vernalization requirements and hence it has been able to establish itself in widely spread geographical areas globally. Subsequent selection, breeding techniques and agronomic practices have enforced the wheat cultivars to produce more and stable yield near it centre of origin. While no significant processor rather approaches for adaptation, domestication, cultivation, optimization of yields have been undertaken for underutilized or orphan crops. Thus, they are constricted to only around the regions of the centre of origin.

The conclusion thus can be summarized as the initial process of crop domestication, adaptation, cultivation of major crop species is very simple in genetic terms. In fact, the farmers of ancient times knew the art of selection and breeding techniques. For example, they selected the phenotypic expression genes in their own field, such as multiple branched teosinte which has single branched mutations. In fact, according to Doebley et al. (1997) and Wang et al. (1999), these phenotypes are known to be caused by molecular changes to the gene promoter of teosinte branched 1 gene. It is proven that until breeders or scientist have understood the science behind genetic analysis as early as the twentieth century, yield improvement was quite restricted. The oil palm (*Elaeis guineensis* Jacq), naturally absices fruit from the ripened bunches often requires a good amount of manpower to harvest the product. However, in wheat, the seed remains on the ear after maturing until threshing is done. Molecular researchers also hint at the frequency of domestication for species. For, example wheat has been domesticated several and innumerable times while for maize the number is only one (Matsuoka et al. 2002).

Modern approaches regarding genetic tools and methodologies regarding genetic control of agronomic traits of principal crops, modern breeders need to replicate the salient features of ancient crops for underutilized crops at ease and rapid pace. Utilizing the knowledge and detailed understandings of gene interactions, introducing a new crop is also feasible. Even, the traits which are acting as a hindrance of the currently underutilized crops for wider adoption may also be altered. Similarly, optimum agronomic practices and management techniques may also contribute significantly to the betterment of yield. It is also noteworthy to mention that in some species genetic manipulation of engineering may be secondary as compared to markets food politics which act as a fuel for better adoption in the local market.

1.4.3.3 Sustainability of Underutilized Crops

The demand–supply chain of the modern market often leads or encourages the farmers to concentrate more and more towards the cultivation of a specific number of crops. They are more indented towards higher profitability, without taking into consideration the local need and demands (Sillitoe and Marzano 2009). These concentric approaches by the farming community have resulted in gradual loss of habitat and agro-biodiversity. The gradual biodiversity losses if not rectified will lead to depletion and irreversible loss of underutilized crop species, which is often very much needful for the well-being of mankind. Overexploitation, climatic hazards, genetic erosions, overgrazing, deforestation, urbanization have resulted in increasing the pace of biodiversity loss (Van de Wouw et al. 2010). Accompanied by loss of species habitat, an equal amount of drainage of local knowledge and traditions is also occurring simultaneously. Thus, to conserve and protect the losses retaining the IK (indigenous knowledge) and encouraging the production of the underutilized crops is an urgent need of the hour keeping in mind the interest of the local people. Apart from this on-farm promotion and production of indigenous breeds need to be up taken for conserving the underutilized crops and side by side maintaining the gene bank consortium (Barbieri et al. 2014). Thus, the sustainability of underutilized orphan crops needs to be ascertained by integrating IK as well as encouraging the local people through gene bank conservations. It is also equally important to undertake breeding programmes to improve the competitiveness among the different crop species and habitat to make it withstand various vagaries of the weather and climate (Ebert 2014; Sthapit et al. 2010). It will be a bigger lie to reject the fact that ages after ages the breeding programme for these underutilized crops has been lacking and more focus has been embedded towards developing genetic resilience. Additionally, value chain, infrastructural improvement strengthening of both local and global market channels and moreover public consciousness should be prioritized (Stamp et al. 2012; Gruère et al. 2009). Lastly, the sustainability of underutilized crop species needs unanimous concentric efforts for betterment and improvement for utilization of the produced accompanied with the conservation of the gene resource bank, genetic upliftment and value chain development.

1.5 Current Status of Research on NUFCs

Agricultural sustainability is a prime concern worldwide. Proper use of the NUS is one of the suitable options to achieve future food and nutritional security (Frison et al. 2011). NUFCs are having enough importance in terms of local significance and their potential for refining diversity in food (Baldermann et al. 2016). During the present time, NUFCs are gaining the attention of persons searching for solutions against malnutrition (Gillespie and van den Bold 2017; Uarrota et al. 2019). The components of the NUS include legumes, small and minor millets, fruits, tuber and root crops which can support biodiversity and ecosystem services in the present context of climate change. Recently, the NUFCs are recognized as climate-smart crops for the future, because they offer enhanced food and nutritional security, climate resilience due to genetic diversity and tolerance to biotic and abiotic stress. These are hardy crops with wider adaptability and used as raw materials for industrial products.

Moreover, the rich nutritional, medicinal and nutraceutical values of the crops ensured their significance for the future. Therefore, there are urgent requirements for making the revised list and mainstreaming of the NUS targeting future needs (Chivenge et al. 2015; Baldermann et al. 2016; Mabhaudhi et al. 2017a, b). In order to popularize the NUFCs, Williams and Haq (2002) have mentioned a detailed list that has been mentioned based on tropical and subtropical regions in India (Table 1.3).

In India, the names of the NUFCs are identified and listed (Arora et al. 2006; Bhag Mal et al. 1997). In the FAO's global plan of action, the NUFCs were taken into consideration for commercialization (Kermali et al. 1997). There is an urgent need for the identification and commercialization of potential NUFCs species. Also, initiatives should be taken to make them popular. In this regard promotional and research activities are to be intensified for wider acceptance and popularity. The concerned authorities need to create support favouring policy considerations and already different international organizations have come forward in this direction with various collaborative countries. Prominent International Organizations like Biodiversity International, Crops for the Future (CFF), International Centre of Underutilized Crops (ICUC) and International Fund for Agricultural Development (IFAD), Global Environment Facility (GEF) and United Nations Environment Program (UNEP) have already launched projects and promotional works for mainstreaming NUFCs for sustaining food and nutritional security (Dandin and Krishna Kumar 2018).

IFAD funded project was implemented from 2001 to 2006 in different developing countries, namely Bolivia, Egypt, Equator, India, Nepal, Peru and Yemen in which crops of culinary, medicinal and nutraceutical importance were promoted and in the project commercialization and conservation of these crops were considered. The UNEP-GEF funded project (2005–2013) on horticultural crops and wild species emphasized on use and conservation of agricultural biodiversity. The countries that implemented the project were China, Kazakhstan, Kyrgyzstan, Tajikistan, Turkmenistan and Uzbekistan where genetic diversity of fruit crops like almond,

Table 1.3 List of the priority under-used and neglected crops of the tropical and subtropical regions of the developing world

Region/ continent	Cereal and pseudocereal	Fruits and nuts	Vegetables and pulses	Roots and tubers
Asia	*Fagopyrum esculentum, Amaranthus* spp., *Panicum miliaceum, Eleusine coracana, Panicum miliare, Triticale, Setaria italica, Chenopodium quinoa, Echinochloa frumentacea/ utilis, Paspalum scrobiculatum*	*Dimocarpus longan, Passiflora edulis, Carya cathayensis, Carica papaya, Hovenia dulcis, Vitis* spp., *Citrus grandis, Annona muricata, Aeglos marmelos, Artocarpus heterophyllus, Litchi chinensis, Salacca zalacca, Anacardium occidentale, Garcinia mangostana, Nephelium lappaceum, Psidium guajava, Averrhoa carambola, Durio zibethinus, Zizyphus mauritiana, Borassus aethiopicum, Canarium* spp., *Terminalia koernbachii Tamarindus indica, Embilica officinalis*	*Ipmoea aquatic, Amaranthus* spp., *Vigna subterranean, Divers local bamboo* spp., *Crambe cordifolia, Corchorus* spp., *Lagenaria* spp., *Rorripa indica, Chenopodium album, Emilia* spp., *Basella rubra, Curcuma* spp., *Momordica* spp., *Vigna umbellate, V. aconitifolia, V. angularis, V. umbellate, Moringa oleifera, Phytolocca acinosa, Macrotyloma uniflorum, Lathyrus* spp., *Psophocarpus tetragonolobus, Canavalia* spp., *Lablab purpureus, Mucuna* spp., *Vegetable Glycine*	*Colocasia esculenta, Xanthosoma sagittifolium, Alocasia* spp., *Pachyrhizus erosus, Dioscorea* spp.
Africa	*Eleusine coracana Panicum miliaceum, Digitaria exilis, Setaria italic, Echinochloa spp Triticale*	*Borassus aethiopicum, Dovyalis spp., Balanites aegyptiaca, Zizyphus mauritiania, Opuntia ficus-indica, Vitellaria paradoxa, Anacardium occidentale, Uapaca* spp., *Psidium guajava,*	*Celosia* spp., *Anaranthus* spp., *Cucurbita* spp., *Lablab purpureus, Lagenaria* spp., *Solanum americanum/ nigrum, S. aethiopicum, S. macrocarpus, Bidens pilosa, Canavalia ensiformis,*	*Dioscorea* spp., *Colocasia, Vigna vexillata, Xanthosoma, Sphenostylis stenocarpa, Dioscorea* spp., *Solenostemon rotundifolius, Plectranthus esculentus, Tylosema fassoglense, T. esculentum,*

(continued)

Table 1.3 (continued)

Region/ continent	Cereal and pseudocereal	Fruits and nuts	Vegetables and pulses	Roots and tubers
		Tamarindus indica, Passiflora edulis, Sclerocarya birrea, Diospyros mespiliforme, Anacardium occidentale, Adansonia digatata Uapaca spp., *Blighia sapida, Strychnos* spp.	*Corchorus* spp., *Mucuna* spp., *Hibiscus sabdariffa, Lablab purpureus, Voandzeia subterranean, Parkia biglobosa, Gymnandropsis synandra, Colocasia/ Xanthosoma leaves, Talinum triangulare, Crotalaria spp., Brassica carinata, Vernonia spp., Basella alba, Citrullus* local spp.	*Harpagophytum procumbens*
Latin America	*Chenopodium quinoa, Triticale, Chenopodium pallidicaule, Amaranthus caudatus*	*Bactris gasipaes, Theobroma grandiflorum, Passiflora edulis/ quadrangularis, P. mollissima, P. ligularis, Cyphomandra betacea, Solanum muricatum, S. quitoense, Annona cherimola/ muricata, Carica pubescens and hybrids, Feijoa sellowiana, Eugenia stipitata, Euterpe oleracea, Manilkara zapota, Couepia, longipendulata Spondias, Psidium angulatum, Paullinia cupana, Juglans—*	*Physalis philadelphicus, Pachyrhizus spp., Lupinus mutabilis, Opuntia* spp., *Bactris hearts, Euterpe hearts, Chamaedorea tepoiilote hearts, Cnidoscolus chayamansa, Cucurbita spp.* espec. squashes	*Dioscorea* spp., *Xanthosoma* spp., *Pachyrhizus erosus/tuberosus, Arracacia xanthorrhiza, Calathea allouia, Ullucus tuberosus, Canna edulis, Tropaeolum tuberosum, Oxalis tuberosa*

(continued)

Table 1.3 (continued)

Region/ continent	Cereal and pseudocereal	Fruits and nuts	Vegetables and pulses	Roots and tubers
		Indigenous species		

alycha, apricot, fig, peach, pear, plum pomegranate, walnut was nurtured. United Nations Environment Program (UNEP) and GEF jointly funded a project on 'Conservation and Sustainable Use of Cultivated and Wild Tropical Fruit Diversity: Promoting Sustainable Livelihood, Food Security and Ecosystem Services' during 2009–2015 spreading over the countries like India, Indonesia, Malaysia and Thailand which focused on fruit crops, namely citrus, mango, mangosteen and rambutan. To promote conservation and use of minor millets and Andean grains, IFAD funded project was implemented (2009–2015) in Bolivia, India and Nepal in the first phase; however, in the second phase (2013–2015) target of the same project was smallholder farmers' food and nutritional security improvement through sustainable use and conservation of agro-biodiversity with a prime focus on abovementioned crops.

Bundesamt für Naturschutz (BFN), the German Federal Agency for Nature Conservation in collaboration with UNEP and GEF worked in Brazil, Kenya, Sri Lanka and Turkey (2012–2017) on 70 native fruits and 40 wild edible grains. The project further considered the nutritional values of the crops with regional significance. In another Germany funded project (2003–2006), the study was carried out in Eastern and Southern Africa on indigenous vegetables. In India, the UNEP supported project (2016–2021) has focused on biodiversity conservation and ecosystem services. Further in another project (2014–2019), the Indian Council of Agricultural Research (ICAR) in collaboration with biodiversity emphasized mainstreaming of minor tropical and subtropical fruits. As the outcome of the different project, 27 potential crops inclusive of minor and pseudocereals, vegetable and fruits of tropical and subtropical regions identified (Table 1.4) were recognized by FAO (Li and Siddique 2018).

1.6 Biotechnological Approaches for Domestication and Improvement of Orphan Crop Species for Climate Resilience

1.6.1 Concept and Importance of Domestication

Domestication is the result of both genotypic and phenotypic alterations of a species from wild to an exclusive extraordinary crop/plant cultivar. The domestication first happened about 12,000 years ago in the Fertile Crescent and the Middle East, followed by China, Mesoamerica, the eastern part of North America, the Andes, Near Oceania, sub-Saharan Africa, then other portions of the world (Purugganan and

Table 1.4 List of potential crops of the tropical and subtropical environment

Sl. No.	Common name	Scientific name
1	Foxtail millet	*Setaria italica*
2	Proso millet	*Panicum miliaceum*
3	Finger millet	*Eleusine coracana*
4	Buckwheat	*Fagopyrum esculentum*
5	Amaranth (leaf and grain)	*Amaranthus species*
6	Drumstick	*Moringa oleifera*
7	Bitter gourd	*Momordica charantia*
8	Fenugreek	*Trigonella foenum-graecum*
9	Rosella	*Hibiscus sabdariffa*
10	Snake guard	*Trichosanthes cucumerina*
11	Bambara groundnut	*Vigna subterranea*
12	Mung bean	*Vigna radiata*
13	Rice bean	*Vigna umbellate*
14	Lentil	*Lens culinaris*
15	Horse gram	*Macrotyloma uniflorum*
16	Taro	*Colocasia esculenta*
17	Elephant foot yam	*Amorphophallus paeoniifolius*
18	Sweet potato	*Ipomoea batatas*
19	Yams	*Dioscorea species*
20	Jackfruit	*Artocarpus heterophyllus*
21	Jamun	*Syzygium cumini*
22	Custard apple	*Annona species*
23	Tamarind	*Tamarindus indica*
24	Ber	*Ziziphus mouratiana*
25	Pickle mango	*Mangifera indica*
26	Kokum butter tree	*Garcinia indica*
27	Rambutan	*Nephelium lappaceum*

Fuller 2009; Meyer and Purugganan 2013). Initially, the most important food and vegetable crops were domesticated from their wild progenitors. Our intimates mostly selected crop plants for their daily requirements as per their likelihood. The modest assortment of crop plants by our intimates simplified the pyramiding of significant genes through recombination and transmutations which eventually stemmed in civilized crop plants. Remarkably, the furthermost of the domesticated crop plant species has several communal characters comprising shattering of seed. For example, the number of the spikelets per spike of rice was unusually improved but the weight of each spikelet persisted virtually unaffected (Si et al. 2016). While Zhu et al. (2018) revealed that the size and weight of fruit in eggplants and tomatoes were improved strangely through the domestication progression.

Out of 5500 crop species, which are generally used as food, 15 crops provide more than 70% calories to human beings, where half or more calories have come from three cereal crops such as rice, wheat and maize (Ross-Ibarra et al. 2007). In the case of 400,000 plant species, 7000 plant species are considered as semi-cultivated orphan crop species for the welfare of human being (Smykal et al. 2018; Lemmon et al. 2018; Li et al. 2018). These crop and plant species could be considered as valuable genetic resources for the development of high-yielding and stress-tolerant crop and plant cultivars under the future changing climate (Lobell and Gourdji 2012; Sedbrook et al. 2014; Zsögön et al. 2018).

1.6.2 Genome Editing Using CRISPR Technology

The utilization of the CRISPR tool for manipulation of gene(s) is a revolutionary technology in biological science (Gasiunas et al. 2012; Jinek et al. 2012). First discovered in 1987, CRISPR in the prokaryote genome has been known as an adaptive immune system in archaea and bacteria within decades (Ishino et al. 1987; Makarova et al. 2011). In 2012, several groups of researchers independently discovered that CRISPR and its associated protein (Cas) as a powerful genome editing technology to generate precise DNA breaks in a targeted location in the living cells of any organisms (Gasiunas et al. 2012; Jinek et al. 2012). During the natural cellular DNA repair process, various changes in DNA can be introduced by using human-designed CRISPR-Cas. The CRISPR-Cas tool was first successfully used in genome editing of plants in 2013 by three independent groups (Shan et al. 2013; Nekrasov et al. 2013; Li et al. 2013). Soon, it has become a powerful toolbox for programmable gene and epigenome editing and transcriptome regulation in plants (Haque et al. 2018; Chen et al. 2019; Schindele et al. 2020). The CRISPR system is also flexible for editing multiple genes (multiplexing), which typically done by a simultaneous delivery and expression of multiple gRNAs (Li et al. 2013). By multiplex gene editing, CRISPR can now not only promote basic research and speed plant breeding but also facilitates plant domestication and germplasm improvement.

At present, the CRISPR-Cas technology is a useful tool for fine-tuning and knockout of master switches in undomesticated potential wild or orphan crops to enhance diversity in the genome and facilitate the de novo process of domestication in a single generation. Therefore, the CRISPR toolkit revolutionizes not only in genomic changes but also accelerates plant research and precise speed breeding. In the context of global climate change the improvement of germplasm through the CRISPR-Cas tool is very important (Haque et al. 2018). The benefits of de novo domestication of orphan crops applying CRISPR-Cas gene editing are: (1) high precision and accuracy, (2) short time required for variety development, (3) product is transgene-free, (4) higher yield and nutritional value and (5) resistance to biotic and abiotic stresses. However, the major limitations or challenges in de novo domestication are: (1) limited genomic information of wild relatives, (2) transformation of wild relatives, (3) optimization of the protocol of tissue culture, (4) fitness cost and (5) public acceptance of genetically modified organisms (GMO).

Recently, the power of CRISPR in de novo plant domestication has been demonstrated by several researchers (Sedbrook et al. 2014; Lemmon et al. 2018;

Li et al. 2018; Zsögön et al. 2018). Furthermore, to stack multiple alleles into one tightly linked locus, the application of CRISPR is extended to the shuffling chromosome. It prevents the segregation of target alleles at the time of plant breeding. By the application of the CRISPR toolkit, plant breeding is expected to elevate to a new era. The breeding will be precise and customized manner. Conventional breeding and genetic engineering are never able to achieve at this level.

Genome editing using CRISPR-Cas is a reliable toolbox for the improvement of targeted plant traits very rapidly and precisely. A good number of success stories in the last 7 years have been published (Zhang et al. 2018; Chen et al. 2019). It has been used for trait change in almost all cultivated food crop plants (Haque et al. 2018; Bhowmik et al. 2019; Islam 2019). Among them, the application of CRISPR-Cas technology in domestication orphan crops is most exciting. At least two success stories, one in tomato and another in wild cherry indicate the promise of the CRISPR-Cas technology for the domestication of a new desired smart crop resilient to the changing environmental change in a single generation.

Tomato is a major cultivated crop as well as a model crop plant for research in genetics. It is one of the top-ranked vegetables produced in the world. It is rich in genetic resources including thousands of landraces, 12 wild related species, modern cultivars and mutants. Recently, it has been demonstrated that the CRISPR toolkit is efficient in the engineering tomato plant. This new tool can be used for developing genetic diversity by gene disruption and gene replacement. It is also effective in the editing of favourable alleles and editing a target base precisely (Van Eck 2018; Yamamoto et al. 2018). Important six loci for key domestication traits in tomato are: (1) general plant growth (SELF PRUNING, *SP*) (Pnueli et al. 1998), (2) fruit shape (Ovate, *O*) (Liu et al. 2002) and (3) size (FASCIATED, *FAS*), (4) fruit weight 2.2 (*FW*2.2) (Xu et al. 2015; Frary et al. 2000), (5) fruit number (MULTIFLORA, *MULT*) (Lippman et al. 2008) and (6) nutritional quality (e.g. LYCOPENE BETA CYCLASE, *CycB*) (Ronen et al. 2000).

Since the early application of CRISPR-Cas9 to tomato, a good number of reviews summarizes the trait engineering in tomato (Scheben et al. 2017; Van Eck 2018; Yamamoto et al. 2018; Rothan et al. 2019). In tomato, somatic embryogenesis mediated genetic transformation gave the high rate of bi-allelic and homozygous heritable mutations (Pan et al. 2016; Ueta et al. 2017). It means that CRISPR-Cas gene editing in tomato plants can be achieved a few months after plant tissue culture. By employing CRISPR, Rodrıguez-Leal et al. (2017) demonstrated a series of mutations in the tomato genome including insertions, deletions and point mutations. Several important plant traits, such as carpel number/fruit, the architecture of plants and production of inflorescence were successfully engineered in tomato that is linked to yield (Rodrıguez-Leal et al. 2017; Park et al. 2014). Based on expanded knowledge generated by many researchers, several researchers have applied the CRISPR-Cas technology for the domestication of wild relative (Lemmon et al. 2018; Li et al. 2018; Zsögön et al. 2018). This section reviews the successful domestication of tomato using CRISPR-Cas. Zsögön et al. (2018) demonstrated multiplex genome editing of several domesticated genes in tomato such as *SP*, *SP5G*, *SlCLV3* and *SlWUS* using CRISPR-Cas, which indicate that de novo domestication can be done within a single generation. In their seminal work, Zsögön et al. (2018) first edited six independent genes that regulate plant architecture, yield and quality in domesticated tomato. Then they aimed to edit them in a wild *Solanum pimpinellifolium* to achieve de novo domestication. Interestingly, the genome-edited lines gave a three-fold increase in fruit size and a ten-fold increase in fruit number in a single generation. Accumulation of fruit lycopene content was also increased by 500% in comparison to the content in the fruits of tomato (*S. lycopersicon*) cultivated widely (Zsögön et al. 2018).

To enrich vitamin C content in the accession of tomato LA1589, a gRNA was designed targeting the uORF (upstream open reading frame) of SlGGP1 which encodes a vitamin C-biosynthetic enzyme (Li et al. 2018). The SlGGP1 uORF mutations in two independent CRISPR-Cas edited lines gave increased foliar ascorbic acid content. As a result, domesticating of wild tomato was achieved by editing of uORF by CRISPR-Cas9 technology for increasing ascorbic acid levels. In a parallel study, Li et al. (2018) applied CRISPR-Cas multiplex genome editing in four accessions of *S. pimpinellifolium*. The accessions were either highly resistant to bacterial spot disease or tolerant to salt (Li et al. 2018). The edited plants retained these abiotic and biotic stress tolerance traits. Therefore, CRISPR-Cas multiplex editing can be used to both introduce and retention of plant traits for the development of an ideal crop (Fernie and Yan 2019).

Similarly, Lemmon et al. (2018) applied CRISPR-Cas multiplex editing in an orphan crop, *Physalis pruinosa* under the Solanaceae family. This orphan plant had no reference genome, data of gene annotation and protocol of transformation. They selected three genes orthologous of *P. pruinosa*, *SP*, *SP5G* and *CLV1* for editing. Although null alleles of Ppr-sp had limitation in the number of fruit production, the Ppr-sp5g mutants gave up to 50% higher fruit number along with each shoot. Furthermore, a Ppr-clv1 null allele took a robust effect on fruit size, with a 24% increase in fruit mass, suggested that CRISPR-Cas technology can efficiently be used for the domestication of semi-domesticated or wild crops. These advancements in genome editing suggest that the domestication process could be accelerated for the creation of an ideal crop for satisfying human demand. Obviously, gene editing-mediated transformation of a wild plant to a cultivated crop in a single generation is exciting evidence of the strength of this revolutionary technology in crop improvement. It is expected that soon this technology is used in the development of novel kinds of crop plants that will fit well with the changing climate (Schindele et al. 2020).

Although CRISPR-Cas genome editing seems a revolutionary tool for the development and re-domestication of smart crop species, several problems currently limit the robust application of this technology. One of the major limitations is that the transformative technologies of many plants are not developed (Lowe et al. 2016). This problem can be solved by using a relatively easier and established pollen-based transformation protocol (Zhao et al. 2017). Nonetheless, regeneration from protoplasts is a challenging approach as it may result in unwanted somaclonal variations in some plants (Li et al. 2019). Ribonucleoproteins (RNPs) are considered an alternative approach that can be delivered into plant tissues. It is a routinely used method for plant regeneration (such as callus) using particle bombardment technique (Liang et al. 2017; Svitashev et al. 2016). RNPs can be delivered into plant zygotes produced by in vitro fertilization. The transformed plant zygote is used to regenerate mature plants without further selection (Toda et al. 2019). Another pre-condition of CRISPR gene editing is that we must know the sequence of the target gene of a trait. Over the last three decades, using various gene silencing and gene complementation strategies, genetic dissection of major crop plants was largely being done (Frary et al. 2000; Fernandez et al. 2009; Grierson 2016). However, the bottlenecks for the realization of an ideal crop using CRISPR include a poor understanding of the structure of the genome and regulatory networks of genes in most of the potential wild or orphan plants. Practically, genome sequences of most of the promising and neglected crops are yet to be done.

Furthermore, the food and nutritional significance of those promising candidates are not thoroughly been explored. The diversity of nutritionally rich root and tuber crops in tropical regions deserve further exploration for improvement (Dahan-Meir et al. 2018). Among the legumes *Vigna* spp. are important promising candidates which have more than 100 species with high human and animal nutritional value. Only ten species of *Vigna* have been domesticated. The rest 90 species of *Vigna* also deserve further exploration for domestication. Regulatory aspects for the acceptance and application of genome-edited domesticated crop in the practical field are still a big challenge as many countries yet to be formulated the regulation or considering the genome-edited non-transgenic crop as non-GMO. However, the US Department of Agriculture gave green passes to several gene-edited plants and considers genome-edited trans-gene free crop plants as non-GMO (Ledford 2013). Considering all existing limitations, it is envisaged that a research-driven genetic architecture dissertation, big-data-driven machine learning and breakthroughs from synthetic biology will help us to realize the goal of designing an idea crop by de novo domestication using the CRISPR toolkit.

A good number of domestication genes spanning the range of plant traits in several domesticated crops such as maize, wheat, rice, tomato and soybean have already been characterized (Fernie and Yan 2019). These domestication genes that can be targeted for CRISPR-Cas genome editing are fruit/seed size and weight of tomato and rice (Xiao et al. 2008; Cong et al. 2008; Frary et al. 2000; Shomura et al. 2008); grain filling in rice and maize (Wang et al. 2008; Sosso et al. 2015); growth habit of soybean (Tian et al. 2010); architecture or inflorescence of barley, maize and rice (Ramsay et al. 2011; Komatsuda et al. 2007; Sigmon and Vollbrecht 2010; Simons et al. 2006; Doebley et al. 1997; Jin et al. 2008; Tan et al. 2008; Asano et al. 2011); seed casing of barley and maize (Taketa et al. 2008; Wang et al. 2005); seed colour of rice (Zhu et al. 2011; Sweeney et al. 2006); seed dormancy of rice, soybean and tomato (Sugimoto et al. 2010; Wang et al. 2018); shattering of rice and sorghum (Konishi et al. 2006; Li et al. 2006; Lin et al. 2012); style length of tomato (Chen et al. 2007) and flowering of wheat and sunflower (Yan et al. 2003, 2004; Blackman et al. 2010). Domestication priority should be given to those plants that have one or more of the following traits: (1) perennial grains; (2) high nutritional value; (3) nitrogen-fixing; (4) high water-use efficiency; and (5) high abiotic stress tolerance.

1.7 Challenges in Production of NUFCs

At present, about 150 plant species are in commercial cultivation, though we have a vast resource of biodiversity (FAO 1996a, b). The geographical distribution of minor crops is found in both the northern and southern hemisphere, but the abundance of these crops is noted in tropical and subtropical regions (Pareek and Sharma 2009a, b). There are some limitations associated with the spread of these crops. The NUFCs are remained neglected because they are economically not much profitable, consumed by local people of a specific region and limited acceptability and preferred by tribal population and considered as poor man's food (Baldermann et al. 2016; Padulosi et al. 2008). Further, it is noted that the distribution of NUFCs is constrained to a confined area especially from where these are originated (Ebert 2014). There is inadequate information in terms of technical know-how related to the post-harvest operation and modern technology on processing and value addition. This leads to an inefficient value chain, poor marketing and improper initiative for popularization. Due to population growth, industrialization and urbanization, normal habitat and ecosystem of neglected and economically less important crops have been disturbed leading to a decline in the area. Sufficient research work has not been carried out on the improvement of production potential and commercialization of the NUFCs. Not only have the abovementioned factors but also over dependence on few crops and their promotion sent the NUS into back foot. But these are having enormous advantages like climate resilience and health benefits including nutraceutical properties and medicinal features (Rodríguez et al. 2020). The minor millets are presently regarded as nutri-cereals and functional foods because of their composition for improving health and recovering ailment (Banerjee and Maitra 2020; Kumar et al. 2016). For alleviation of hunger and achieving Sustainable Development Goals, the focus should be given to the use and prioritization of NUFCs. Despite the enormous advantages of NUFCs, these crops are still not properly considered by farmers, consumers and policymakers due to some reasons which can be removed to popularize these.

Traditional farmers already switched to the growing of high-yielding food crops, because of consumers' demand, better marketing facility and remunerative price. But recently productivity declining trend of major staple food crops has been noted. Conversely, population growth increased the enhancement of food production target. But promotional activities by the government and other agencies and even contract farming motivated the farmers to grow major cereals and cash crops avoiding NUFCs. Industrialization and urbanization have changed the food habit and directed the consumers to purchase limited major cereals and their value-added products. But in the developing countries where hunger and malnutrition are great issues, NUFCs can alleviate all. Moreover, NUFCs are cost-effective and within the purchasing capacity of the economically underprivileged section of the society (Khalid and Kaushik 2008; Mal et al. 2010).

The consciousness about the nutritional value of foods among common people is also lacking which leads to consuming unhealthy food. The loss of Indigenous Technical Knowledge (ITK) is also another constraint. The present generation of farmers is lacking ITKs related to the cultivation of NUFCs and the nutritional value of these crops (Johns and Sthapit 2004). Hidden hunger for micronutrient deficiency is a common issue among the economically backward classes (Osei et al. 2010). Another major constraint was policy issues.

The present international policies, legal frameworks on biodiversity and plant genetic resources have so far provided limited stimulus and funding for the conservation and sustainable use of the genetic resources of NUFCs (Galluzzi and Noriega 2014). But most of the governments of the concerned countries are not considering it with top priority (Hagen 2004). Required strategies to support NUFCs are still neglected in different countries. Policies of different countries on farm subsidy, minimum support price, insurance like supportive initiatives are confined to major crops. Similarly, market policy and trade rarely consider the NUFCs and their nutritional and ecological value of NUFCs. The Food and Agriculture Organization's (FAO's) Global Plan of Action for Plant Genetic Resources duly emphasized the importance of the NUFCs (FAO 2009). But, NUFCs are still not been adequately mainstreamed in the global food and nutrition security agenda, leading to negligible demand for these in global as well as national markets (Williams and Haq 2002).

1.8 Framework for Mainstreaming NUFCs for Sustainable Nutrition Security

Nature gifted biodiversity as well as diversified food options for all living being on the earth for healthy nourishment. The early civilizations witnessed the nurture of a wide range of plant varieties (40,000–100,000 plant species) for the fulfilment of daily needs (Brown 2010; Eyzaguirre and Padulosi 1999), but gradual change over time in terms of modification of food habit gave more importance of only few food crops and thus rest become neglected and underutilized (Chivenge et al. 2015; Popoola et al. 2019). Further, malnutrition is a common problem in the developing world. In the present scenario with the threat of climate change, food basket diversity and a diversified diet are authoritative to address the issues of malnutrition and undernutrition (Mbow et al. 2019; Palai et al. 2019). The factors affecting malnutrition and hunger are comprised of population growth in the developing countries, ever-dwindling natural resources inclusive of biodiversity (UN 2008). The fulfilment of the requirement of foods for the ever-growing population directed to search beyond the few fine cereals like rice, wheat and maize for food and nutritional security (Shah and Wu 2019). At present, the three major cereals, namely, rice, wheat and maize, vegetables like tomato and potato and meat from five animals are providing food and energy to 60% of the global population. The improper distribution of specific food-stuff caused malnutrition in many ways where alternatives can be looked out for nutrition that may be sufficient in the surroundings (Weaver et al. 2014). But due to a lack of knowledge and ideas still, some nutritionally valuable food is remaining untouched and underutilized and hence focus should be given to those food crops for food and nutritional security. The declining variety in the food basket and repetitious diet has caused micronutrient deficiencies like iron, zinc and iodine (Govender et al. 2016; Vijaya Bhaskar et al. 2017), along with different vitamins, which are now highly prevalent, among women and children among the underprivileged population of the developing countries (Müller and Krawinkel 2005). Mainstreaming variety for food and nutrition security with climate resilience was accentuated by the Convention on Biological Diversity (CBD) in 1992 (FAO 1996a, b) and Biodiversity International (2017) also advocated in favour of the same. In the present circumstances, there is an urgent need for mainstreaming potential NUS as Future Smart Food (FSF) targeting food and nutritional security.

Conservation of biodiversity has enough potential not only in establishing greater ecosystem services but also in nurturing the people with food and nutritional security. These were actually neglected once for the pretty interest of short-term achievement of food sufficiency, but later the importance of the NUFCs was recognized for agricultural sustainability. A huge population of Asia and African countries is nutritionally undernourished, where the NUFCs can play a great role as these are promising to combat malnutrition with a superior livelihood option. In this regard, policy support and integration in the local food system are very much important. These would not be a problem, because once the NUFCs were an integral part of the social and cultural life of a locality. The policy support is much more required for the creation of awareness, making inputs and planting materials available for cultivation, the arrangement of processing and on-farm value addition and creation of market demand.

The developing countries have their policies on food and nutritional security and the NUS are to be included in the programme. Equal importance is to be given to NUFCs like major crops. Generally, traditional food crops do not need many of high energy inputs and grown in marginal and degraded lands with less investment with a comparatively low carbon footprint (Padulosi et al. 2009). Even different millets can be grown in degraded and erosion-prone lands as intercrop as well as in pure stand for soil conservation and soil fertility improvement. The availability of genotypes and cultivars is to be assured as the mountain and tribal farmers get easy access. Incentives can also be given to the farmers adopting the cultivation of NUFCs in the cropping system as it is practised in crop diversification program in different countries (Adhikari et al. 2017. Further, the institutional setup has to be made targeting credit, extension services, value addition and marketing. A media campaign can be done for awareness creation. The NUFCs based food-stuffs can be included in the mid-day meal program and public delivery system. Presently, the nutritional value of different small millets has been re-evaluated by health-conscious people and increasing demand is observed (Handschuch and Wollni 2016a; Maitra 2020). In Kenya, increased demand of finger millet has made this course cereal costlier than maize (Handschuch and Wollni 2016b). Further from consumers' point of view, there should be holistic support to favour the local food industry (Oduori 2005) which may provide better nutrition compared to unhealthy and instant junk foods. Interestingly, locally made value-added food products like cookies, biscuits, flakes like snacks from small millets are made in different countries. Additional benefits of the inclusion of NUFCs are enriched biodiversity, food security, adaptive measure against climate change and greater ecosystem services. Ultimately, a sound and healthy ecosystem leads to agricultural sustainability.

1.9 Conclusion

To meet the food and nutritional security of nine billion people by the year 2050, agricultural productivity has to be increased by 70%. While 90% of the increased demand can be achieved through the improvement of agricultural productivity. The remaining demand can be achieved by adding unutilized land for crop production. Besides these, the cultivation of neglected and underutilized species (NUFCs) along with modern crop varieties may be an innovative strategy for managing future agricultural production under the changing environmental conditions. The NUFCs are generally well-known as 'minor crops' in terms of global production and market value. Wider use of today's NUFCs provides more options to build temporal and spatial heterogeneity into uniform cropping systems and will enhance resilience to both biotic and abiotic stress. Many traditional vegetables and underutilized legume crops are an essential source of vitamins, micronutrients and protein and, thus, a valuable component to attain nutritional security. The current chapter has been focused on the potentiality of NUFCs for fighting poverty, hunger, malnutrition and also genetic resources for the development of crop cultivars suitable to grow under the future changing environment.

References

Aboagye LM, Obirih-Opareh N, Amissah L, Adu-Dapaah H (2007) Underutilized species policies and strategies: analysis of existing national policies and legislations that enable or inhibit the wider use of underutilized plant species for food and agriculture in Ghana

Adhikari L, Hussain A, Rasul G (2017) Tapping the potential of neglected and underutilized food crops for sustainable nutrition security in the mountains of Pakistan and Nepal. Sustainability 9:291. https://doi.org/10.3390/su9020291

Amagloh FK, Hardacre A, Mutukumira AN, Weber JL, Brough L, Coad J (2012) A household-level sweet potato-based infant food to complement vitamin a supplementation initiative. Matern Child Nutr 2012(8):512–521

Arora RK (2014) Diversity in underutilized plant species—an Asia-Pacific perspective. Bioversity International, New Delhi, p 203

Arora D, Chandel KPS, Joshi BS, Pent KC (1980) Rice bean: tribal pulse of eastern India. Econ Bot 34:260–263

Arora RK, Sharma GD, Joshi V, Phogat BS, Bhatt KC, Rana JC (2006) Underutilized crops. In: Dhillon BS, Saxena S, Agrawal A, Tyagi RS (eds) Plant genetic resources: food grain crops. Narosa Publishing House, New Delhi, pp 320–342

Asano K, Yamasaki M, Takuno S, Miura K, Katagiri S, Ito T, Doi K, Wu J, Ebana K, Matsumoto T et al (2011) Artificial selection for a green revolution gene during japonica rice domestication. Proc Natl Acad Sci U S A 108:11034–11039

AVRDC (2008) The World Vegetable Center. Point of impact: healthy urban fast food: a new Maasai Enterprise. AVRDC—The World Vegetable Center, Tainan

Baldermann S, Blagojević L, Frede K, Klopsch R, Neugart S, Neumann A, Ngwene B, Norkeweit J, Schröter D, Schröter A, Schweigert FJ, Wiesner M, Schreiner M (2016) Are neglected plants the food for the future. Crit Rev Plant Sci 35:106–119. https://doi.org/10.1080/07352689.2016.1201399

Baloda S, Sehrawat SK, Yadav BS, Ahlawat VP, Singh S (2012) Present status of ber production and future thrusts in India—a review. Agric Rev 33(3):256–264

Banerjee P, Maitra S (2020) The role of small millets as functional food to combat malnutrition in developing countries. Indian J Nat Sci 10(60):20412–20417

Barbieri RL, Gomes JCC, Alercia A, Padulosi S (2014) Agricultural biodiversity in Southern Brazil: integrating efforts for conservation and use of neglected and underutilized species. Sustainability 6:741–757

Beluhan S, Ranogajec A (2011) Chemical composition and non-volatile components of Croatian wild edible mushrooms. Food Chem 124(3):1076–1082

Bhag Mal M, Paroda RS, Kochhar S (1997) Underutilized crops and their implications in farming systems in India. In: Smartt J, Haq N (eds) Domestication, production and utilization of new crops. ICUC (International Centre for Underutilized Crops), Southampton, pp 30–45

Bhowmik P, Hassan MM, Molla K, Rahman M, Islam MT (2019) Application of CRISPR-Cas genome editing tools for the improvement of plant abiotic stress tolerance. In: Hasanuzzaman M, Nahar K, Fujita M, Oku H, Islam MT (eds) Approaches for enhancing abiotic stress tolerance in plants. CRC Press, Boca Raton, pp 459–472

Biodiversity International (2017) Mainstreaming agrobiodiversity in sustainable food systems: scientific foundations for an agrobiodiversity index. Bioversity International, Rome

Blackman BK, Strasburg JL, Raduski AR, Michaels SD, Rieseberg LH (2010) The role of recently derived FT paralogs in sunflower domestication. Curr Biol 20:629–635

Blagbrough IS, Bayoumi SAL, Rowan MG, Beeching JR (2010) Cassava: an appraisal of its phytochemistry and its biotechnological prospects—review. Phytochemistry 71 (17–18):1940–1951

Bobojonov I, Lamers JP, Bekchanov M, Djanibekov N, Franz-Vasdeki J, Ruzimov J, Martius C (2013) Options and constraints for crop diversification: a case study in sustainable agriculture in Uzbekistan. Agroecol Sustain Food Syst 37:788–811

Bosch CH (2004) Moringa oleifera Lam. In: Grubben GJH, Denton OA (eds) Plant resources of tropical Africa, vol 2: vegetables. Backhuys Publishers, Kerkwerve, pp 392–395

Bovell-Benjamin C (2007) Sweet potato: a review of its past, present, and future role in human nutrition. Adv Food Nutr Res 52:1–59

Bressani R (1985) Nutritive value of cowpea. In: Singh SR, Rachie KO (eds) Cowpea research, production and utilization. Wiley, Chichester, pp 353–359

Brown AH (2010) Variation under domestication in plants: 1859 and today. Philos Trans R Soc Lond Ser B Biol Sci 365(1552):2523–2530. https://doi.org/10.1098/rstb.2010.0006

Bruinsma J (2009) The resource outlook to 2050: by how much do land, water and crop yields need to increase by 2050? In: Proceedings of the technical meeting of experts on how to feed the world in 2050, Rome, Italy, 24–26 June 2009. Food and Agriculture Organization (FAO), Rome, pp 1–33

Burton GW, Wallace AT, Rachie KO (1972) Chemical composition and nutritive value of pearl millet [*Pennisetum typhoides* (burm.) stapf and E. C. Hubbard] grain. Crop Sci 12:187–188

Burton W, Pymer S, Salisbury P, Kirk J, Oram R (1999) Performance of Australian canola quality Indian mustard breeding lines. In: Proceedings of 10th international rapeseed congress, Canberra, Australia, 26–29 September 1999 [CD ROM]. The Regional Institute, Canberra

Chandrasekara A, Shahidi F (2011) Determination of antioxidant activity in free and hydrolyzed fractions of millet grains and characterization of their phenolic profiles by HPLC-DAD-ESI-MSn. J Funct Foods 3:144–158

Chandrasekara A, Josheph Kumar T (2016) Roots and tuber crops as functional foods: a review on phytochemical constituents and their potential health benefits. Int J Food Sci:3631647. https://doi.org/10.1155/2016/3631647

Chen KY, Cong B, Wing R, Vrebalov J, Tanksley SD (2007) Changes in regulation of a transcription factor lead to autogamy in cultivated tomatoes. Science 318:643–645
Chen K, Wang Y, Zhang R, Zhang H, Gao C (2019) CRISPR/Cas genome editing and precision plant breeding in agriculture. Annu Rev Plant Biol 70:28.1–28.31
Cheng A, Mayes S, Dalle G, Demissew S, Massawe F (2017) Diversifying crops for food and nutrition security – a case of teff. Biol Rev 92(1):188–198
Chethan S, Malleshi NG (2007) Finger millet polyphenols: optimization of extraction and the effect of pH on their stability. Food Chem 105:862–870
Chibarabada T, Modi A, Mabhaudhi T (2017) Expounding the value of grain legumes in the semi- and arid tropics. Sustainability 9:60. https://doi.org/10.3390/su9010060
Chivenge P, Mabhaudhi T, Modi A, Mafongoya P (2015) The potential role of neglected and underutilised crop species as future crops under water scarce conditions in sub-Saharan Africa. Int J Environ Res Public Health 12:5685–5711. https://doi.org/10.3390/ijerph120605685
Chweya JA, Eyzaguirre PB (1999) The biodiversity of traditional leafy vegetables. IPGRI, Rome
Cong B, Barrero LS, Tanksley SD (2008) Regulatory change in YABBY-like transcription factor led to evolution of extreme fruit size during tomato domestication. Nat Genet 40:800–804
Da Silva JG (2014) Don't just produce more food—produce better food. Natl Geograph Spec Ser Future Food:1–2
da Silveira Vasconcelos M, Gomes-Rochette NF, de Oliveira ML, Nunes-Pinheiro DC, Tomé AR, Maia de Sousa FY, Pinheiro FG, Moura CF, Miranda MR, Mota EF, de Melo DF (2015) Anti-inflammatory and wound healing potential of cashew apple juice (*Anacardium occidentale* l.) in mice. Exp Biol Med 240:1648–1655
Dahan-Meir T, Filler-Hayut S, Melamed-Bessudo C, Bocobza S, Czosnek H, Aharoni A, Levy AA (2018) Efficient in planta gene targeting in tomato using geminiviral replicons and the CRISPR/Cas9 system. Plant J 95:5–16
Dahlberg JA, Wilson JP, Snyder T (2003) Sorghum and pearl millet: health foods and industrial products in developed countries. In: Alternative uses of sorghum and pearl millet in Asia. Proceedings of expert meeting. ICRISAT, Patancheru, pp 42–59
Dandin SB, Krishna Kumar NK (2018) Mainstreaming future smart food for improving nutrition security. In: Li X, Siddique KHM (eds) Future smart foods, rediscovering hidden treasures of neglected and underutilized species for Zero Hunger in Asia, Bangkok, pp 79–88
de Abreu FP, Dornier M, Dionisio AP, Carail M, Caris-Veyrat C, Dhuique-Mayer C (2013) Cashew apple (*Anacardium occidentale* L.) extract from by-product of juice processing: a focus on carotenoids. Food Chem 138:25–31
De la Peña RC, Ebert AW, Gniffke P, Hanson P, Symonds RC (2011) Genetic adjustment to changing climates: vegetables. In: Yadav SS, Redden RJ, Hatfield JL, Lotze-Campen H, Hall AE (eds) Crop adaptation to climate change, 1st edn. Wiley, Chichester, pp 396–410
Dehmer KJ (2003) Molecular diversity in the genus Amaranthus. In: Knüpffer H, Ochsmann J (eds) Rudolf Mansfeld and plant genetic resources. ZADI, Bonn, pp 208–215
Doebley J, Stec A, Hubbard L (1997) The evolution of apical dominance in maize. Nature 386:485–488
Doebley J, Gaut BS, Smith BD (2006) The molecular genetics of crop domestication. Cell 127:1309–1321
Duke JA (2013) Moringa oleifera Lam. http://www.hort.purdue.edu/newcrop/duke_energy/Moringa_oleifera.html. Accessed 21 Nov 2013
Ebert AW (2014) Potential of underutilized traditional vegetables and legume crops to contribute to food and nutritional security, income and more sustainable production systems. Sustainability 6 (1):319–335. https://doi.org/10.3390/su6010319
Ebert AW, Wu TH, Wang ST (2011) International cooperators' guide—vegetable Amaranth (Amaranthus L.). AVRDC—The World Vegetable Center, Tainan, p 8
El-Jasser AS (2011) Chemical and biological properties of local cowpea seed protein grown in Gizan region. Int J Agric Res Rev 1:68–75

Esfeld K, Uauy C, Tadele Z (2013) Application of TILLING for orphan crop improvement. In: Biotechnology of neglected and underutilized crops. Springer, Dordrecht, pp 83–113

Espinosa JR (1996) Dominant two-loop corrections to the MSSM finite-temperature effective potential. Nucl Phys 475(1–2):273–292

Eyzaguirre PB, Padulosi S, Hodgkin T (1999) IPGRI's strategy for neglected and underutilized species and the human dimension of agrobiodiversity. In: Padulosi S (ed) Priority setting for underutilized and neglected species of the Mediterranean region. Report of the IPGRI Conference 9-11Feb. 1998, ICARDA, Aleppo, Syria IPGRI, Rome Italy. pp 1-19

Fahey JW (2005) *Moringa oleifera*: a review of the medical evidence for its nutritional, therapeutic, and prophylactic properties. Part 1. Phytochemistry 47:123–157

Falco SD, Chavas JP (2009) On crop biodiversity, risk exposure, and food security in the highlands of Ethiopia. Am J Agric Econ 91(3):599–611

FAO (1996a) Global plan of action for the conservation and sustainable utilization of world's plant genetic resources for food and agriculture and Leipzig declaration. In: International technical conference on plant genetic resources, Leipzig, Germany, 17–23 June 1996. Food and Agriculture Organization of the United Nations, Rome. www.fao.org/pgrfa-gpa-archive/docs/gpa_en.pdf

FAO (1996b) Report on the state of the world's plant genetic resources for food and agriculture. Prepared for the international technical conference on plant genetic resources, Leipzig, Germany, 17–23 June 1996. Food and Agriculture Organization of the United Nations, Rome, Italy. p 511

FAO (2009) FAO and traditional knowledge: the linkages with sustainability, food security and climate change impacts. Food and Agriculture Organization (FAO), Rome

FAO (2014) The state of food insecurity in the world: strengthening the enabling environment for food security and nutrition. Food and Agriculture Organization (FAO), Rome

FAO (Food and Agricultural Organization) (1998) The state of the world's plant genetic resources for food and agriculture. FAO, Rome

FAO (Food and agriculture organization) (1999) Fermented cereals. A global perspective. FAO Agricultural Services Bulletin 138 Food and Agriculture Organization of the United Nations, Rome (1999). http://www.fao.org/docrep/x2184E/x2184e00.HTM. Accessed 30 June 2009

FAOSTAT (2013). http://faostat3.fao.org

Fernandez AI, Viron N, Alhagdow M et al (2009) Flexible tools for gene expression and silencing in tomato. Plant Physiol 151:1729–1740

Fernie AR, Yan J (2019) De novo domestication: an alternative route toward new crops for the future. Mol Plant 12:615–631

Frary A, Nesbitt TC, Grandillo S et al (2000) fw2.2: a quantitative trait locus key to the evolution of tomato fruit size. Science 289:85–88

Friedman M (1997) Chemistry, biochemistry and dietary role of potato polyphenols—a review. J Agric Food Chem 45:1523–1540

Frison EA, Cherfas J, Hodgkin T (2011) Agricultural biodiversity is essential for a sustainable improvement in food and nutrition security. Sustainability 3:238–253

Galluzzi G, Noriega I (2014) Conservation and use of genetic resources of underutilized crops in the Americas—a continental analysis. Sustainability 6:980–1017. https://doi.org/10.3390/su6020980

Ganry J (2006) The nutritional value of fruits and vegetables. Fruits 61(4):223–224. https://doi.org/10.1051/fruits:2006034

Garai S, Mondal M, Mukherjee S (2020) Resource conservation technologies for achieving sustenance in agricultural production system. In: Maitra S, Pramanick B (eds) Advanced agriculture, 1st edn. New Delhi Publishers, New Delhi, pp 325–349

Gasiunas G, Barrangou R, Horvath P, Siksnys V (2012) Cas9–crRNA ribonucleoprotein complex mediates specific DNA cleavage for adaptive immunity in bacteria. Proc Natl Acad Sci 109 (39):2579–2586

Gillespie S, van den Bold M (2017) Agriculture, food systems, and nutrition: meeting the challenge. Global Chall 1:1600002. https://doi.org/10.1002/gch2.201600002

Godfray HC, Beddington JR, Crute IR, Haddad L, Lawrence D, Muir JF, Pretty J, Robinson S, Thomas SM, Toulmin C (2010) Food security: the challenge of feeding 9 billion people. Science 327:812–818

Govender L, Pillay K, Siwela M, Modi A, Mabhaudhi T (2016) Food and nutrition insecurity in selected rural communities of KwaZulu-Natal, South Africa-linking human nutrition and agriculture. Int J Environ Res Public Health 14(1):17. https://doi.org/10.3390/ijerph14010017

Grierson D (2016) Identifying and silencing tomato ripening genes with antisense genes. Plant Biotechnol J 14:835–838

Gruère G, Nagarajan L, King EO (2009) The role of collective action in the marketing of underutilized plant species: lessons from a case study on minor millets in South India. Food Policy 34(1):39–45

Hadebe S, Mabhaudhi T, Modi AT (2017) Drought tolerance and water use of cereal crops: a focus on sorghum as a food security crop in sub-Saharan Africa. J Agron Crop Sci 203:177–199. https://doi.org/10.1111/jac.12191

Hagen T (2004) Traditional framing practices and farmers' rights in the HKH Region, Policy Brief. No. 6. SWATEE, Lalitpur

Handschuch C, Wollni M (2016a) Improved production systems for traditional food crops: the case of finger millet in Western Kenya. Food Secur 2016(8):783–797

Handschuch C, Wollni M (2016b) Traditional food crop marketing in sub-Saharan Africa: does gender matter? J Dev Stud 52:343–359

Haque E, Taniguchi H, Hassan MM, Bhowmik P, Karim MR, Smiech M, Zhao K, Rahman M, Islam T (2018) Application of CRISPR/Cas9 genome editing technology for the improvement of crops cultivated in tropical climates: recent progress, prospects, and challenges. Front Plant Sci 9:617

Hasan MA, Singh SR, Majhi D, Devi HL, Somi Singh Y, Biswas S, Das BC (2010) Significance of minor fruits in health care. In: Proc botanicals in integrated health care, pp 162–166

Hazell P (2003) The green revolution. In: Mokyr J (ed) Oxford encyclopedia of economic history. Oxford University Press, Oxford, UK

Hernández-Bermejo JE (2013) Cultivos infrautilizados en España: pasado, presente y futuro. Ambienta. http://www.revistaambienta.es/WebAmbienta/marm/Dinamicas/pdfs/versionpdf/Esteban13.pdf

Hughes JDA, Ebert AW (2013) Research and development of underutilized plant species: the role of vegetables in assuring food and nutritional security. In: Massawe F, Mayes S, Alderson P (eds) Proceedings of the 2nd international symposium on underutilized plant species: crops for the future—beyond food security; International Society for Horticultural Sciences (ISHS). Korbeek-Lo, Belgium, pp 79–91

IPGRI (1998) IPGRI's strategy for neglected and under-utilized species and the human dimension of agrobiodiversity. Priority setting for under-utilized and neglected plant species of the Mediterranean region. IPGRI-Aleppo, Syria

IPGRI (International Plant Genetic Resources Institute) (2002) Neglected and underutilized plant species: strategic action plan of the International Plant Genetic Resources Institute. International Plant Genetic Resources Institute, Rome

Ishino Y, Shinagawa H, Makino K, Amemura M, Nakata A (1987) Nucleotide sequence of the iap gene, responsible for alkaline phosphatase isozyme conversion in Escherichia coli, and identification of the gene product. J Bacteriol 169:5429–5433

Islam T (2019) CRISPR-Cas technology in modifying food crops. CAB Rev 14(50):1–16

Jackson LE, Pascual U, Hodgkin T (2007) Utilizing and conserving agrobiodiversity in agricultural landscapes. Agric Ecosyst Environ 121:196–210

Jahan FN, Rahim MA, Bokhtiar SM, Samanta AK (2020) Potentiality of underutilized crop Dioscorea spp.: a source of nutraceutical. SAARC J Agric 17:1–13. https://doi.org/10.3329/sja.v17i2.45290

Jain SM, Gupta SD (2013) Biotechnology of neglected and underutilized crops. Springer, Berlin. https://doi.org/10.1007/978-94-007-5500-0
Jayaweera DMA (1982) Medicinal plants (indigenous and exotic) used in Ceylon. Part 11:1982
Jin J, Huang W, Gao JP, Yang J, Shi M, Zhu MZ, Luo D, Lin HX (2008) Genetic control of rice plant architecture under domestication. Nat Genet 40:1365–1369
Jinek M, Chylinski K, Fonfara I, Hauer M, Doudna JA, Charpentier E (2012) A programmable dual-RNA-guided DNA endonuclease in adaptive bacterial immunity. Science 337:816–821
Johns T, Sthapit BR (2004) Biocultural diversity in the sustainability of developing-country food systems. Food Nutr Bull 25:143–155
Jones JM, Engleson J (2010) Whole grains: benefits and challenges. Annu Rev Food Sci Technol 40
Joshi BK, Shrestha R (2018) Nepal. In: Li X, Siddique KHM (eds) Future smart food—rediscovering hidden treasures of neglected and underutilized species for Zero Hunger in Asia. FAO, Bangkok, pp 161–178
Joshi KD, Bhanduri B, Gautam R, Bajracharya J, Hollington PB (2008) Rice bean: a multi-purpose underutilized legume. In: Smart JH (ed) New crops and uses: their role in a rapidly changing world. CUC, UK, pp 234–248
Kahane R, Hodgkin T, Jaenicke H, Hoogendoorn C, Hermann M, Keatinge JDH, Hughes JA, Padulosi S, Looney N (2013) Agrobiodiversity for food security, health and income. Agron Sustain Dev 33:671–693
Kannan SM, Thooyavathy RA, Kariyapa RT, Subramanian K, Vijayalakshmi K (2013) Seed production techniques for cereals and millets. In: Vijayalakshmi K (ed) Seed node of the revitalizing rainfed agriculture network Centre for Indian knowledge systems (CIICS), pp 1–39. http://www.ciks.org/downloads/seeds/5.%20Seed%20Production%20Techniques%20for%20Cereals%20and%20Millets.pdf. Accessed 29 Dec 2017
Kaul T, Eswaran M, Thangaraj A, Meyyazhagan A, Nehra M, Raman NM, Bharti J, Gayacharan, Badapanda C, Balamurali B (2019) Rice Bean (*Vigna umbellata*) draft genome sequence: unravelling the late flowering and unpalatability related genomic resources for efficient domestication of this underutilized crop. bioRxiv. https://doi.org/10.1101/816595
Keatinge JDH, Waliyar F, Jamnadass RH, Moustafa A, Andrade M, Drechsel P, Hughes J'A, Palchamy K, Luther K (2010) Re-learning old lessons for the future of food—by bread alone no longer: diversifying diets with fruit and vegetables. Crop Sci 50:51–62
Keatinge JDH, Easdown W, Yang R-Y, Chadha M, Shanmugasundaram S (2011) Overcoming chronic malnutrition in a future warming world: the key importance of mungbean and vegetable soybean. Euphytica 180:129–141
Keatinge JD, Chadha ML, Hughes JDA, Easdown WJ, Holmer RJ, Tenkouano A, Yang RY, Mavlyanova R, Neave S, Afari-Sefa V, Luther G (2012) Vegetable gardens and their impact on the attainment of the Millennium Development Goals. Biol Agric Hortic 28(2):71–85
Kermali SR, Anishetty NM, Cooper HD (1997) Promoting development and commercialization of underutilized crops and species in the FAO global plan of action. In: Smartt J, Haq N (eds) Domestication, production and utilization of new crops. ICUC, Southampton, pp 19–27
Khalid MA, Kaushik G (2008) Food security in mountains: challenges and sustainable strategies. http://www.mtnforum.org/sites/default/files/publication/files/1850.pdf. Accessed 25 Aug 2020
Khoury CK, Bjorkman AD, Dempewolf H, Ramirez-Villegas J, Guarino L, Jarvis A, Rieseberg LH, Struik PC (2014) Increasing homogeneity in global food supplies and the implications for food security. Proc Natl Acad Sci 111(11):4001–4006
Komatsuda T, Pourkheirandish M, He C, Azhaguvel P, Kanamori H, Perovic D, Stein N, Graner A, Wicker T, Tagiri A et al (2007) Six-rowed barley originated from a mutation in a homeodomain-leucine zipper I-class homeobox gene. Proc Natl Acad Sci U S A 104:1424–1429
Koneri R, Balaraman R, Saraswati CD (2006) Antiovulatory and abortifacient potential of the ethanolic extract of roots of Momordica cymbalaria Fenzl in rats. Indian J Pharmacol 38:111–114
Konishi S, Izawa T, Lin SY, Ebana K, Fukuta Y, Sasaki T, Yano M (2006) An SNP caused loss of seed shattering during rice domestication. Science 312:1392–1396

Kour S, Bakshi P, Sharma A, Wali VK, Jasrotia A, Kumari S (2018) Strategies on conservation, improvement and utilization of underutilized fruit crops. Int J Curr Microbiol App Sci 7:638–650. https://doi.org/10.20546/ijcmas.2018.703.075

Kubo J, Lee JR, Kubo I (1999) Anti-helicobacter pylori agents from the cashew apple. J Agric Food Chem 47:533–537

Kumar A, Metwal M, Kaur S, Gupta AK, Puranik S, Singh S, Singh M, Gupta S, Babu BK, Sood S, Yadav R (2016) Nutraceutical value of finger millet [Eleusine coracana (L.) Gaertn.], and their improvement using omics approaches. Front. Plant Sci 7:934. https://doi.org/10.3389/fpls.2016.00934

Kumar A, Tomer V, Kaur A, Kumar V, Gupta K (2018) Millets: a solution to agrarian and nutritional challenges. Agric Food Secur 7(1):31. https://doi.org/10.1186/s40066-018-0183-3

Kuo CG, Chen HM, Sun HC (1992) Membrane thermostability and heat tolerance of vegetable leaves. In: Adaptation of food crops to temperature and water stress. AVRDC—The World Vegetable Center, Tainan, pp 160–168

Laurie S, Van Heerden S (2012) Consumer acceptability of four products made from beta-carotene-rich sweet potato. Afr J Food Sci 6:96–103

Lebot V (2009) Tropical root and tuber crops: cassava, sweet potato, yams and aroids. CABI, Oxfordshire

Leder I (2004) Sorghum and millets. Cultivated plants, primarily as food sources. In: Gyargy F (ed) Encyclopaedia of life support systems. UNESCO, Eolss Publishers, Oxford

Ledford H (2013) US regulation misses some GM crops. Nature 500:389–390

Lemmon ZH, Reem NT, Dalrymple J, Soyk S, Swartwood KE, Rodriguez-Leal D, van Eck J, Lippman ZB (2018) Rapid improvement of domestication traits in an orphan crop by genome editing. Nat Plants 4:766–770

Li X, Siddique KHM (2018) Future smart food—rediscovering hidden treasures of neglected and underutilized species for Zero Hunger in Asia. FAO, Bangkok. 242 pp. http://www.fao.org/documents/card/en/c/I8907EN/. Accessed 15 Sept 2019

Li C, Zhou A, Sang T (2006) Rice domestication by reducing shattering. Science 311:1936–1939

Li J-F et al (2013) Multiplex and homologous recombination-mediated genome editing in Arabidopsis and Nicotiana benthamiana using guide RNA and Cas9. Nat Biotechnol 31:688–691

Li T, Yang X, Yu Y, Si X, Zhai X, Zhang H, Dong W, Gao C, Xu C (2018) Domestication of wild tomato is accelerated by genome editing. Nat Biotechnol. https://doi.org/10.1038/nbt.4273

Li J et al (2019) Whole genome sequencing reveals rare off-target mutations and considerable inherent genetic or/and somaclonal variations in CRISPR/Cas9-edited cotton plants. Plant Biotechnol J 17:858–868

Liang Z et al (2017) Efficient DNA-free genome editing of bread wheat using CRISPR/Cas9 ribonucleoprotein complexes. Nat Commun 8:14261

Liebman M, Helmers MJ, Schulte LA, Chase CA (2013) Using biodiversity to link agricultural productivity with environmental quality: results from three field experiments in Iowa. Renew Agric Food Syst 28:115–128

Lim TK (2015) Arracacia xanthorrhiza. In: Edible medicinal and non medicinal plants. Springer, Dordrecht, pp 361–366

Lin BB (2011) Resilience in agriculture through crop diversification: adaptive management for environmental change. Bioscience 61(3):183–193. https://doi.org/10.1525/bio.2011.61.3.4

Lin Z, Li X, Shannon LM, Yeh CT, Wang ML, Bai G, Peng Z, Li J, Trick HN, Clemente TE et al (2012) Parallel domestication of the Shattering1 genes in cereals. Nat Genet 44:720–724

Lippman ZB et al (2008) The making of a compound inflorescence in tomato and related nightshades. PLoS Biol 6:e288

Liu J, Van Eck J, Cong B, Tanksley SD (2002) A new class of regulatory genes underlying the cause of pear-shaped tomato fruit. Proc Natl Acad Sci U S A 99:13302–13306

Liu YW, Shang HF, Wang CK, Hsu FL, Hou WC (2007) Immunomodulatory activity of dioscorin, the storage protein of yam (*Dioscorea alata* cv. Tainong No. 1) tuber. Food Chem Toxicol 45 (11):2312–2318

Lobell DB, Gourdji SM (2012) The influence of climate change on global crop productivity. Plant Physiol 160:1686–1697

Low JW, Arimond M, Osman N, Cunguara B, Zano F, Tschirley D (2007) A food-based approach introducing orange-fleshed sweet potatoes increased vitamin A intake and serum retinol concentrations in young children in rural Mozambique. J Nutr 2007(137):1320–1327

Lowe K, Wu E, Wang N, Hoerster G, Hastings C, Cho M-J, Scelonge C, Lenderts B, Chamberlin M, Cushatt J, Wang L, Ryan L, Khan T, Chow-Yiu J, Hua W, Yu M, Banh J, Bao Z, Brink K, Igo E, Rudrappa B, Shamseer PM, Bruce W, Newman L, Shen B, Zheng P, Bidney D, Falco C, Register J, Zhao Z-Y, Xu D, Jones T, Gordon-Kamm W (2016) Morphogenic regulators *Baby boom* and *Wuschel* improve monocot transformation. Plant Cell 28 (9):1998–2015

Mabhaudhi T, O'Reilly P, Walker S, Mwale S (2016) Opportunities for underutilised crops in Southern Africa's post–2015 development agenda. Sustainability 8:302

Mabhaudhi T, Chimonyo VGP, Chibarabada TP, Modi AT (2017a) Developing a roadmap for improving neglected and underutilized crops: a case study of South Africa. Front Plant Sci 8:2143. https://doi.org/10.3389/fpls.2017.02143

Mabhaudhi T, Chimonyo VGP, Modi AT (2017b) Status of underutilised crops in South Africa: opportunities for developing research capacity. Sustainability 9:1569. https://doi.org/10.3390/su9091569

Magbagbeola JAO, Adetoso JA, Owolabi OA (2010) Neglected and underutilized species (NUS): a panacea for community focused development to poverty alleviation/poverty reduction in Nigeria. J Econ Int Finance 2:208–211

Maitra S (2020) Potential horizon of brown-top millet cultivation in drylands: a review. Crop Res 55 (1–2):57–63. https://doi.org/10.31830/2454-1761.2020.012

Makarova KS, Haft DH, Barrangou R, Brouns SJJ, Charpentier E, Horvath P, Moineau S, Mojica FJM, Wolf YI, Yakunin AF, van der Oost J, Koonin EV (2011) Evolution and classification of the CRISPR–Cas systems. Nat Rev Microbiol 9:467–477

Makate C, Makate M, Mango N (2017) Smallholder farmers' perceptions on climate change and the use of sustainable agricultural practices in the Chinyanja Triangle, Southern Africa. Soc Sci 6 (1):30. https://doi.org/10.3390/socsci6010030

Mal B (2007) Neglected and underutilized crop genetic resources for sustainable agriculture. Indian J Plant Genet Resour 20(1):1–4

Mal B, Padulosi S, Ravi SB (2010) Minor millets in South Asia: Learnings from IFAD-NUS project in India and Nepal, Biodiversity International, Rome, Italy. The M.S. Swaminathan Research Foundation, Chennai

Malalavidhane S, Wickramasinghe SM, Jansz ER (2001) An aqueous extract of the of the green leafy vegetable Ipomoea aquatica is as effective as the oral hypoglycaemic drug tolbutamide in reducing the blood sugar levels of wistar rats. Phytother Res 15:635–637

Malkanthi SHP (2017) Importance of underutilized crops in Thanamalwila Divisional Secretariat Division in Monaragala District in Sri Lanka. J Agric Sci 12(3):197–206

Matsuoka Y, Vigouroux Y, Goodman MM, Sanchez JG, Buckler E, Doebley J (2002) A single domestication for maize shown by multilocus microsatellite genotyping. Proc Natl Acad Sci U S A 99:6080–6084

Mayes S, Massawe FJ, Alderson PG, Roberts JA, Azam-Ali SN, Hermann M (2012) The potential for underutilized crops to improve security of food production. J Exp Bot 63(3):1075–1079

Mbow C, Rosenzweig C, Barioni LG, Benton TG, Herrero M, Krishnapillai M, Liwenga E, Pradhan P, Rivera-Ferre MG, Sapkota T, Tubiello FN, Xu Y (2019) Food security. In: Shukla PR, Skea J, Calvo Buendia E, Masson-Delmotte V, Pörtner HO, Roberts DC, Zhai P, Slade R, Connors S, van Diemen R, Ferrat M, Haughey E, Luz S, Neogi S, Pathak M, Petzold J, Portugal Pereira J, Vyas P, Huntley E, Kissick K, Belkacemi M, Malley J (eds) Climate Change and

Land: an IPCC special report on climate change, desertification, land degradation, sustainable land management, food security, and greenhouse gas fluxes in terrestrial ecosystems. In Press. https://www.ipcc.ch/site/assets/uploads/2019/11/08_Chapter-5.pdf. Accessed 26 Aug 2020

McCouch S (2013) Feeding the future. Nature 2013(499):23–24

Meyer RS, Purugganan MD (2013) Evolution of crop species: genetics of domestication and diversification. Nat Rev Genet 14:840–852

Michaelraj PSJ, Shanmugam A (2013) A study on millets based cultivation and consumption in India. Int J Market Finance Manage 2(4):49–58

Modi AT, Mabhaudhi T (2013) Water use and drought tolerance of selected traditional and indigenous crops; Final report of Water Research Commission Project K5/1771//4; WRC Report No. 1771/1/13, ISBN 978-1-4312-0434-2. Water Research Commission, Pretoria

Mondal M, Garai S, Banerjee H (2020) Smart practices and adaptive technologies for climate resilient agriculture. In: Maitra S, Pramanick B (eds) Advanced agriculture, 1st edn. New Delhi Publishers, New Delhi, pp 3–35

Morris JB (2007) Swordbean (Canavalia ensiformis (L.) DC.) genetic resources regenerated for potential medical, nutraceutical and agricultural traits. Genet Resour Crop Evol 54:585–592

Morton JF (1987) Fruits of warm climates. Julia F. Morton, Miami

Motsa N (2015) Agronomic and physiological approaches to improving productivity of selected sweet potato (*Ipomoea batatas* l.) cultivars in KwaZulu–Natal: a focus on drought tolerance. PhD thesis, University of KwaZulu-Natal, Pietermaritzburg, South Africa

Mukhopadhyay S, Sen H, Jana P (1990) Effect of planting materials on growth and yield of sweet potato. J Root Crops 1990(16):119–122

Mukhopadhyay SK, Chattopadhyay A, Chakraborty I, Bhattacharya I (2011) Crops that feed the world 5. Sweetpotato. Sweetpotatoes for income and food security. Food Secur 3(3):283

Müller O, Krawinkel M (2005) Malnutrition and health in developing countries. CMAJ 173 (3):279–286. https://doi.org/10.1503/cmaj.050342

Murphy K, Matanguihan J (2015) Quinoa: improvement and sustainable production. Wiley, Hoboken

Muthamilarasan M, Singh NK, Prasad M (2019) Multi-omics approaches for strategic improvement of stress tolerance in underutilized crop species: a climate change perspective. Adv Genet 103:1–38

Nair RM, Schafleitner R, Kenyon L, Srinivasan R, Easdown W, Ebert AW, Hanson P (2012) Genetic improvement of mungbean. SABRAO J Breed Genet 44:177–190

Nalawadi UG, Jayasheela N (1975) Progress Hortic 7:37–38

Nekrasov V, Staskawicz B, Weigel D, Jones JDG, Kamoun S (2013) Targeted mutagenesis in the model plant Nicotiana benthamiana using Cas9 RNA-guided endonuclease. Nat Biotechnol 31:691–693

Newton AC, Johnson SN, Gregory PJ (2011) Implications of climate change for diseases, crop yields and food security. Euphytica 179:3–18

Njeru EM (2013) Crop diversification: a potential strategy to mitigate food insecurity by smallholders in sub-Saharan Africa. J Agric Food Syst Community Dev 3(4):63–69. https://doi.org/10.5304/jafscd.2013.034.006

Ochatt S, Jain SM (2007) Breeding of neglected and under-utilized crops, spices and herbs. Science Publishers Inc., Enfield

Oduori C (2005) The importance and research status of finger millet in Africa. Presented at the McKnight Foundation Collaborative Crop Research Program workshop on Tef & Finger Millet: comparative genomics of the Chloridoid cereals at the Biosciences for East and Central Africa (BECA), Nairobi, Kenya, 28–30 June 2005

Olson M (2013) The home page of the plant family Moringaceae. http://www.mobot.org/gradstudents/olson/moringahome.html. Accessed 21 Nov 2013

Osei A, Pandey P, Spiro D, Nielson J, Shrestha R, Talukder Z, Quinn V, Haselow N (2010) Household food insecurity and nutritional status of children aged 6 to 23 months in Kailali district of Nepal. Food Nut Bull 31:483–494

Padulosi S (2006) Finger millet. Crop diversity, priority crops. The global crop diversity trust. http://www.croptrust.org/main/priority.php

Padulosi S, Eyzaquirre P, Hodgkin T (1999) Challenges and strategies in promoting conservation and use of neglected and underutilized crop species. In: Alexandria VA (ed) Perspectives on new crops and new uses. ASHS Press, Alexandria, pp 140–145

Padulosi S, Irmgard H, Bordoni P (2008) Minor crops and underutilized species: lessons and prospects. In: Maxted N, Ford-Lloyd BV, Kell SP, Iriondo JM, Dulloo ME, Turok J (eds) Crop wild relatives, conservation and use. CAB International, Wallingford

Padulosi S, Mal B, Bala Ravi S, Gowda J, Gowda KTK, Shanthakumar G, Yenagi N, Dutta M (2009) Food security and climate change: role of plant genetic resources of minor millets. Indian J Plant Genet Resour 22:1–16

Padulosi S, Amaya K, Jäger M, Gotor E, Rojas W, Valdivia RA (2014) Holistic approach to enhance the use of neglected and underutilized species: the case of Andean grains in Bolivia and Peru. Sustainability 6:1283–1312

Padulosi S, Mal B, King OI, Gotor E (2015) Minor millets as a central element for sustainably enhanced incomes, empowerment, and nutrition in rural India. Sustainability 7:8904–8933

Padulosi S, Cawthorn DM, Meldrum G, Flore R, Halloran A, Mattei F (2019a) Leveraging neglected and underutilized plant, fungi, and animal species for more nutrition sensitive and sustainable food systems. Encyclopaedia Food Secur Sustain 3:361–370

Padulosi S, Roy P, May FJR (2019b) Supporting nutrition sensitive agriculture through neglected and underutilized species—operational framework. Bioversity International and IFAD, Rome

Palai JB, Jena J, Maitra S (2019) Prospects of underutilized food legumes in sustaining pulse needs in India—a review. Crop Res 54:82–88. https://doi.org/10.31830/2454-1761.2019.014

Pan C, Ye L, Qin L, Liu X, He Y, Wang J et al (2016) CRISPR/Cas9-mediated efficient and heritable targeted mutagenesis in tomato plants in the first and later generations. Sci Rep 6:1–9. https://doi.org/10.1038/srep24765

Pareek OP, Sharma S (2009a) Underutilized fruits and nuts, vol 1. Diversity and utilization and fruits of tropical and temperate region. Aavishkar Publishers and Distributors, p 366

Pareek OP, Sharma S (2009b) Underutilized fruits and nuts, vol 2. Fruits of tropical region. Aavishkar Publishers and Distributors, p 342

Park SJ, Jiang K, Tal L, Yichie Y, Gar O, Zamir D, Eshed Y, Lippman ZB (2014) Optimization of crop productivity in tomato using induced mutations in the florigen pathway. Nat Genet 46:1337–1342

Patricio HG, Palada MC, Ebert AW (2012) Adaptability and horticultural characterization of Moringa accessions under Central Philippines conditions. In: Holmer R, Linwattana G, Nath P, Keatinge JDH (eds) High value vegetables in Southeast Asia: production, supply and demand; proceedings of the SEAVEG 2012 regional symposium. AVRDC—The World Vegetable Center, Shanhua, pp 61–70

Pedersen B, Kalinowski LS, Eggum BO (1987) The nutritive value of amaranth grain (*Amaranthus caudatus*). Plant Foods Hum Nutr 36(4):309–324

Pnueli L et al (1998) The SELF-PRUNING gene of tomato regulates vegetative to reproductive switching of sympodial meristems and is the ortholog of CEN and TFL1. Development 125:1979–1989

Polprasid P (1996) Moringa oleifera Lamk. In: Siemonsma JS, Piluek K (eds) PROSEA—plant resources of South-East Asia No. 8: vegetables. Prosea Foundation, Bogor, pp 213–215

Popoola J, Ojuederie O, Omonhinmin C, Adegbite A (2019) Neglected and underutilized legume crops: improvement and future prospects. In: Shah F, Khan Z, Iqbal A, Turan M, Olgun M (eds) Recent advances in grain crops research. IntechOpen. https://doi.org/10.5772/intechopen.87069

Pudasaini R, Sthapit S, Suwal R, Sthapit B (2013) The role of integrated home gardens and local, neglected and underutilized plant species in food security in Nepal and meeting the millennium development goal 1 (MDG). In: Diversifying food and diets: using agricultural biodiversity to improve nutrition and health, pp 242–256

Purugganan MD, Fuller DQ (2009) The nature of selection during plant domestication. Nature 457:843–848

Rahmatullah M, Noman A, Hossan MS, Rashid MH, Rahman T, Chowdhury MH, Jahan R (2009) A survey of medicinal plants in two areas of Dinajpur district, Bangladesh including plants which can be used as functional foods. Am Eurasian J Sustain Agric 3(4):862–876

Rai M, Pandey S, Ram D, Rai N, Pandey AK, Yadav DS (2007) Plant genetic resources of legumes and underutilized vegetable crops in India. Acta Hort 752:225–230

Ramachandran C, Peter KV, Gopalakrishnan PK (1980) Drumstick (*Moringa oleifera*): a multipurpose Indian vegetable. Econ Bot 34:276–283

Ramsay L, Comadran J, Druka A, Marshall DF, Thomas WT, Macaulay M, MacKenzie K, Simpson C, Fuller J, Bonar N et al (2011) INTERMEDIUM-C, a modifier of lateral spikelet fertility in barley, is an ortholog of the maize domestication gene TEOSINTE BRANCHED 1. Nat Genet 43:169–172

Ransom EI, Elder LK (2003) Nutrition of women and adolescent girls: why it matters. Population Reference Bureau, Washington, DC

Rathore SK, Bhatt SH, Dhyani S, Jain A (2012) Preliminary phytochemical screening of medicinal plant Ziziphus mauritiana Lam. fruits. Int J Curr Pharm Res 4(3):160–162

Ravi SB, Hrideek TK, Kumar AK, Prabhakaran TR, Mal B, Padulosi S (2010) Mobilizing neglected and underutilized crops to strengthen food security and alleviate poverty in India. Indian J Plant Genet Resour 1(23):110–116

Reddy RVSK, Subbaiah YPV, Reddy MGDM (2007) Exploit kaasara kaya (Momordica tuberosa) for diversification of vegetables. Acta Hort 752:577–579

Rodríguez JP, Rahman H, Thushar S, Singh RK (2020) Healthy and resilient cereals and pseudo-cereals for marginal agriculture: molecular advances for improving nutrient bioavailability. Front Genet 11:49. https://doi.org/10.3389/fgene.2020.00049

Rodrıguez-Leal D, Lemmon ZH, Man J, Bartlett ME, Lippman ZB (2017) Engineering quantitative trait variation for crop improvement by genome editing. Cell 171:470–480

Ronen G, Carmel-Goren L, Zamir D, Hirschberg J (2000) An alternative pathway to β-carotene formation in plant chromoplasts discovered by map-based cloning of beta and old-gold color mutations in tomato. Proc Natl Acad Sci U S A 97:11102–11107

Ross-Ibarra J, Morrell PL, Gaut BS (2007) Plant domestication, a unique opportunity to identify the genetic basis of adaptation. Proc Natl Acad Sci U S A 104:8641–8648

Rothan C, Diouf I, Causse M (2019) Trait discovery and editing in tomato. Plant J 97:73–90

Ruiz KB, Biondi S, Oses R, Acuña-Rodríguez IS, Antognoni F, Martinez-Mosqueira EA, Coulibaly A, Canahua-Murillo A, Pinto M, Zurita-Silva A, Bazile D (2014) Quinoa biodiversity and sustainability for food security under climate change. A review. Agron Sustain Dev 34 (2):349–359

Saleh ASM, Zhang Q, Chen J, Shen Q (2013) Millet grains: nutritional quality, processing, and potential health benefits. Compr Rev Food Sci Food Saf 12:281–295

Scheben A, Wolter F, Batley J, Puchta H, Edwards D (2017) Towards CRISPR/Cas crops—bringing together genomics and genome editing. New Phytol 216:682–698

Schindele A, Dorn A, Puchta H (2020) CRISPR/Cas brings plant biology and breeding into the fast lane. Curr Opin Biotechnol 61:7–14

Schmidt M, Lam NT, Hoanh MT, Padulosi S (2010) Promoting neglected and underutilized tuberous plant species in Vietnam. In: Haas R, Canavari M, Slee B, Tong C, Anurugsa B (eds) Looking east looking west: organic and quality food marketing in Asia and Europe. Wageningen Academic Publishers, Wageningen, pp 183–194

Sebetha ET, Ayodele VI, Kutu FR, Mariga IK (2010) Yields and protein content of two cowpea varieties grown under different production practices in Limpopo province. S Afr J Biotechnol 9:628–634

Sedbrook JC, Phippen WB, Marks MD (2014) New approaches to facilitate rapid domestication of a wild plant to an oilseed crop: example pennycress (Thlaspi arvense L.). Plant Sci 227:122–132

Shah F, Wu W (2019) Soil and crop management strategies to ensure higher crop productivity within sustainable environments. Sustainability 11:1485. https://doi.org/10.3390/su11051485

Shan Q et al (2013) Targeted genome modification of crop plants using a CRISPR-Cas system. Nat Biotechnol 31:686–688

Shanmugasundaram S, Keatinge JDH, Hughes JA (2009) The mungbean transformation: diversifying crops, defeating malnutrition. In: Proven successes in agricultural development. A technical compendium to millions fed. IFPRI discussion paper 00922. International Food Policy Research Institute (IFPRI), Washington

Shobana S, Usha Kumari SR, Malleshi NG, Ali SZ (2007) Glycemic response of rice, wheat and finger millet based diabetic food formulations in normoglycemic subjects. Int J Food Sci Nutr 58 (5):363–372

Shomura A, Izawa T, Ebana K, Ebitani T, Kanegae H, Konishi S, Yano M (2008) Deletion in a gene associated with grain size increased yields during rice domestication. Nat Genet 40:1023–1028

Si L, Chen J, Huang X, Gong H, Luo J, Hou Q, Zhou T, Lu T, Zhu J, Shangguan Y et al (2016) OsSPL13 controls grain size in cultivated rice. Nat Genet 48:447–456

Sigmon B, Vollbrecht E (2010) Evidence of selection at the ramosa1 locus during maize domestication. Mol Ecol 19:1296–1311

Sillitoe P, Marzano M (2009) Future of indigenous knowledge research in development. Futures 41:13–23

Silva Ramos MÁ (2002) Elrol de los cultivos marginales en la seguridad alimentaria. Instituto Interamericano de Cooperación para la Agricultura,Oficina IICA, Bolivia, La Paz, Bolivia

Simons KJ, Fellers JP, Trick HN, Zhang Z, Tai YS, Gill BS, Faris JD (2006) Molecular characterization of the major wheat domestication gene Q. Genetics 172:547–555

Singh SP, Mishra BK, Chandel KPS, Pant KC (1980) Major food constituents of rice bean (Vigna umbellata). J Food Sci Technol 17:23

Siqueira MVBM (2011) Yam: a neglected and underutilized crop in Brazil. Hortic Bras 29:16–20. https://doi.org/10.1590/S0102-05362011000100003

Smykal P, Nelson MN, Berger JD, von Wettberg EJB (2018) The impact of genetic changes during crop domestication. Agronomy 8:119

Solh ME (2016) Importance, challenges and potential of neglected and underutilized crops. FAO Regional Initiative on Zero Hunger. Regional consultation on scoping and mapping of neglected and underutilized crop species in Asia. FAO, Bangkok

Sosso D, Luo D, Li QB, Sasse J, Yang J, Gendrot G, Suzuki M, Koch KE, McCarty DR, Chourey PS et al (2015) Seed filling in domesticated maize and rice depends on SWEET-mediated hexose transport. Nat Genet 47:1489–1493

Stamp P, Messmer R, Walter A (2012) Competitive underutilized crops will depend on the state funding of breeding programmes: an opinion on the example of Europe. Plant Breed 131:461–464

Sthapit B, Padulosi S, Mal B (2010) Role of on-farm/*in situ* conservation and underutilized crops in the wake of climate change. Indian J Plant Genet Resour 23:145–156

Sugimoto K, Takeuchi Y, Ebana K, Miyao A, Hirochika H, Hara N, Ishiyama K, Kobayashi M, Ban Y, Hattori T et al (2010) Molecular cloning of Sdr4, a regulator involved in seed dormancy and domestication of rice. Proc Natl Acad Sci U S A 107:5792–5797

Svitashev S, Schwartz C, Lenderts B, Young JK, Cigan AM (2016) Genome editing in maize directed by CRISPR–Cas9 ribonucleoprotein complexes. Nat Commun 7:13274

Sweeney MT, Thomson MJ, Pfeil BE, McCouch S (2006) Caught red-handed: Rc encodes a basic helix-loop-helix protein conditioning red pericarp in rice. Plant Cell 18:283–294

Tadele Z (2018) African orphan crops under abiotic stresses: challenges and opportunities. Scientifica (Cairo) 2018:1–19. https://doi.org/10.1155/2018/1451894

Taketa S, Amano S, Tsujino Y, Sato T, Saisho D, Kakeda K, Nomura M, Suzuki T, Matsumoto T, Sato K et al (2008) Barley grain with adhering hulls is controlled by an ERF family transcription factor gene regulating a lipid biosynthesis pathway. Proc Natl Acad Sci U S A 105:4062–4067

Tan L, Li X, Liu F, Sun X, Li C, Zhu Z, Fu Y, Cai H, Wang X, Xie D et al (2008) Control of a key transition from prostrate to erect growth in rice domestication. Nat Genet 40:1360–1364
Thies E (2000) Promising and underutilized species crops and breed. Eschborn, Germany. https://cgspace.cgiar.org/bitstream/handle/10568/3682/thies
Thompson J, Hodgkin T, Atta-Krah K, Jarvis D, Hoogendoorn C, Padulosi S (2007) Biodiversity in agroecosystems. In: Scherr SJ, McNeely JA (eds) Farming with nature: the science and practice of ecoagriculture. Island Press, New York, pp 46–60
Tian Z, Wang X, Lee R, Li Y, Specht JE, Nelson RL, McClean PE, Qiu L, Ma J (2010) Artificial selection for determinate growth habit in soybean. Proc Natl Acad Sci U S A 107:8563–8568
Tiwari RJ, Banafar RNS (1995) Studies on the nutritive constituents, yield and yield attributing characters in some ber (*Zizyphus jujuba*) genotypes. Indian J Plant Physiol 38(1995):88–89
Tiwari S, Kumar S (2013) Neglected oil crop biotechnology. In: Biotechnology of neglected and underutilized crops. Springer, Dordrecht, pp 117–171
Toda E et al (2019) An efficient DNA- and selectable-marker-free genome-editing system using zygotes in rice. Nat Plants 5:363–368
Triantaphylides C, Havaux M (2009) Singlet oxygen in plants: production, detoxification and signaling. Trends Plant Sci 14:219–228
Uarrota VG, de Bairros A d FM, Gindri D, Leolato LS, de Andrade GC, Nerling D, Stefen D, Arijama M, Razão EH, Rocha M (2019) From neglected and underutilized crops to powerful sources of vitamin A: three case studies of Mozambican cultivated Tacca leontopetaloides, cowpea, and cassava. In: Zepka LQ, de Rosso VV, Jacob-Lopes E (eds) Vitamin A. IntechOpen. https://doi.org/10.5772/intechopen.84829. https://www.intechopen.com/books/vitamin-a/from-neglected-and-underutilized-crops-to-powerful-sources-of-vitamin-a-three-case-studies-of-mozamb
Uawonggul N, Chaveerach A, Thammasirirak S, Arkaravichien T, Chuachan C, Daduang S (2006) Screening of plants acting against *Heterometrus laoticus* scorpion venom activity on fibroblast cell lysis. J Ethnopharmacol 103:201–207
Ueta R, Abe C, Watanabe T, Sugano SS, Ishihara R, Ezura H et al (2017) Rapid breeding of parthenocarpic tomato plants using CRISPR/Cas9. Sci Rep 7:1–8. https://doi.org/10.1038/s41598-017-00501-4
UN (2008) Achieving sustainable development and promoting development cooperation, dialogues at the economic and social council. Department of Economic and Social Affairs Office for ECOSOC Support and Coordination, United Nations Publications, New York, p 295. ISBN: 978-92-1-104587-1
van de Wouw M, Kik C, van Hintum T, van Treuren R, Visser B (2010) Genetic erosion in crops: concept, research results and challenges. Plant Genet Resour 8:1–15
Van Eck J (2018) Genome editing and plant transformation of solanaceous food crops. Curr Opin Biotechnol 49:35–41
Varshney RK, Nayak SN, May GD, Jackson SA (2009) Nextgeneration sequencing technologies and their implications for crop genetics and breeding. Trends Biotechnol 27:522–530
Vaughan DA, Balazs E, Heslop-Harrison JS (2007) From crop domestication to super-domestication. Ann Bot 100:893–901
Verdcourt B (1970) Studies in the Leguminosae-Papilionoïdeae for the "Flora of tropical East Africa": IV. Kew Bull 24:507–569
Vijaya Bhaskar AV, Nithya DJ, Raju S, Bhavani RV (2017) Establishing integrated agriculture-nutrition programmes to diversify household food and diets in rural India. Food Secur 9:981–999. https://doi.org/10.1007/s12571-017-0721-z
Virmani R, Virmani T, Singh C, Sorout G, Gupta J (2017) Hidden potential of natural herb Carissa carandas (Karonda). Res Pharm Health Sci 3:294–302
Wang R-L, Stec A, Hey J, Lukens L, Doebley J (1999) The limits of selection during maize domestication. Nature 398:236–239
Wang H, Nussbaum-Wagler T, Li B, Zhao Q, Vigouroux Y, Faller M, Bomblies K, Lukens L, Doebley JF (2005) The origin of the naked grains of maize. Nature 436:714–719

Wang E, Wang J, Zhu X, Hao W, Wang L, Li Q, Zhang L, He W, Lu B, Lin H et al (2008) Control of rice grain-filling and yield by a gene with a potential signature of domestication. Nat Genet 40:1370–1374

Wang M, Li W, Fang C, Xu F, Liu Y, Wang Z, Yang R, Zhang M, Liu S, Lu S et al (2018) Parallel selection on a dormancy gene during domestication of crops from multiple families. Nat Genet 50:1435–1441

Weaver CM, Dwyer J, Fulgoni VL, King JC, Leveille GA, MacDonald RS, Ordovas J, Schnakenberg D (2014) Processed foods: contributions to nutrition. Am J Clin Nutri 99 (6):1525–1542. https://doi.org/10.3945/ajcn.114.089284

Weinberger K (2003) Impact analysis of mungbean research in south and Southeast Asia. AVRDC—The World Vegetable Center, Tainan

Weinberger K (2007) Are indigenous vegetables underutilized crops? Some evidence from Eastern Africa and Southeast Asia. Acta Hortic 2007(752):29–34

Williams JT, Haq N (2002) Global research on underutilized crops. An assessment of current activities and proposals for enhanced cooperation. ICUC, Southampton

Woods D, Capcara J, Downey R (1991) The potential of mustard (Brassica juncea (L.) Coss) as an edible oil crop on the Canadian Prairies. Can J Plant Sci 71:195–198

Wu W, Yu Q, You L, Chen K, Tang H, Liu J (2018) Global cropping intensity gaps: increasing food production without cropland expansion. Land Use Policy 76:515–525. https://doi.org/10.1016/j.landusepol.2018.02.032

Xiao H, Jiang N, Schaffner E, Stockinger EJ, van der Knaap E (2008) A retrotransposon-mediated gene duplication underlies morphological variation of tomato fruit. Science 319:1527–1530

Xu C et al (2015) A cascade of arabinosyltransferases controls shoot meristem size in tomato. Nat Genet 47:784–792

Yamamoto T, Kashojiya S, Kamimura S, Kameyama T, Ariizumi T, Ezura H, Miura K (2018) Application and development of genome editing technologies to the Solanaceae plants. Plant Physiol Biochem 131:37–46

Yan L, Loukoianov A, Tranquilli G, Helguera M, Fahima T, Dubcovsky J (2003) Positional cloning of the wheat vernalization gene VRN1. Proc Natl Acad Sci U S A 100:6263–6268

Yan L, Loukoianov A, Blechl A, Tranquilli G, Ramakrishna W, SanMiguel P, Bennetzen JL, Echenique V, Dubcovsky J (2004) The wheat VRN2 gene is a flowering repressor downregulated by vernalization. Science 303:1640–1644

Yang RY, Keding GB (2009) Nutritional contributions of important African indigenous vegetables. African indigenous vegetables in urban agriculture. Earthscan, London, pp 105–144

Zhang F, Batley J (2020) Exploring the application of wild species for crop improvement in a changing climate. Curr Opin Plant Biol 56:218–222. https://doi.org/10.1016/j.pbi.2019.12.013

Zhang Y, Massel K, Godwin ID, Gao C (2018) Applications and potential of genome editing in crop improvement. Genome Biol 19:210

Zhao X, Lu X, Yin Z, Wang D, Wang J, Fan W, Wang S, Zhang T, Ye W (2017) Genome-wide identification and structural analysis of pyrophosphatase gene family in cotton. Genomics Mol Genet Biotechnol 56(4):1831–1840

Zhu BF, Si L, Wang Z, Zhou Y, Zhu J, Shangguan Y, Lu D, Fan D, Li C, Lin H et al (2011) Genetic control of a transition from black to straw-white seed hull in rice domestication. Plant Physiol 155:1301–1311

Zhu G, Wang S, Huang Z, Zhang S, Liao Q, Zhang C, Lin T, Qin M, Peng M, Yang C et al (2018) Rewiring of the fruit metabolome in tomato breeding. Cell 172:249–261.e12

Zsögön A, Cermak T, Naves ER, Notini MM, Edel KH, Weinl S, Freschi L, Voytas DF, Kudla J, Peres LEP (2018) De novo domestication of wild tomato using genome editing. Nat Biotechnol. https://doi.org/10.1038/nbt.4272

Neglected and Underutilized Crops: Present Status and Future Prospectus

2

Amir B. Wani, Biaza, and Irtiza

Abstract

Neglected and underutilized species (NUS) are ancient crop species indigenous to a community, which are usually wild or semi-domesticated. Such crops are often referred to as orphan crops. These crops are predominantly grown in their centre of origin, where they are still considered significant and used for the sustenance of the local population. NUS are usually grown under harsh and extreme agro-climatic conditions. Overtime, they have become more resistant and resilient to stresses, biotic as well as abiotic. Even though these crops may be less yielders as compared to major crops, they have been favourably relied upon whenever major staple crops have failed due to threats or catastrophe. Further, NUS also ensures elevated income for small-scale farmers. NUS plays major role in improving the food security; however, such crops are poorly/less documented and, hence, are neglected in the fields of research and conservation. Bioversity International is taking a considerable interest along with FAO in curbing years of monumental negligence of NUS and to smoothly roll these species out of the say. At the global level, many organizations have put their efforts forward to ensure the improvement and utilization of neglected and underutilized crops. As these crops are rich in essential amino acids, vitamins, etc. and thereby, such crops have a

A. B. Wani (✉)
Division of Plant Biotechnology, Sher-e-Kashmir University of Agricultural Science and Technology, Srinagar, Jammu and Kashmir, India

Biaza
Division of Fruit Science, Sher-e-Kashmir University of Agricultural Science and Technology, Srinagar, Jammu and Kashmir, India

Irtiza
Division of Basic Sciences and Humanities, Sher-e-Kashmir University of Agricultural Science and Technology, Srinagar, Jammu and Kashmir, India

S. M. Zargar et al. (eds.), *Neglected and Underutilized Crops - Towards Nutritional Security and Sustainability*, https://doi.org/10.1007/978-981-16-3876-3_2

characteristic of mitigating the 'hidden hunger' in the poorest communities. This chapter highlights the present status and prospectus of NUS around the world, with a brief outlook across countries, potential of these crops in mitigating hidden hunger and ensuring global food security and role of various organizations in conservation, preservation and elevating the status of these species.

Keywords

Neglected and underutilized crops · Food security · Hidden hunger · Medicinal values · Economic values

2.1 Introduction

Neglected and underutilized species (NUS) are wild or semi-domesticated species, predominantly grown by the local farmers in places where these crops are still considered significant for the sustenance. These crops are still being used within the local, national and even international communities. Even though neglected by the breeders and researchers, these crops have the potential to contribute to minimize the food security issues in the world. NUS crops are overlooked and respond with a biased tinge when it comes to their importance and utilization in research and extension services. Such species are least promoted or never promoted by the government, therefore, are neglected in the fields of research and conservation (IPGRI 2002; Mayes et al. 2012; Padulosi et al. 2013; FAO 2020).

Since ages these underutilized crops were more of importance but after the Green Revolution, the significance of these indigenous crops was lost and became neglected and underutilized. Ultimately such crops lost their value and were completely forgotten and least demanded, which is one of the possible reasons as to why many such species have now fallen into undermined eminence. This is more appropriately attributed to various economic, agronomic, genetic and cultural reasons. Various other reasons which profoundly led to the drift of various traditional crops into NUS category include prevalence of mono-cropping systems, mechanized cultivation, standardized agricultural practices, standardized marketing, changed food habits and incorporation of fewer species into dominant food systems at local, national and global level. All these trends were marked off by the start of 'Green revolution' and are still in continuance due to globalization of agricultural markets. NUS are also labelled as 'minor crops' or 'orphan crops' and the label is generally tagged to the indigenous species rather than the exotic or non-native crops. With 'neglected' clearly referring to the 'lack of attention', apparently reveals the position of these species, such species are rarely attended by the researchers, policymakers. Rather they are majorly cultivated and utilized by traditional farmers in the native regions. There are, although, many contributory factors that have played a role in leading to a possible neglect and consequent insignificant use of NUS, however, important amongst these factors are the lack of awareness about their nutritional and economic value. Thus, their potential remains underutilized and poorly exploited, which has resulted in their categorization under NUS. It is for this reason NUS are subjected to number of risks like genetic erosion, erosion of cultural diversity, loss of competitiveness in the market, sometimes such crops are abandoned by the farmers and ultimately are escalated to an endangered condition. It has also been of keen observance that such crops are mostly grown and tended by women in the traditional communities (IPGRI 2002; Dansi et al. 2012; Mayes et al. 2012; Padulosi et al. 2011, 2013, 2019; Chivenge et al. 2015).

However, the scenario is now changing to identify NUS as having economic potential as well as potential to reduce food and nutritional insecurity. The latter advantage of NUS is due to their considerable nutritional composition, adaptability to low input agricultural systems, survival ability in harsh environmental conditions and the essential role they play in resilience of food production system. The rich nutritional profile of NUS allows their use to fight the malnutrition problems observed in poor sections across the globe and to ensure wider and better nutrition availability to the increasing population. Slowly but steadily NUS are gaining back their significance as they have been found out to have greater reliability in curbing the global food insecurity. Breeders and researchers are now understanding their prominence and thus tagging their major role in missions to inhibit their complete loss from the biodiversity. There are several global organizations constantly involved in designing strategies to stabilize and improve the status of NUS as well as deeply indulged with its research and development, to name a few CFF, AOCC, ICUC, etc. Bioversity International is taking a considerable interest along with FAO to spread awareness through various means and ensure active participation of the breeders, farmers, extension workers, government, etc. in curbing negligence of NUS and to smoothly roll these species out of the say. Nevertheless, the process is not as easy as it seems, as there are a number of obstacles and constraints that have generated over the years, which are to be focused upon. The steps would include policymaking, increasing NUS's cultural acceptability, creating awareness amongst the farmers and consumers, improving genetic management of NUS, optimizing their marketability (Mayes et al. 2012; Chivenge et al. 2015).

2.2 Present Status of Neglected and Underutilized Species

Green Revolution is considered to be the most important event in modernizing the agricultural system to an exceptional level. Since green revolution, some primary major crops like rice, wheat, barley, etc. became a prime focus, while rest of the local crops were started to being neglected. This has ultimately led to their decrease at an alarming rate, the depletion and narrowing of gene pool has followed thereby.

The government customarily provides subsidies for major crops, which is a pure stance of political discount corresponding to corrosion in the development of NUS. Although there have been enormous campaigning activities and prominent studies by researchers over the importance of preservation of NUS, nevertheless, efforts being put forward are still bleak, but optimistically not the least. There are several organizations currently promoting conservation of NUS. Such organizations are slowly but actively elevating the status of NUS from a down trodden state to a new prominent one. Some organizations working day in and day out are referral in the chapter along with few of the participatory work.

2.2.1 Status of NUS Around the World

Around 100,000 crop species have been frequently needed for food, fibres, industrial, cultural and medicinal purposes. More than 7000 cultivated crops are still being used globally. Due to the increased population from last few centuries, around 30 crops around the world have become important and commonly grown. These crops now form the backbone of the world's agriculture, leading to the decrease in cultivation and use of NUS. As per IPGRI the management (both ex situ and in situ) of the genetic diversity of NUS is in awful conditions: moreover, 80% of these 'minor species' preserved in gene banks globally are pictured by just 1–10 accessions. In the Second Report of the State of the World's Plant Genetic Resources for Food and Agriculture, SoWPGR-2, by FAO in 2010, Bioversity International studied the status of Neglected and Underutilized Crops from around the world. It was found out in the analysis that a medium to high importance is accorded to NUS in most of the countries of Latin America, sub-Saharan Africa, South Asia and East Asia, while 15% of the countries considered NUS of low importance. Asian Countries shown to have increased use of NUS are Nepal, India, Malaysia and Philippines. Whereas countries such as Indonesia, Sri Lanka, Papua New Guinea show a moderate trend in usage of NUS (Padulosi et al. 2013).

2.2.2 Global Scenario of NUS

To get a better understanding about the status of NUS around the world, here is the list of some of such crops, which underlie in different categories, viz. fruits, vegetables, spices and condiments, roots and tubers, etc. Examples of some NUS from around the world are represented in Table 2.1.

Bambara Groundnut (*Vigna subterranean* L.), one of the neglected and underutilized crop is grown primarily in sub-Saharan Africa as a traditional diet, rich in proteins, carbohydrates, fats, calcium, phosphorus, iron, vitamins and amino acids. Being highly nutritious, it is also considered as complete food. Besides Bambara Groundnut has the capability to mitigate the hunger and thus secure the food population of Africa. The seeds are having potential therapeutic properties. It has the ability to fix atmospheric nitrogen and the legume is highly drought tolerant. Despite the vast range of contribution in human and environmental health, Bambara groundnut has been ironically neglected in modern research fields. Also, the research is mainly confined to testing to evaluate the potential yield and other characters related to yield or the objectives are mainly confined on characterization of landraces.

Table 2.1 NUS grown in different parts of the world

Species	Common name
Artocarpus altilis	Breadfruit
Adansonia digitata	Baobab
Bertholletia excelsa	Brazil nut
Physalis peruviana	Cape gooseberry
Ziziphus mauritiana	Jujube
Artocarpus heterophyllus	Jackfruit
Brosimum alicastrum	Maya nut
Annona cherimola	Cherimoya
Punica granatum	Pomegranate
Couepia longipendula	Egg nut
Hylocereus sp.	Dragon fruit
Duguetia lepidota	Yara Yara
Pachira aquatica	Malabar chestnut
Myrciaria dubia	Camu camu
Eugenia stipitate	Araza
Solanum quitoense	Naranjilla
Annona sp.	Annona
Morinda citrifolia	Noni
Sclerocarya birrea	Marula
Lansium domesticum	Duku
Synsepalum dulcificum	Miracle fruit
Lucuma obovate	Lúcuma
Eugenia uniflora	Pitanga
Dacryodes edulis	Safou
Salacca sp.	Salak
Carissa edulis	Carissa
Acacia tortilis	Acacia
Dovyalis caffra	Kei apple
Vanguera spp	Medlars
Celosia argentea	Celosia
Thrichosantes cucumerina var. anguina	Snake gourd
Luffa acutangula	Angle gourd
Amaranthus sp.	Leaf amaranth
Brassica rapa	Brassica
Psophocarpus tetragonolobus	Winged bean
Basella rubra	Ceylon spinach
Vernonia amygdalina	Bitter leaf
Moringa oleifera	Moringa
Parkia biglobosa	Locust bean
Solanum aethiopicum	African eggplant
Coccinia grandis	Ivy gourd
Cynara cardunculus	Cardoon
Portulaca oleracea	Purslane

(continued)

Table 2.1 (continued)

Species	Common name
Sechium edule	Chayote
Irvingia sp.	Dika
Tylosema esculentum	Marama
Solanum nigrum	Black nightshade
Crambe sp.	Crambe
Gnetum africanum	Eru
Scolymus hispanicus	Golden thistle
Cyrtosperma merkusii	Taro
Gundelia tournefortii	Akoub
Momordica charantia	Bitter gourd
Vigna angularis	Adzuki bean
Vigna umbellate	Rice bean
Cajanus cajan	Pigeon pea
Vigna subterranean	Bambara groundnut
Lathyrus sativus	Grasspea
Sphenostylis stenocarpa	African yam bean
Canavalia ensiformis	Jack bean
Lupinus mutabilis	Lupin
Macrotyloma geocarpum	Kersting's groundnut
Lablab purpureus	Lablab
Vigna radiate	Mung bean
Fagopyrum sp.	Buckwheat
Eleusine coracana	Finger millet
Setaria italic	Foxtail millet
Panicum miliaceum	Proso millet
Coix lacryma-jobi	Job's tears
Digitaria exilis	Fonio
Panicum sumatrense	Little millet
Chenopodium pallidicaule	Cañihua
Amaranthus caudatus	Amaranth
Triticum spelta	Spelt
Eragrostis tef	Tef
Triticum monococcum	Einkorn
Triticum dicoccon	Emmer
Xanthosoma sagittifolium	Cocoyam
Dioscorea sp.	Yams
Ullucus tuberosus	Ulluco
Pachyrhizus sp.	American yam bean
Colocasia esculenta	Taro
Lepidium meyenii	Maca
Oxalis tuberose	Oca
Pastinaca sativa	Parsnip
Arracacia xanthorrhiza	Arracacha

(continued)

Table 2.1 (continued)

Species	Common name
Smallanthus sonchifolius	Yacon
Brassica juncea	Mustard seed
Nigella sativa	Nigella
Trigonella foenum-graecum	Fenugreek
Polygonum odoratum	Polygonum
Hippophae rhamnoides Le	Sea buckthorn
Pandanus amaryllifolius	Pandan
Diplotaxis sp.	Rocket
Antidesma venosum	Antidesma
Cola lateritia	Monkey cola
Capparis spinosa	Caper
Lippia carviodora	Uer
Bixa Orellana	Annatto
Fadogia ancylantha	Makoni

Padulosi et al. (2013)

Peach palm (*Bactris gasipaes*) is a palm tree endemic to South and Central America, it has been in use for centuries, but is a limited exported crop. With many health benefits, because of its many bioactive compounds, it is used in treating diabetes (due to the low glycemic index), preventing macular degeneration (due to the high content of vitamin A) and increasing the high-density lipoprotein cholesterol (Jimenez et al. 2012; Carvalho et al. 2013; Graefe et al. 2013; Espinosa-Pardo et al. 2014; Yumbya et al. 2014).

Nance or wild cherry (*Byrsonima crassifolia*) is grown mainly in Central and South America in tropical and sub-tropical climates. The fruits are taken as raw, or cooked, and used to make a beer-like drink known as chichi (Orwa et al. 2009). It also has some medicinal properties like treating skin infections and gastrointestinal disorders, used as bactericide, fungicide or antipyretic, also, acts as an antidote for snakebites (Caceres et al. 1991, 1993; Bejar and Malone 1993; Martinez-Vazquez et al. 1999).

Sesamum indicum, an edible oilseed, is mainly used for salad oil. Its seeds are used as condiments, while as its vegetative parts are used as fodder, also used in medicines and cosmetics. Another legume crop Lathyrus is mostly used as fodder in Turkey, while as in South Asia it is mostly used as a food, also now being used as a new crop for forage, feed and green manure in Australia (Williams and Haq 2002).

Date palms and bitter gourds substantially help in iron uptake in diets (Arora 2014). Seaweed provides micronutrients, proteins, fibre and is commonly used in the Asian-Pacific region (examples are given in Cardozo et al. 2007; El Gamal 2010). *Gnetum africanum* cultivars known as Okazi, Ukazi or Afang in different parts of Nigeria, are having hard and bitter leaves, used for medicinal properties by the Efik people in South-eastern Nigeria (Ali et al. 2011; Ingram et al. 2012). Hulled wheat (*Triticum monococcum*, *T. dicoccum*) is used for making biscuits and pasta cultivated in Italy and now being also cultivated by many countries including Australia. Similarly, the spicy vegetable rocket (*Erucasativa* and *Diplotaxis species*) grown mostly in whole Mediterranean is now being cultivated and used in Italy. Roselle (*Hibiscus sabdariffa*) mainly grown in Sub-Sahara Africa for many decades is now being used as beverage in Europe.

2.2.3 Status of NUS in Europe

The importance of NUS in some of the countries is not clear, which includes Europe as well. As compared to other countries, Europe shows a decreasing trend in usage and conservation of NUS. Amongst the European countries, NUS are used for domestic and commercial purposes as well as in improvement programmes in Ukraine and has been reported to be used for multiple purposes in Romania. While NUS is primarily used for medicinal purpose and aromatic use in Poland and for food in Germany. In Georgia, such crops receive importance based on the use and their potential to generate the income (Padulosi et al. 2013).

2.2.4 Status in Sub-Saharan Africa (SSA)

NUS that are generally found in the cropping systems of sub-Saharan Africa are Sweet potatoes (*Ipomoea batata*), taro (*Colocasia esculenta*), wild melon (*Cucurbita spp.*), wild watermelon (*Citrullus lanatus* L.), bambara groundnut (*Vigna subterranea*), wild mustard (*Brassica* spp.), Cowpea (*Vigna unguiculata*), *Amaranthus* spp. and various other leafy vegetables, which are edible (Chivenge et al. 2015). The South Africa's Miombo ecosystem is a shelter and a land of about 200 species of fruits, out of which about 167 species are edible. The species like *Adansonia digitata, Uapaca kirkiana, Parinari curatellifolia, Strychnos cocculoides,* which are native to Miombo ecosystem represent a diverse nutritional status as they are found to be rich in proteins, carbohydrates, sugars, oils, vitamins, etc., thus contribute to an important part of dietary systems of humans. Miombo woodlands comprise a vast range of the tropical grasslands, shrub lands, savannas, which covers much of the southern and the central Africa and the ecosystem is named after the oak like Miombo trees, *Brachustegia* spp. (Saka and Msonthi 1994; Kwesiga et al. 2000; World Wide fund for Nature, WWF 2020).

In Ivory Coast, the mostly utilized NUS include the groups of legumes, minor leafy vegetables and vegetable crops, owing to their immensely rich nutritional content. The use of NUS in SSA is mostly found in Ghana, Uganda, Malawi, Nigeria, Zimbabwe, while East and West Africa also finds utilization of NUS for various purposes. For instance, in Ghana, sword beans (*Canavalia ensiformis*) are commonly cultivated NUS for food, whereas the minor forest and timber species are collected and utilized for developmental purposes like construction. In Malawi, the cropping system mostly notices Bambara groundnut, pearl millets and sorghum, in Nigeria NUS, such as, Bambara groundnut, Ethiopian teff (*Eragrostis tef*) originating from Ethiopia and Eritrea, pigeon peas (*Cajanus cajan*), lima beans (*Phaseolus lunatus*), fonio (*Digitaria exilis*), *Solanum nigrum*, *Crotalaria spp.* finds significant use. NUS are also used as a mean of income generation in West Africa apart from the use in medicinal and ornamental purposes.

Uganda shows a neglecting trend in research and development programmes of NUS and lack of documentation in identifying the NUS varieties; however, local communities are consistently involved in conservation and utilization of various neglected crop species. There are the local groups, which are preparing and creating public awareness about the values and potential of NUS in human and environmental health. CONVOL is an American NGO, which especially comprehends promotion of commercial cultivation of Shea butter (*Vitellaria paradoxa* subsp. *nilotica*) in the northern, north-western and north-eastern districts of Ghana, plus lays the importance on women employment for a number of tasks such as processing and fruit gathering (Padulosi et al. 2013).

2.2.5 Status of NUS in West Asia

In most of the West Asian countries, the government is recognizing the dire need to conserve, use and improve neglected and underutilized crops in response to the opportunities and potentials allowed by such species. The government is thus executing a number of policies and activities for the protection and promotion of such crop species. The attention is being laid upon various aspects and advantages served by minor medicinal plants, having the status of NUS, in combating human health problems. Also, the government is recognizing the capability of other NUS in adding to surplus income of the farmers along with fighting against the environmental deterioration. In Turkey, NUS have profound importance and are recognized as having the potential of providing for food security and are also part of healthy and balanced diet. A very little use of NUS has been evident in Uzbekistan, moderate in Yemen, while the information about the utilization of NUS in rest of the West Asian countries is quite diminished.

2.2.6 Status of NUS in Asia

Due to varying climatic conditions in Asia humongous diversity of plant genetic resources is found, but their potential is not utilized in present cropping systems. As per Arora (2014), there are about 778 NUS in the Asian-Pacific region that include 261 fruits, 213 vegetables, 55 root/tuber types, 34 nuts, 28 pseudocereals and millets, 14 grain legumes/pulses, 25 industrial crops and 148 others.

2.3 China, Cambodia, Thailand, Vietnam and Southeast Asia

China being an extensive agriculture country yet manages and preserves immense range of NUS crops. The most encouraging NUS are buckwheat (*Fagopyrum esculentum*), followed by fleece flower root (*Fallopia multiflora*), mulberry tree (*Morus alba*) and cashew (*Anacardium occidentale).* Albeit having no definitive policy in dealing with the utilization and protection of NUS, few policies are there for the protection NUS in some minority regions.

Cambodia's agriculture and food supply are grossly dependant on rice. This makes virtually rest of the crops as NUS in Cambodia. Some of the important NUS in Cambodia are lemon grass (*Cymbopogon citratus*), different varieties (yellow, sugar and green) of banana (Musa sp.), mung bean (*Vigna radiata*), sweet potato (*Ipomoea batatas*), taro (*Colocasia esculenta*), groundnut (*Arachis hypogaea*), ginger (*Zingiber officinale*) and Durian (*Durio zibethinus*). There is no policy on NUS protection in Cambodia. Only the *Kbalkoh* vegetable research station manages to produce good quality seeds of some NUS vegetables. Vietnam is one of the countries in the world with the highest agro-biodiversity. Rice being the dominant crop protection and cultivation of minor crops such as NUS is of less importance for government. Few NUS crops are as litchi (*Litchi chinensis*), mustard (*Brassica juncea, Brassica campestris*), edible canna (*Canna edulis*), taro (*Colocasia esculenta*) and lablab (*Lablab purpureus*).

Majority of people in Thailand practise farming. Farming is traditional crop-based with only few major crops being mostly cultivated, such as sugarcane, cassava, maize, rice and rubber. Some important NUS crops grown there are as kenaf (*Hibiscus sabdariffa*), safflower (*Carthamus tinctorius*), rice bean (*Vigna umbellata*) and a second kenaf species (*Hibiscus cannabinus*). With no research policy for NUS crops, Thailand mostly imports NUS crops from abroad (Polthanee 2001; Schmidt et al. 2008).

2.3.1 India

The lack of proper documentation limits the development and creation of policies for NUS in India. However, Indian Agricultural Research Institute, New Delhi (IARI) initiated the recognition of underutilized crops in 1960. Later in 1982, the research was extended by All India Co-ordinated Research Project (AICRP) on underutilized crop plants (UUP). The research aims at conservation, collection, evaluation and utilization of UUC. The work was later on extended to various parts of India. Also, Ministry of Agriculture has launched a national project for the purpose of carrying out an extensive research on UUC. However, the threat to underutilized crops remains persistent in India, even though there are many ongoing activities aiming at promotion of NUS, but the activities are still lacking the proportion, ultimately, we may lose some of the important species to disappearance. Table 2.2 provides details of some commonly grown NUS.

Table 2.2 List of few of the neglected and underutilized species in India

Crop	Common name	References
Aegle marmelos	Bael, bel, wood apple	Kumari et al. (2011)
Averrhoa carambola	Carambola, star fruit	Kapoor (1990)
Cordia sp.	Indian cherry, Gonda	Patil et al. (2010)
Eleusine sp.	Finger millet	Ravi (2010)
Grewia subinaequalis	Phalsa	Ali and Rab (2000)
Mushrooms	Fungi	Tibuhwa (2013)
Rhododendron sp.	Buraansh	Scott (2012)
Simarouba glauca	Paradise tree	Joshi and Joshi (2008)
Pennisetum sp.	Pearl millet	Ravi (2010)
Syzygium cumini	Jamun, black plum	Ved (1991)
Tamarindus indica	Tamarind	Maiti et al. (2004)
Ziziphus mauritiana	Ber, Indian jujube	Kumari et al. (2011)

2.3.2 Sri Lanka and Bangladesh

Sri Lanka and Bangladesh being dominant agricultural countries of South Asia with huge agricultural biodiversity despite small size. More than 60 NUS fruit crops are grown in Sri Lanka, with most importantly *Limonia acidissima* L., *Aegle marmelos* (L.) Correa, *Annona muricata* L. and *Tamarindus indica* L. Bangladesh grows most of underutilized vegetables in the homestead, fallow and forest areas with no policy for protection of these NUS (Hossain and Hasan 2018; Ratnayake et al. 2020).

2.3.3 Pakistan and Nepal

Pakistan and Nepal also mainly being agricultural countries with enormous biodiversity potential. Some of NUS crop species grown in these countries are presented in Table 2.3.

2.4 Global Research Organizations and Events Promoting NUS

At global level, many organizations and events have put their efforts forward to ensure the improvement and utilization of neglected and underutilized crops. Some of the names are renowned (mentioned further in the chapter) and have contributed a lot in the development of NUS and are still working upon the objectives. Some of the names are:

The Genetic Resources Policy Committee of the Consultative Group on International Agricultural Research (CGIAR) 1971, International Centre for Underutilized Crops (ICUC) 1988, International Symposia on New Crops, 1992 and 1996, The FAO Global Plan for Action for Plant Genetic Resources for Food and Agriculture 1996, Global Facilitation Unit for Underutilized crops (GFU) 2002, Crops for Future

Table 2.3 NUS grown in Himalayas of Pakistan and Nepal

Crop	Common name
Amaranthus caudatus	Amaranthus
Hordeum vulgare	Naked barley
Vigna radiate	Black gram
Macrotyloma uniflorum	Horse crop
Amorphophallus campanulatus	Olarum
Dioscorea spp.	Yam
Brassica juncea	Rayo
Sesamum indicum	Sesame
Guizotia abyssinica	Niger
Myrica esculenta	Kaphal
Aesandra butyracea	Chiuri
Phyllanthus emblica	Amla
Citrus maxima	Pumello
Syzygium cumini	Jamun

Adhikari et al. (2017)

(CFF) 2008, The African Orphan Crop Consortium (AOCC) 2011, etc. Global Forum for Agricultural Research's Commodity Chain agenda also gives a high priority to underutilized crops (UUC). US National Academy of Sciences (NAS) in 1975 issued a report on 'Underutilized Tropical Plants with Promising Economic Value'. The issuance of report was prompted in response to the repeated threats of global food shortage. FAO has issued several reports as FAO 1988. IPGRI issued 24 monographs on individual species, which covers a diversity of crops and is issuing a further of 8 monographs on individual species (Williams and Haq 2002).

2.5 The Genetic Resources Policy Committee of the Consultative Group on International Agricultural Research (CGIAR)

CGIAR was founded in the year 1971, with its headquarters in Montpellier, France, with a global partnership to unite the international organizations, which are involved in research about food security. CGIAR works on the purpose of Nutrition, Partnership, Leadership and concisely aims at enhancing the food security, human health, abate poverty and sustainably manage the natural resources. With over 50 years of experience and an expanded presence in about 70 countries, CGIAR culminates deeper understanding of customs, market operations and traditions. CGIAR has a built-up of more than 7,770,000 germplasm accessions. The 15 research centres of CGIAR are autonomous, non-profit centres with a goal of conducting useful and innovative research, welcoming and favouring the approach of CGIAR to design a better and food and health secure world for the poor. These research centres have facilitated and organized many important events for betterment of UUC (Thakur 2014).

2.6 International Centre for Underutilized Crops (ICUC)

Established in 1988, ICUC is an independent, non-profit making research centre, mainly focused on research, training, development and use of UUC for food, industrial and medicinal purposes and for improvement of environmental health. The range of research and technology incorporates many disciplines such as agriculture, agroforestry and processing (CTA 1991). The organization has recently expanded its objectives towards including processing and marketing assessments of UUC. With professional networks in over 21 countries as UTVAPNET (Underutilized Tropical Vegetables for Asia and the Pacific Network), UTFANET (Underutilized Tropical Fruits in Asia Network), ACUC (Asian Centre for Underutilized Crops), etc., the organization provides know-how and supports researches on national priorities for germplasm collection of UUC and training events and the conferences associated with it (Thakur 2014).

2.7 Global Facilitation Unit for Underutilized Species (GFU)

Established in 2002, under Global Forum on Agriculture Research (GFAR), hosted by IPGRI, Rome, Italy, which is currently its headquarter, it is a multi-institutional initiative aiming at encouraging and facilitating the work on and use of underutilized species at different levels by agencies, networks and organizations, around the world. The mission has some pre-fixed goals of enabling an environment for discussing strategies for facilitating sustainable use of UUC, providing the stakeholders an access to financial resources and information and developing such an independent mechanism to encourage work on UUC.

2.8 Crops for the Future (CFF)

CFF, known as crops for the future is an independent global organization, located in Malaysia, Seminyeh, aiming at ensuring, facilitating and promoting research, development, conservation and use of neglected and underutilized crop species for securing the agricultural, traditional and dietary diversity. CFF was found in 2008, it works with its partners (ICUC and GFU) to advocate its aim to diversify NUS. CFF is funded by the Government of Malaysia and the University of Nottingham for its research, also the funds are collectively generated from its various competitive contracts which include UK Newton Fund, the EU, FAO. Crops for the Future Centre (CFRC) is the research centre of CFF focusing on research and development of underutilized crops for non-food and food uses. CFF has proved to be quite a successful foundation in vitalizing its aims and objectives; the recent achievements include the database development, which consists of agronomic, phenological and nutritional information about various underutilized crops.

CFF initiated the research work with Bambara Groundnut as model crops for developing improved genotypes. The improved genotypes were derived from its wild types and landraces. University of Nottingham in collaboration with the international partners studied its physiology and identified two limiting traits, which hindered their wide adoption. The traits studied were its short-day photoperiod requirement for pod-filling and the second trait as its hard-to-cook nature. The short-day photoperiod restricts the timing of planting and the hard-to-cook trait aggravates the fuel demands. CFF upon studying the given limited traits, which are still under improvisation, has developed through controlled crossing, a range of lines with source of lower photoperiod sensitivity, and such lines are currently being tested in African countries and the testing shall as well be conducted in North Africa and Southern Europe (Brough and Azam-Ali 1992; Adeniji et al. 2008; Mongomaké et al. 2011; Hillocks et al. 2012; Kendabie et al. 2015; Mubaiwa et al. 2017; Chang et al. 2018; Gregory et al. 2019; Halimi et al. 2019). CFF is working on Winged Bean (*Psophocarpus tetragonolobus*) for its development and characterization in association with various seed companies and researchers. CFF currently possesses about 400 accessions of winged beans, the 200 of which they have secured from East-West Seeds breeding programme (Wong et al. 2017; Gregory et al. 2019).

There has been a lot of research and improvement of a number of NUS by CFF in various ways. CFF is readily indulged in using these NUS as ingredients in popular foods or using the crops as food and beverages or to prepare snacks and instant foods from them, just as Bambara groundnut, Bambara groundnut noodles, Moringa instant soup, Moringa and Bambara groundnut granola bar, Dragon fruit tortellini, etc. This is one of the popular ways of increasing the status of NUS by adding appeal to them, optimizing their marketability and demand amongst the consumers, by which these become known to the people and enhanced demand ultimately ensures enhanced production, improvement and research. CFF in association with African Orphan Crops Consortium published a draft genome of five of the Africa's underutilized crops. CFF has done a considerable work on various NUS and the work is still undergoing (Mayes et al. 2012; Chang et al. 2018; Marahaini et al. 2019; Gregory et al. 2019).

2.9 The African Orphan Crop Consortium (AOCC)

The African Orphan Crop Consortium (AOCC) is an international organization, which was established in the year 2011. The organization started with the aim of providing nutritional security to sub-Saharan Africa by means of curbing the malnutrition, food insecurity issues and concerns through improvement of locally adapted, traditional orphan crops characterized with high nutritional content to meet the standards. AOCC has tasked the sequencing of genome of 101 traditional, underutilized crops of Africa through a network of private, public, regional and international collaborations, to make them available to the public domain. Initiative was taken in order to facilitate its use by the breeders and hence lead to their genetic improvement. AOCC holds training programme for plant breeders of Africa, with teaching hosted by World Agroforestry Centre (ICRAF) to train them about the advanced breeding methods. At AOCC, scientists are actively involved in incorporating breeding techniques to improve the nutritional status as well as resilience of underutilized crops (www.cgiar.org/changing-worlds-food-systems-can-help-protect-planet/). For about 60 crops sequence data has been published, while 19 are still lying in progress and 6 have almost reached completion (Hendre et al. 2019; FAO 2020).

2.10 Potential and Future Prospects of NUS

More than 800 million people around the world particularly from the developing countries go through food and nutrition insecurity as reported by Food and Agriculture Organization (FAO) of the United Nations (Dansi et al. 2012). More than 50% of the world population is residing in cities and have less access for crop cultivation. An approach to address this challenging situation is to promote crop diversification and production, consumption and utilization of the NUS crops in the diets (Dansi et al. 2012; Padulosi et al. 2014). NUS are rich in nutrients and are having health benefits, mostly being used as food, fodder and for medicinal purposes for the local communities. With changing environmental conditions round the globe, an impact on natural resources is predicted. NUS adapt to the environmental conditions of their regions of origin. The researchers have now come to a deeper understanding of the fact that the reliance only upon fewer crops, as has been the result of green revolution, to feed the exponentially rising global population will never be an attribute of security and credibility. Therefore, the sustenance of this dynamic population would have to call upon the need for diversity of crops and species to be incorporated in our agricultural systems. The understanding led to further investigation and a glance of researchers towards potentially fit crops to name for the means of food security. Owing to the exceptionally huge capabilities and characteristics possessed by NUS, researchers have now realized their immense potential to account for the coming enhanced food needs. The nutritional richness and potential of many neglected and underutilized species has been underlined and their consequent advantage in ensuring food security and health of the poor has been well proven. There are now evidently many such NUS known, which are being identified as eminent elements in sustainable agriculture. These crops have a vast potential to be utilized to overcome food insecurity at local, national as well as global level. Neglected or underutilized crops play major roles in improving the food security by reducing the risk of over-dependence on some limited number of major crops. In turn, NUS may help poor in ascending their income and subsistence, contributing to quality food, conserving and cherishing dietary and cultural diversity. In addition, they help to increase sustainability of agriculture by means of reducing inputs, such as nitrogenous fertilizers derived from fossil fuels and fuel for agriculture, knowing the risks of the carbon footprint of agriculture on the climate change and the transition to a post peak-oil world.

Considering the fact that many NUS are hardy and are adapted to comparatively harsh environmental conditions, they possess greater resilience, resistance and/or tolerance to a number of biotic and abiotic stresses. Hence, NUS have been found to play a pivotal role in nature conservation and there are many NUS which are found to be resistant to drought, low water requirements, poor soil conditions, etc. Owing to their strong adapting potential, NUS such as finger millet (*Eleusine coracana*), Bambara groundnut (*Vigna subterranean*), fonio (*Digitaria exilis*), etc. may account for an active choice in times like now, when climate changes are quite evident. Such climate changes have become a major reason for increased risk of today's world to natural calamities and catastrophe like tsunamis, floods, famines, etc. Such natural threats and disasters jeopardize the existence of population rendering them food-less, which is further aggravated as we have been relying more and more upon fewer crops. Even though these may yield less as compared to major crops, but favourably have been relied upon whenever major staple crops have failed due to disease epidemics or catastrophe. NUS have consequently proved to provide an outright safety from economic and food crisis following the emergencies and disasters. Another important potential of NUS is that it ensures elevated income for small-scale farmers. In India, a threefold enhanced income could be generated by adding value to little millet, which has also led to employment generation, especially for women, thereupon, boosting their independence and social status. One of the specific attributes of NUS is the dietary diversity, such crops have high levels of micronutrients, macronutrients, also are rich in essential amino acids, vitamins, etc., and thereby, such crops have a characteristic of mitigating the 'hidden hunger' and nutritional deficiencies in the poorest communities. NUS can help in curbing the problems of malnutrition amongst the poor people and serve as a cheap and balanced diet which they can easily afford and thus enjoy the relishes of a healthy body. For example, the use of leaves from *Moringa oleifera,* rich in proteins could be a healthy substitute for animal protein scarce diets. Moringa leaf powder is added to paste of soybean or peanut, which forms a supplemental food, rich in energy and is more appropriately known as ready to use food (RUF), the supplement helps to treat acute malnutrition. Moringa has already reached the status of an economically important crop in India, the Philippines, Ethiopia and Sudan. Amaranth is a leafy vegetable, which is highly nutritious in both cooked as well as raw form, its nutritional status is as rich as spinach. Given the amount of nutrients, some NUS serve as 'Super foods', for instance; goji berry (*Lycium* sp.), chia seeds (*Salvia hispanica*), açaíberry (*Euterpe oleracea*) being highly nutritious foods and are really appealable to the health-conscious consumers. Such NUS share an ascended consumer preference plus a high market demand. There are other NUS gaining recognition and demand being harvested in the countries, where such crops were not known or grown previously like quinoa, which is becoming a global food commodity, Salep tea made from the tubers of *Orchis sp,* which is an orchid (IPGRI 2002; Jilcott et al. 2010; Vijayalakshmi et al. 2010; Ebert et al. 2011; Padulosi et al. 2011, 2013, 2019; Mayes et al. 2012; Johannes 2015; Gregory et al. 2019).

2.11 Conclusion

NUS, with their versatility to extreme climatic conditions, resistance to various biotic and abiotic stresses, low input requirements, are very much effective for curbing food insecurity, high adaptation to agro-ecological niches, also rich in carbohydrates and fats. They have a hidden potential to encourage small-holder farmers and rural communities by bettering their earnings. NUS are also known to encourage food and nutritional security. Making adequate use of these NUS crops, however, will engage making them more commercially competitive with improved 'modern' varieties. An understanding of morpho-anatomical responses would commit undoubtedly towards breeding for stress tolerance and more developed varieties of these NUS. The demand is for a unified push across all levels in all aspects, ranging from basic genetics, use of crop modelling technique in the interpretation of agronomic field data, to more investment in continuing research, breeding programs and upgraded seed supply of these NUS crops in order to safeguard their competitiveness in the marketplace.

References

Adeniji OT, Peter JM, Bake I (2008) Variation and interrelationships for pod and seed yield characters in Bambara groundnut (*Vigna subterranea*) in Adamawa state, Nigeria. Afr J Agric Soc Res 3:617–621

Adhikari L, Hussain A, Rasul G (2017) Tapping the potential of neglected and underutilized food crops for sustainable nutrition security in the mountains of Pakistan and Nepal. Sustainability 9:29

Ali R, Rab F (2000) Research needs and new products development from underutilized tropical fruits. Acta Hort (ISHS) 518:241–248. http://www.actahort.org/books/518/518_33.htm

Ali F, Assanta MA, Robert C (2011) *Gnetum africanum*: a wild food plant from the African forest with many nutritional and medicinal properties. J Med Food 14:1289–1297

Arora RK (2014) Diversity in underutilized plant species—an Asia-Pacific perspective. Bioversity International, New Delhi, India, p 203

Bejar E, Malone MH (1993) Pharmacological and chemical-screening of *Byrsonima crassifolia*, a medicinal tree from Mexico. J Ethnopharmacol 39:141–158

Brough SH, Azam-Ali SN (1992) The effect of soil moisture on proximate composition of Bambara groundnut (*Vigna subterranean* L.). J Sci Food Agric 60:197–203

Caceres A, Lopez BR, Giron MA, Logemann H (1991) Plants used in Guatemala for the treatment of dermatophytic infections. 1. Screening for antimycotic activity of 44 plant extracts. J Ethnopharmacol 31:263–276

Caceres A, Lopez B, Juarez X, del Aguila J, Garcia S (1993) Plants used in Guatemala for the treatment of dermatophytic infections. 2. Evaluation of antifungal activity of seven American plants. J Ethnopharmacol 40:207–213

Cardozo KH, Guaratini T, Barros MP, Falcao VR, Tonon AP, Lopes NP, Campos S, Torres MA, Souza AO, Colepicolo P, Pinto E (2007) Metabolites from algae with economical impact. Comp Biochem Physiol C Toxicol Pharmacol 146:60–78

Carvalho RP, Lemos JR, de Aquino SRS, Martins MG, Nascimento CH, Bayona M, Marcon JL, Monteiro JB (2013) The consumption of red pupunha (*Bactris gasipaes* Kunth) increases HDL cholesterol and reduces weight gain of lactating and post-lactating wistar rats. J Aging Res Clin Pract 2:257–260

Chang Y, Liu H, Liu M, Liu M, Liao X, Sahu SK, Fu Y, Song B et al (2018) The draft genome of five agriculturally important African orphan crops. Giga Sci 8:1–16

Chivenge P, Mabhaudhi T, Modi AT, Mafongoya P (2015) The potential role of neglected and underutilised crop species as future crops under water scarce conditions in sub-Saharan Africa. Int J Environ Res Public Health 12:5685–5711

CTA (1991) ICUC—International Centre for Underutilized Crops Spore 36. CTA, Wageningen, The Netherlands

Dansi A, Vodouhe R, Azokpota P, Yedomonhan H, Assogba P, Adjatin A et al (2012, 2012) Diversity of the neglected and underutilized crop species of importance in Benin. Sci World J:1–19

Ebert AW, Wu TH, Wang ST (2011) International cooperators' guide—vegetable Amaranth (*Amaranthus* L.). AVRDC—The World Vegetable Center, Tainan, Taiwan, p 8

El Gamal AA (2010) Biological importance of marine algae. Saudi Pharm J 18:1–25

Espinosa-Pardo FA, Martinez J, Martinez-Correa HA (2014) Extraction of bioactive compounds from peach palm pulp (*Bactris gasipaes*) using supercritical CO2. J Supercrit Fluid 93:2–6

FAO (2020) Promoting neglected and underutilized crop species

Graefe S, Dufour D, van Zonneveld M, Rodriguez F, Gonzalez A (2013) Peach palm (*Bactris gasipaes*) in tropical Latin-America: implications for biodiversity conservation, natural resource management and human nutrition. Biodivers Conserv 22:269–300

Gregory PJ, Mayes S, Hui Hui C, Jahanshiri E, Julkifle A, Kuppusamy G, Wai Kua H, Lin TX, Massawe F, Suhari TASTM, Azam Ali SN (2019) Crops for the future (CFF): its research and role in the adoption of underutilised species. Planta. ISSN 00320935

Halimi RA, Barkla B, Mayes S, King G (2019) The potential of the underutilized pulse Bambara groundnut (*Vigna subterranean* L. Verdc.) for nutritional food security 177:47–59

Hendre PS, Muthemba S, Kariba R et al (2019) African Orphan Crops Consortium (AOCC): status of developing genomic resources for African orphan crops. Planta 250:989–1003

Hillocks RJ, Bennett C, Mponda OM (2012) Bambara nut: a review of utilisation, market potential and crop improvement. Afr Crop Sci J 20:1–16

Hossain MA, Hasan SS (2018) Potentiality of underutilized vegetables for contribution to sustainable development goals (SDGs) in Bangladesh. Asian J Agric Extension Econ Sociol 26(2):1–9

Ingram V, Ndumbe LN, Ewane ME (2012) Small scale, high value: *Gnetum africanum* and buch holzianum value chains in Cameroon. Small-Scale Forest 11:539–556

IPGRI (International Plant Genetic Resources Institute) (2002) Neglected and underutilized plant species: strategic action plan of the International Plant Genetic Resources Institute. International Plant Genetic Resources Institute, Rome, Italy

Jilcott SB, Ickes SB, Ammerman AS, Myhre JA (2010) Iterative design, implementation and evaluation of a supplemental feeding program for underweight children ages 6–59 months in Western Uganda. Matern Child Health J 14:299–306

Jimenez G, Gomez G, Perez AM, Blanco-Metzler A (2012) Estimation of glycaemic index of peach palm (*Bactris gasipaes*) cooked fruits and chips, and pitahaya (Hylocereus spp.) pulp. Arch Latinoam Nutr 62:242–248

Johannes MME (2015) The importance of neglected and underutilized species for sustainable agriculture, their conservation and their contribution to food security. Conference paper

Joshi S, Joshi S (2008) *Simarouba glauca* D.C. (paradise tree) (brochure)—ideas. https://ideas.repec.org/p/ags/uasbcp/43624.html

Kapoor LD (1990) CRC handbook of Ayurvedic medicinal plants. CRC Press, Boca Raton, FL. 58 pp. ISBN 0-8493-0559-4

Kendabie P, Massawe F, Mayes S (2015) Developing genetic mapping resources from landrace-derived genotypes that differ for photoperiod sensitivity in Bambara groundnut (*Vigna subterranean* L.). Asp Appl Biol 124:1–8

Kumari P, Joshi GC, Tiwari LM (2011) Diversity and status of ethno-medicinal plants of Almora district in Uttarakhand, India. Int J Biodivers Conserv 3(7):298–326

Kwesiga F, Akinnifesi F K, Ramadhani T, Kadzere I, Saka J (2000) Domestication of indigenous fruit trees of the Miombo in southern Africa. In: Proceedings of a SADC Tree Seed Centre Network technical meeting, Windhoek, Namibia, 14 March 2000

Maiti R, Jana D, Das UK, Gosh H (2004) Antidiabetic effect of aqueous extract of seed of *Tamarindus indica* in streptozotocin-induced diabetic rats. J Ethnopharmacol 92(1):85–91

Marahaini N, Jahanshiri E, Salama A, Tengku-Adhwa-Syaherah TMS (2019) Linking data across the value chain of underutilised crops—a multidisciplinary approach. Food Res 3:108–116

Martinez-Vazquez M, Gonzalez-Esquinca AR, Cazares-Luna L, Moreno Gutierrez MN, Garcia-Argaez AN (1999) Antimicrobial activity of *Byrsonima crassifolia* (l.). J Ethnopharmacol 66:79–82

Mayes S, Massawe FJ, Alderson PG, Roberts JA, Azam-Ali SN, Hermann M (2012) The potential for underutilized crops to improve security of food production. J Exp Bot 63(3):1075–1079

Mongomaké K, Alistair GP, Yaya T (2011) Bambara groundnut [*Vigna subterranea* (L.) Verdc. (Fabaceae)] usage in human health. Nuts Seeds Health Dis Prev 22:189–196

Mubaiwa J, Fogliano V, Chidewe C, Linnemann AR (2017) Hard-to-cook phenomenon in bambara groundnut (*Vigna subterranean* (L.) Verdc.) processing: options to improve its role in providing food security. Food Rev Int 33:167–194

Orwa C, Mutua A, Kindt R, Jamnadass R, Anthony S (2009) Agroforestry database: a tree reference and selection guide version 4.0

Padulosi S, Heywood V, Hunter D, Jarvis A (2011) Underutilized species and climate change: current status and outlook. In: Yadav Shyam S, Redden RJ, Hatfield JL (eds) Crop adaptation to climate change, pp 507–521

Padulosi S, Thompson J, Rudebjer P (2013) Fighting poverty, hunger and malnutrition with neglected and underutilized species needs, challenges and way forward. Bioversity International, Rome

Padulosi S, Amaya K, Jäger M, Gotor E, Rojas W, Valdivia R (2014) A holistic approach to enhance the use of neglected and underutilized species: the case of Andean grains in Bolivia and Peru. Sustainability 6(3):1283–1312

Padulosi S, Cawthornb D, Meldruma G, Florec R, Hallor A, Mattei F (2019) Leveraging neglected and underutilized plant, fungi, and animal species for more nutrition sensitive and sustainable food systems

Patil DN, Kulkarni AR, Shahapurkar AA, Patil BS (2010) Release of drug neomycin from *Cordia dichotoma* transdermal film. Int J Pharma Bio Sci 2:1–5

Polthanee A (2001) Indigenous farming practices and knowledge in north-east Thailand. Khon Kaen Printing Limited, Khon Kaen

Ratnayake SS, Kumar L, Kariyawasam CS (2020) Neglected and underutilized fruit species in Sri Lanka: prioritisation and understanding the potential distribution under climate change. Agronomy 10(34):1–19

Ravi BS (2010) Neglected millets that save the poor from starvation. LEISA India (1):34–36

Saka JK, Msonthi JD (1994) Nutritional value of edible fruits of indigenous wild trees in Malawi. Forest Ecol Manage 64:245–248

Schmidt M, Wei W, Polthanee A, Zhou H (2008) Ambiguity in a trans-disciplinary stakeholder assessment of neglected and underutilized species in China, Cambodia, Thailand and Vietnam. Biodivers Conserv 17:1645–1666

Scott S (2012) Benefits of *Rhododendron* herb. www.livestrong.com/article/120355-benefits-rhododendronherb/

Thakur M (2014) Underutilized food crops: treasure for the future India. Food Sci Res J 5 (2):174–183

Tibuhwa DD (2013) Wild mushroom—an underutilized healthy food resource and income generator: experience from Tanzania rural areas. J Ethnobiol Ethnomed 9:49

Ved P (1991) Indian medicinal plant. In: Samant SS, Dhar U, Palni LMS (eds) Current status in Himalayan medicinal plants: potential and prospects. Gramodaya Prakashan, Nainital, pp 45–63

Vijayalakshmi D, Geetha K, Gowda J, Ravi B, Padulosi S, Bhag Mal S (2010) Empowerment of women farmers through value addition on minor millets genetic resources: a case study in Karnataka. Indian J Plant Genet Resour 23(1):132–135

Williams JT, Haq N (2002) Global research on underutilized crops. An assessment of current activities and proposals for enhanced cooperation. ICUC, Southampton, UK

Wong QN, Tanzi AS, Ho WK, Malla S, Blythe M, Karunaratne A, Massawe F, Mayes S (2017) Development of gene-based SSR markers in winged bean (*Psophocarpus tetragonolobus* (L.) DC.) for diversity assessment. Genes 8:100

World Wide fund for Nature, WWF (2020) Miombo, Africa. wwf.panda.org

Yumbya P, Ambuko J, Shibairo S, Owino W (2014) Effect of modified atmosphere packaging (map) on the shelf life and postharvest quality of purple passion fruit (*Passiflora edulis* sims). J Post-Harvest Technol 2:25–36

Germplasm Resources of Major Underutilized Crops

3

Surabhi Jasrotia and Romesh Kumar Salgotra

Abstract

The cultivation of a very limited number of crops does not fulfil the nutritional requirements of the growing population. For this, underutilized crops can play a vital role to meet food and nutrition security globally. They are of significant importance in localized areas, as they are highly adapted to marginal lands, do not require high inputs and is resilient to climate variability. The underutilized crops include food crops, such as cereals, vegetables, legumes, oil seeds, root, and tubers, mainly produced as a source of income for livehood of poor farmers in developing countries. Limited germplasm resource availability, lack of information on production, nutritional quality of many of the underutilized plant products, and the lack of improved quality material are the main constraints impacting the availability and productivity of these crops. Plant germplasm resources are the materials required for initiating any crop improvement programme with the help of advances in genomic techniques and various programmes for the improvement of major underutilized crops. At national and international level, genebanks are concerned with the collection, maintenance, ex situ conservation, regional and global germplasm exchange. The chapter highlights the potential of underutilized crops in context to nutrition, food security, and germplasm resources of major underutilized crops for further crop improvement studies.

Keywords

Germplasm resources · Underutilized crops · Food and nutrition security · Gene banks · Ex situ conservation

S. Jasrotia · R. K. Salgotra (✉)
School of Biotechnology, Sher-e-Kashmir University of Agricultural Sciences and Technology of Jammu, Srinagar, Jammu and Kashmir, India

S. M. Zargar et al. (eds.), *Neglected and Underutilized Crops - Towards Nutritional Security and Sustainability*, https://doi.org/10.1007/978-981-16-3876-3_3

3.1 Introduction

Increasing population pressure and fast depletion of natural resources necessitate diversification of present day agriculture in order to meet global food security (WCMC 1992; UNEP 1995). Plant biodiversity contributes as a source for food, feed, shelter and is important for ecosystem functioning, sustainable agricultural output, and achieving food and nutritional security (Thrupp 2000; Toledo and Burlingame 2006). Total 7000 plant species found useful in agriculture (Wilson 1992), only about 150 species are commonly used and out of which less than 39 plant species fulfils about 90 per cent of world's food requirement (Ravi et al. 2010). The primary cultivated crops—wheat, rice and maize, account for over half of the world's caloric needs, and a diet consisting solely of these foods fails to offer nutritional security to the malnourished. With the selection of these limited crops for cultivation, the United Nations objective of achieving zero hunger by 2030 will fail to achieve its target (Malik and Chaudhary 2018).

The surge of 'Green Revolution' from 1960 to 80, laid stress on narrowing of species and improvements to major staple crops in order to meet food requirements of the population. Furthermore, modern plant breeding efforts aided in the development and cultivation of high-yielding crop species, resulting in the increase of a limited number of crop species. This dominance of monoculture cultivation is the root cause of genetic erosion of biodiversity in the soil (pests, diseases, etc.). As a result, crop diversity in agricultural fields around the world is declining. In addition, the lack of genetic diversity within the gene pools of these few crops makes our agricultural systems vulnerable to pests, diseases, and abiotic stresses (Salgotra and Gupta 2015). Essential nutrients cannot be obtained just from a plentiful supply of few basic crops. The cultivation of underutilized crops must be stressed as they provide greater genetic biodiversity, which can help to increase food security (Chivenge et al. 2015).

The underutilized crops commonly called orphan, minor, forgotten crops include (pseudo) cereals, legumes, vegetables, oilseeds, roots and tubers, aromatic and medicinal plants, fruits and nuts, and others (Padulosi and Hoeschle-Zeledon 2004, 2008). They are mainly indigenous and traditional crops, whose distribution, biology, cultivation, and applications are poorly documented. They are recognized for their traditional applications in specific places and are sourced from the wild or produced in traditional production systems with little or no external inputs. They have a deep connection to their place of origin's cultural legacy, as they have been a vital part of cultural variety linked to traditions, culinary habits, and health practices (Mayes et al. 2012). Globally, traditional crops that offer greater genetic biodiversity have dropped in cultivation around the world, owing to a lack of interest from farmers, researchers, extension services, policy and decision makers, technology providers and consumers.

Underutilized crops have high nutritional benefits, as they are key source of calories, proteins, micronutrients, and fibres, thus they have a lot of potential to lessen the risk of over-reliance on a very limited number of crops. As they are highly nutritious and have therapeutic properties, they hold a good candidate for the 'hidden hunger' caused by micronutrient deficiencies amongst the poor (Salvi and Katewa 2016). They have enormous potential to assist the needy with subsistence farming as a source of income. In addition to superior nutritional qualities, majority of these crops do not require larger inputs, such as nitrogen fertilizers. Also, they can be produced on marginal lands and intercropped or rotated with staple crops with ease. They are more adaptable to climate stresses than major cereals and cash crops, because they are adapted to marginal conditions and many have the unique ability to tolerate or withstand stresses.

The criteria for identifying and evaluating underutilized species/crops include (1) local importance in consumption and production, (2) highly suited to marginal areas, (3) receive little focus from national agricultural and biodiversity conservation programmes on research and development, (4) largely represented by landraces, native diversity and dependent on local knowledge, and (5) poorly represented in ex situ (Williams and Haq 2002).

3.2 Categories of Underutilized Crops

Underutilized crops are known by different names as Abandoned crops, Alternative crops, Niche crops, Local crops, Minor crops, Traditional crops. Underdeveloped crops by Padulosi (2017), Disadvantaged crops (Massawe et al. 2015), Underused crops (EcoBusiness 2015), Forgotten crops and Promising crops by FAO (2017a, b), Future smart food (Li and Siddique 2020), Neglected crops (Bermejo and León 1994), and Orphan crops (AOCC 2018) are some of the terms used to describe these crops. These crops include cereals, fruits, legumes, vegetables, etc. (Padulosi et al. 2013) (Fig. 3.1).

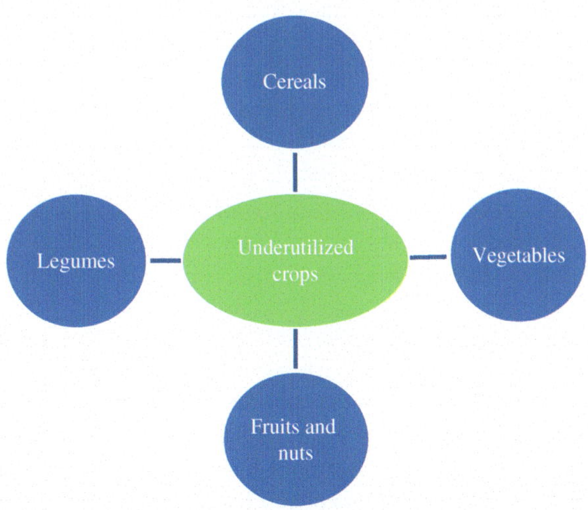

Fig. 3.1 Important underutilized crops (Padulosi et al. 2013)

3.2.1 Cereals

The cultivation of (peudo) cereals has been emphasized as a key priority as they are rich sources of nourishment for both humans and other animals. Pseudocereals claim their close relationship to the 'true cereals' by their nutritional content particularly in terms of carbohydrate composition. These crops have traditionally been utilized as cattle feed (Cheng 2018). Zhu (2020) discussed the importance of cultivating these neglected crops as a source of 'novel' starches for a variety of food and non-food applications. The underutilized cereals include small millets, such as finger millet (*Eleusine coracana*), barnyard millet (*Echinochloa crus-galli*), foxtail millet (*Setaria italica*), little millet (*Panicum sumatrense*), proso millet (*Panicum miliaceum*), and kodo millet (*Paspalum scrobiculatum*). Buckwheat (*Fagopyrum esculentum*), Grain amaranths (*Amaranthus* spp.), and quinoa (*Chenopodium quinoa*) are also included in this category (Bekkering and Tian 2019). These grains have a low glycaemic index and high nutritional content, making them attractive healthy foods. Besides teff is another small cereal, which is widely grown in the Horn of Africa due to its resistance to abiotic stresses, particularly to poorly drained soil. In addition, the lack of gluten in these crops, which is the cause for celiac disease, has boosted its demand in the recent years (Spaenij-Dekking et al. 2005). These crops also provide a range of health-related benefits such as oxidative stress reduction, prevention of cardiovascular diseases, inflammation, and diabetes (Mir et al. 2018).

3.2.2 Vegetables

Vegetables are a good source of income for farmers with small farming fields (Ashraf et al. 2018). There are various indigenous vegetables that have benefits in agronomic and nutritional traits (Table 3.1). For example, *Adansonia digitata* (Baobab), a multi-purpose tree, with leaves rich in iron and vitamin C enriched fruits (Sanchez 2018). The African spider flower plant, *Cleome gynandra*, eaten in rural Africa, is much higher in iron, zinc, and β-carotene. The World Vegetable Center, situated in Taiwan, includes six regional centres in Asia and Africa where it conducts research and promotes vegetables. In this centre, over 60,000 accessions of 440 species from 151 nations have been conserved, with 10,000 of them being traditional vegetables (AVRDC 2019).

3.2.3 Legumes

Legumes are important underutilized crops for soil improvement as they are able to fix atmospheric nitrogen. In Africa, bambara groundnut (*Vigna subterranean*) and cowpea (*Vigna unguiculata*) are widely grown. Horse gram (*Macrotyloma uniforum*) is mainly grown in Asia (FAOSTAT 2018). Grass pea (*Lathyrus sativus*), cultivated in Asia, Africa, and Europe, is highly resistant to drought. The crop is recognized as an insurance crop because of this trait, as it delivers consistent yields

Table 3.1 Some underutilized vegetables and leguminous crops with potential applications

Name of the crop (scientific name)	Common name	Importance	References
Solanum aethiopicum	African eggplant	Edible as source of nutrients and antioxidants	Gürbüz et al. (2018)
Parkia biglobosa	Locust bean	Edible as fermented food condiment. Making non-timber forest products. Leaves, bark, roots, and flowers have medicinal properties.	Houndonougbo et al. (2020)
Sechium edule	Chayote	Stems, leaves, and tubers as food and feed. Controls diabetes and obesity	Coronel et al. (2017)
Luffa acutangula	Angle gourd	Appetizer, low calorie diet, pulp used in making cuisines	Kandoliya et al. (2016), Shendge and Belemkar (2018)
Trichosanthes cucumerina	Snake gourd	Fresh root has anti-convulsant activity. Leaves and stems are used for bilious disorders and skin diseases	Devi (2017)
Cleome gynandra	Spider plant	Leaf rich in minerals like potassium, calcium, magnesium, phosphorus, etc.	Omondi et al. (2017)
Solanum nigrum	Black nightshade	Leaves and seeds are used as vegetables in soup. Fruit extract have anti-tumour properties	Akubugwo et al. (2007)
Celosia argentea	Celosia	Edible leaves, tender stems, and young flower spikes as soups, sauces, etc.	Fayaz et al. (2019)
Tylosema esculentum	Marama	Edible beans used as traditional medicines against diarrhoea. Root and bean extracts are effective against rota virus infection	Chingwaru et al. (2007, 2011)
Cyrtosperma chamissonis	Giant swamp taro	Edible leaves and starchy corms. Petiole yield a fibre for weaving	Plucknett (1976)
Scolymus hispanicus	Golden thistle	As dietary supplements. Roots as a coffee substitute and its flowers as a colouring alternative to saffron	Polo et al. (2009)
Basella rubra	Ceylon spinach	Leaves edible as boiled, fried or mix vegetable items, laxative, alleviate labour pain, blood purifier, bruise, ringworm, acne freckle, anti-pruritus, treats diarrhoea and inflammation	Singh et al. (2019c)
V. angularis	Adzuki bean	Rich source of carbohydrates, fats, proteins, minerals, antioxidants and is gluten free	Liu et al. (2017)
V. umbellate	Rice bean	High-protein content, essential amino acids, vitamins, minerals, and lower fat content	Katoch (2013)
Lupinus mutabilis	Lupin	High-protein content comparable to soybean, seed oil rich in polyunsaturated fatty acids	Abraham et al. (2019)

(continued)

Table 3.1 (continued)

Name of the crop (scientific name)	Common name	Importance	References
Canavalia ensiformis	Jack bean	As gelling agents and energy supplements in livestock feeds	Akande (2016)
Lathyrus sativus	Grass pea	Seeds edible as boiled, roasted, sauce, drink, snacks, and animal feed	Lambein et al. (2019)
Cajanus cajan	Pigeon pea	Source of aminoacids, proteins, vitamins, minerals, ten times more fat, five times more vitamin A, and three times more vitamin C than ordinary peas	Sharma et al. (2011a)
Macrotyloma geocarpum	Kersting's groundnut	High-protein content and richer in amino acids, such as lysine and methionine than most of the other legume crops	Ayenan and Ezin (2016)

when all other crops fail owing to significant moisture deficit. Cassava, cowpea, and arrowroot are crops of the future. Their use must be maximized to aid in the eradication of malnutrition caused by the deficiency of vitamin A in Mozambique, in Africa, and in rest of the world (Uarrota et al. 2019).

3.2.4 Roots and Tuber Crops

After potato, sweet potato, and cassava, yam (*Dioscorea* spp.) is considered as the fourth most significant root crop and it has nutritionally advantages over other root crops (Table 3.2). Yam is exclusively cultivated in Africa. Enset (*Ensete ventricosum*), commonly known as the 'false banana' because of its resemblance to the domesticated banana plant, is a staple food for more than 20 million Ethiopians living in densely populated areas. Unlike bananas, which have edible fruit, the pseudo-stem and the underground corm of the enset are edible. Vetiver grass (*Vetiveria zizanioides*) is used to stabilize soil, and is cultivated in contours along a slope because of its dense root structure, has an average tensile strength of 75 MPa, and contributes to ecological safety. In addition, vetiver is used for thatching, rope manufacturing, basket weaving, etc. and its roots contain a valuable oil that is used as a base for aftershaves in the perfume industry.

3.2.5 Fruits

Essential nutrients, such as, proteins, minerals, carbohydrates, dietary fibres, phytochemicals, and antioxidants can be found in underutilized fruits (Table 3.3). Baobab (*Adansonia digitata*), kei apple (*Dovyalis caffra*), and Marula (*Sclerocarya birrea*) are the most cultivable underutilized fruits in the developing nations and are

Table 3.2 Examples of underutilized roots and tubers and their importance

Name of the crop (scientific name)	Common name	Importance	References
Smallanthus sonchifolius	Yacon	Starchy fruit like roots contain fructooligosacharides, soluble dietary fibres that exhibit prebiotic activity, regulate glucose homeostasis	Caetano et al. (2016), Cao et al. (2018)
Colocasia esculenta	Taro	Tubers as rich source of carbohydrates, nutrients, phytate, and tannin	Rashmi et al. (2018)
Lepidium meyenii	Maca	As ayurvedic medicine in anaemia, memory, fertility, weak bones, and boost immunity	Li et al. (2017)
Oxalis tuberose	Oca	Roots as vegetable, dietary fibre, contain phenolic compounds and have antioxidant activity	Jimenez et al. (2015)
Xanthosoma sagittifolium	Cocoyam	All plant parts are edible as food/feed and also as antipoison agents	Boakye et al. (2018)
Myrciaria dubia	Camu camu	Fruit processed into juices and Vitamin C capsules	Langley et al. (2014)

considered to be healthy foods despite their limited genetic resources and study and their usage is yet to be realized. A variety of Myriad tropical fruits, such as *Annona* spp. (soursop, sweetsop, cherimoya, etc.), *Aegle marmelos* (Beli), *Feronia limonia* (Wood apple), and *Choerospondias axillaris* (Lapsi), can be processed into juices, jams, candy, etc. by small-scale entrepreneurs and sold in shops or supermarkets to generate income.

3.2.6 Oilseeds

Sesame (*Sesamum indicum* L.), an oilseed, cultivated on over ten million hectares of land each year, primarily in India, Sudan, Myanmar, and Tanzania (FAOSTAT 2018). In developing countries, the important oilseeds are castor bean (*Ricinus communis*), linseed (*Linum usitatissimum*), and noug (*Guizotia abyssinica*). Some plant species, such as *Artemisia annua,* have anti-malarial properties, and have been used in Chinese traditional medicine for over two millennia, particularly by the rural poor. As a result, it has grabbed much interest amongst research institutions in the West and international agencies such as WHO. Amaranth (*Amaranthus* spp.) has a significant cultural value in Peru, where women dancers often use the scarlet amaranth flower to paint their cheeks during carnival festivities.

Table 3.3 Some underutilized fruits, nuts species and their importance

Name of the crop (scientific name)	Common name	Importance	References
Brosimum alicastrum	Maya nut	Edible seed, caffeine free drink, nutrient dense, and free of gluten	Subiria-Cueto et al. (2019)
Artocarpus altilis	Breadfruit	Energy booster, rich in fibre, and vitamins	Badrie and Broomes (2010), Li et al. (2017)
Artocarpus heterophyllus	Jack fruit	Rich in starch, dietary fibres, and antioxidants and has anti-cancer properties	Ranasinghe et al. (2019)
Adansonia digitata L.	Baobab	Edible (leaves, seeds, fruit pulp). Non-food parts are used as timber and fodder	Gebauer et al. (2016), Rahul et al. (2015)
Ziziphus jujube	Jujuba	Medicinal plant used in Iranian folk medicine as antitussive, laxative agent, and blood pressure reducer. Powders of stem bark and leaves of jujube to cure wounds and oral wounds	Jin (2018)
Physalis peruviana	Cape gooseberry	Consumed as fruit, juices, salads, jams, snacks, oil, etc.	Singh et al. (2019b), Olivares-Tenorio et al. (2016)
Morinda citrifolia	Noni	Leaves, fruits, barks, and roots are used as noni juice, capsule powder, and tea	Ali et al. (2016)
Sclerocarya birrea	Marula	All parts are used as food and medicine. Seeds are rich in protein and oils, such as oleic acid, palmitic acid, etc.	Mariod and Abdelwahab (2012)
Tamarindus indica	Tamarind	Medicinal properties and rich source of most of the essential amino acids and phytochemicals	Kuru (2014), Ebifa-Othieno et al. (2020)
Garcinia mangostana	Mangosteen	Fruits as antioxidant and anti-inflammatory and treats diarrhoea, wounds, and fever.	Aizat et al. (2019a, b)
Dovyalis caffra	Kei apple	Furniture construction, animal fodder, and bee production as it provides good source of nectar for bees	Omotayo et al. (2019)
Cydonia oblonga	Quince	Shrub with edible fruit after cooking or baking. Effective against cough, diabetes, and urinary problems	Sharma et al. (2011b), Hegedus et al. (2013)

3.3 Major Underutilized Crops and Their Germplasm Resources

Germplasm are the materials that are used for the improvement of a cultivated plant species. The present chapter highlights cultivation, nutritional characteristics, and germplasm collections of major underutilized crops, such as millets, hulled wheat, buckwheat, quinoa, teff, etc. Legumes are rich in protein and are discussed as bambara groundnut, horsebean, winged bean, etc. Similarly vegetables are represented by moringa, amaranthus spp., etc. In recent times, these are used as breakfast cereals, snacks, in processed foods, flour, beverages, etc., but they are not a considerable source of calories and are currently underutilized.

3.3.1 Millets

Millets include six cereal crops—finger millet (*Eleusine coracana*), foxtail millet (*Setaria italica*), kodo millet (*Paspalum scrobiculatum*), barnyard millet (*Echinochloa* spp.), little millet (*Panicum sumatrense*), and proso millet (*Panicum miliaceum*) (Table 3.4). They are grown on a total of 32.2 million hectares of land each year, primarily in India, Sudan, Niger, and Nigeria (FAOSTAT 2018). They outperform commercially farmed cereal crops like rice and wheat in terms of nutrition as they contain less endosperm and a greater proportion of embryos accumulating proteins and lipids (Valdivia-López and Tecante 2015). In addition to micronutrients, these underutilized grains contain significant amount of phenolic acids and flavonoids with antioxidant properties (Zhang et al. 2014; Singh and Sharma 2017; Tang and Tsao 2017).

Ragi or Finger millet is a well known millet that belongs to the Poaceae family. It was domesticated in Ethiopian and Western Uganda after being imported in India. It is a short duration crop that can produce seed in 75–90 days. Due to its adaptability

Table 3.4 Areas of cultivation different types of millets around the world

Types of millets	Scientific name	Cultivated regions	References
Finger millet	*Eleusine coracana*	Parts of Asia and Africa	Dwivedi et al. (2012)
Foxtail millet	*Setaria italic*	Europe, India, China, Indonesia	Vetriventhan et al. (2012)
Kudo millet	*Paspalum scrobiculatum*	Tropical and sub-tropical regions	Dwivedi et al. (2012)
Proso millet	*Panicum miliaceum*	China, India, Russia, Middle East	Zarnkow et al. (2009)
Barnyard millet	*Echinochloa* spp.	India, China, Japan, Pakistan, Nepal, Africa	Gomashe (2016)
Little millet	*Panicum sumatrense*	India, Sri Lanka, Pakistan, Myanmar, and South East Asia	Hiremath et al. (1990)

to unfavourable climatic and soil conditions, particularly drought, finger millet is widely cultivated in the semi-arid regions of Asia and Africa. It is now grown in numerous nations, like India, Nepal, China, Myanmar, Sri Lanka, Kenya, Uganda, Eritrea, Zimbabwe, Zambia, Madagascar, Malawi, Rwanda, Sudan, Burundi, etc. The United States National Academies have recognized it as a potential 'super cereal' crop due to its nutritive properties amongst the major grain crops (National Research Council 1996). It forms a primary staple food for people living on marginal lands with limited economic means. Its seeds are rich in protein (12.3%) and minerals (3.3%) (Rai 2002). It is the richest source of iron, calcium, and fibre than other cereals like rice, wheat or maize as well as vitamins, such as thiamine, riboflavin, niacin, tocopherols, etc. It also contains aminoacids like lysine, methionine, and tryptophan, which are missing from starch rich diets, and are an essential nutritional requirement for growth and health (Bhatt et al. 2011). Finger millet contains beneficial polyunsaturated fatty acids (PUFAs) such as linoleic acid, α-linolenic acid as well as phytochemicals, like tannins, steroids, alkaloids, terpenoids, cardiac glycosides, lignans, phyto-oestrogens, phytocyanins, polyphenols, that have detoxifying, immune modulating, and antioxidant properties (Devi et al. 2014) that facilitate the proper functioning of nervous system. The dietary fibre present in finger millet promotes slow digestion and lowers blood glucose levels (American Diabetes Association 2010; Shukla and Srivastava 2014). Iron and calcium are naturally fortified in this plant, making it a nutrient dense food crop (Ramprasad 2005).

Cultivation of finger millet is cost-effective as it requires very low amount of nitrogen between 20 and 60 kg/ha (Pradhan et al. 2011). It constitutes the largest germplasm collection of all the millets in the world. More than 46,000 foxtail millet and 37,000 finger millet germplasm accessions have been conserved worldwide, with India and China having the largest collections of finger and foxtail millets germplasm accessions, respectively (Vetriventhan et al. 2015). Many organizations maintain their own germplasm banks of which India holds the largest collection (10,507) of finger millet accessions at ICAR-National Bureau of Plant Genetic Resources, New Delhi (Sood et al. 2019) (Table 3.5). ICRISAT maintains a collection of 6804 finger millet germplasm accessions from 25 different countries (Goron and Raizada 2015). Other germplasm collection organizations include the All India Coordinated Small millets Improvement Project (AICSMP), the Kenya Agricultural Research Institute (KARI), the Institute of Biodiversity Conservation (IBC), the United States Department of Agriculture-Agricultural Research Service (USDA-ARS), the Central Plant Breeding and Biotechnology Division Agricultural Research Council (CPBBD), the Central Plant Breeding and Biotechnology Division, the Nepal Agricultural Research Council (CPBBD), the National Institute of Agrobiological Sciences (NIAS), the Institute of Crop Germplasm Resources, and the Chinese Academy of Agricultural Sciences (ICGR-CAAS).

Foxtail millet (*Setaria italica*) popularly known as Kangni has been farmed as a food crop in Asia for nearly 5000 years and as animal feed in the USA and Europe. It is called foxtail millet as its panicles resemble fox tail like appearance. It was first domesticated in Central China (Ho 1975). India, China, and Japan are the main

Table 3.5 Germplasm resources of small millets around the world

Common name	Institution/organization	Headquarters	Number of accessions
Finger millets	ICAR-National Bureau of Plant Genetic Resources, New Delhi	India	10,507
	International Crops Research Institute for the SemiArid Tropics (ICRISAT), Patancheru	India	6804
	All India Coordinated Small Millets Improvement Project, Bengaluru	India	6257
	Kenya Agricultural and Livestock Research Organization (KALRO), Muguga	Kenya	2857
	Ethiopian Biodiversity Institute (EBI), Addis Ababa	Ethiopia	2156
	USDA Agricultural Research Service (USDA-ARS)	USA	1452
	Serene Agricultural and Animal Production Research Institute, Soroti	Uganda	1231
	SADC Plant genetic Resources Centre, Lusaka	Zimbawe	1158
	Nepal Agricultural Research Council, Kathmandu	Zambia	1037
	National Centre for Genetic Resources Preservation, Fort Collins	Nepal	869
	National Institute of Agrobiological Sciences, Kannondai	USA	702
	Mt. Makulu Central Research Station, Chilanga	Japan	565
	Institute of Crop Germplasm Resources, Chinese Academy of Agricultural Sciences, Beijing	China	300
Foxtail millet	Chinese National Genebank (CNGB)	China	26,670
	National Bureau of Plant Genetic Resources (NBPGR)	India	4330
	ORSTOM-MONTP	France	3500
	All India Coordinated Minor Millet Project (AICMMP)	India	2512
	International Crops Research Institute for the Semi-Arid Tropics (ICRISAT)	India	1535
	National Institute of Agrobiological Sciences (NIAS)	Japan	1299
	North Central Regional Plant Introduction Station, USDA-ARS	USA	1000
	Biologie Végétale Appliquée, Institut Louis Pasteur (IUT)	France	850
	Kenya Agricultural Research Institute (KARI)	Kenya	772
	USDA Agricultural Research Service (USDA-ARS)	USA	762
	Estación de Iguala, Instituto Nacional de Investigaciones Agrícolas (INIA)	Mexico	350
Kudo millet	National Bureau of Plant Genetic Resources (NBPGR)	India	2170

(continued)

Table 3.5 (continued)

Common name	Institution/organization	Headquarters	Number of accessions
	All India Coordinated Minor Millet Project (AICMMP)	India	1111
	International Crops Research Institute for the Semi-Arid Tropics (ICRISAT)	India	656
	USDA Agricultural Research Service (USDA-ARS)	India	336
Proso millet	N.I. Vavilov All-Russian Scientific Research Institute of Plant Industry	Russia	8778
	Institute of Crop Germplasm Resources, Chinese Academy of Agricultural Sciences (ICGR-CAAS)	China	6517
	Ustymivka Experimental Station of Plant Production	Ukraine	3976
	V Y Yuriev Plant Production Institute UAAS	Ukraine	1046
	International Crops Research Institute for the Semi-Arid Tropics (ICRISAT)	India	842
	Botanical Garden of the Plant Breeding and Acclimatization Institute	Poland	721
	USDA Agricultural Research Service (USDA-ARS)	USA	719
	North Central Reg. Plant Introd. Station, USDA-ARS	USA	713
	Estación de Iguala, Instituto Nacional de Investigaciones Agrícolas (INIA)	Mexico	400
	National Institute of Agrobiological Sciences (NIAS)	Japan	302
Barnyard millet	National Bureau of Plant Genetic Resources (NBPGR)	India	1888
	University of Agricultural Sciences, Bangalore	India	985
	International Crops Research Institute for the Semi-Arid Tropics (ICRISAT)	India	749
	Vivekananda Parvatiya Krishi Anusandhan Sansthan	India	300
	Indian Institute of Millets Research	India	1561
	National Institute of Agrobiological Sciences	Japan	3671
	Plant Germplasm Institute	Kyoto university	65
	USDA Agricultural Research Service (USDA-ARS)		232
	National Centre for Genetic Resources Conservation	Collins	306
	North Central Regional Plant Introduction Station	Ames	304
Little millets	Institute of Crop science, Chinese Academy of Agricultural Sciences	China	717
	All India Coordinated Minor Millet Project (AICMMP)	India	544

(continued)

Table 3.5 (continued)

Common name	Institution/organization	Headquarters	Number of accessions
	International Crops Research Institute for the Semi-Arid Tropics (ICRISAT)	India	466
	USDA Agricultural Research Service (USDA-ARS)	USA	211

producers of this crop. Taxonomically, two subspecies of foxtail millet, *S. italica* subsp. *italica* and *Setaria viridis* subsp. *viridis*, are considered the progenitor of foxtail millet. There are three races of foxtail millet—'Moharia' is common in Europe, Southeast Russia, Afghanistan, and Pakistan, 'Maxima' is common in Eastern China, Georgia (Eurasia), Japan, Korea, Nepal, Northern India and for the purpose of animal feed in the USA, and 'Indica' is found in the remaining parts of India and Sri Lanka (Prasada Rao et al. 1987). It is tolerant to salt and drought stress as well as insect, pest and disease resistant, just like other millet crops (Jayaraman et al. 2008).

The Chinese National Genebank (CNGB) has the largest germplasm collection of this crop, with over 26,670 accessions as of 2012 (Wang et al. 2012). International Crop Research Institute for the Semi-Arid Tropics (ICRISAT) has germplasm collections from 26 countries. Genebanks in the USA (USDA, Plant Genetic Resources Conservation Unit, PGRCU) and Japan (National Institute of Agrobiological Sciences, NIAS) provide access to a wide range of foxtail millet cultivars (Table 3.5).

Kodo millet (*Paspalum scrobiculatum*) is known by various names such as rice grass, African bastard millet, creeping paspalum, ditch millet, Indian paspalum, kodo grass, varagu, and so on in different parts of the world. It was domesticated in India around 3000 years ago mostly as a cereal crop. It was once considered a weedy crop that grew alongside major cereal crops like rice. Its wild varieties can be found in Africa and it can be grown as a pure crop or intercropped with pulses, mainly pigeon pea, soybean, and mung bean. As compared to other millets, this grain is enriched with high quality protein (Kulkarni and Naik 2000). It is rich in fibre and has antioxidant activity, just like finger millet (Geervani and Eggum 1989). Its grains are enriched in B vitamins, especially niacin, B_6, and folic acid, as well as minerals, like calcium, iron, potassium, magnesium, and zinc. Due to limited research in crop improvement methods, the area under cultivation of kudo millet is declining, resulting in making it neglected or underutilized crop.

More than 8000 germplasm accessions of kodo millet have been maintained in genebanks around the world, majority of them are in India (Upadhyaya et al. 2015). The ICRISAT genebank in India conserves 656 landrace accessions of all three races of kodo millet (http://genebank.icrisat.org/). The National Plant Germplasm System (GRIN) of the US Department of Agriculture (USDA) holds 336 accessions. Other major germplasm conservation centres include National Bureau of Plant Genetic

Resources (NBPGR), USDA Agricultural Research Service (USDA-ARS) (Table 3.5).

Broomcorn millet, common millet, hog millet, and Russian millet are all names for Proso millet (*Panicum miliaceum*). It was first domesticated in China about 10,000 years ago. Northern China, Mongolia, Republic of Korea, South-eastern Russia, Afghanistan, Pakistan, India, and Southern Europe are currently cultivating it. It is used to feed birds, animals and fodder in developed nations and for food in some parts of Asia (Rajput et al. 2014). The majority of the proso millet harvest is utilized for birdfeed and cattle fattening diets in the USA (Habiyaremye et al. 2017). In comparison to other staple cereals, proso millet grains are high in protein, vitamins, minerals, and micronutrients like iron, zinc, copper, and manganese. The protein content of proso millet is 12.5%, which is comparable with that of wheat and its grains are richer in essential amino acids like leucine, isoleucine, and methionine than those of wheat (Kalinova and Moudry 2006; Saleh et al. 2013).

Over 29,000 germplasm accessions of proso millets have been conserved in genebanks around the world, with the largest collections of proso millet germplasm accessions being found in Russia, China, Ukraine, and India (Reddy et al. 2007). The N.I. Vavilov All-Russian Scientific Research Institute of Plant Industry in St. Petersburg has the largest collection of this millet, with over 8778 accessions as of 2012 (Dwivedi et al. 2012). ICRISAT holds 842 accessions covering each of the five races (Upadhyaya et al. 2011). Flowering period, plant height, panicle length, and inflorescence length have all been used to define the diversity of this collection (Reddy et al. 2007) (Table 3.5).

Barnyard millet (*Echinochloa* spp.) is cultivated in countries like India, China, Japan, and Korea. In terms of area and production, India is the largest producer of barnyard millet, with an average productivity of 1034 kg/ha during the last 3 years (IIMR 2018). *Echinochloa frumentacea* (Indian barnyard millet) and *Echinochloa esculenta* (Japanese barnyard millet) are the two species of Barnyard millet (Sood et al. 2015). The richness of protein, carbohydrates, fibre, and micronutrients such as iron (Fe) and zinc (Zn) all contribute to a wide range of its health benefits (Saleh et al. 2013; Chandel et al. 2014; Ugare et al. 2014). All these characteristics make it an ideal supplementary crop for poor farmers and as a backup crop during the failure of monsoons in major crop cultivating areas (Gupta et al. 2009). Despite its great nutritional and agronomic attributes, the lack of awareness has led this crop to be considered as a neglected and underutilized crop (Renganathan et al. 2020).

India and Japan have the largest collections of barnyard millet germplasm accessions. Currently, 8000 barnyard millet germplasms are being conserved at different centres throughout the world. The National Bureau of Plant Genetic Resources (NBPGR) in New Delhi has the highest germplasm accessions (1888). ICRISAT currently holds 749 accessions of these barnyard millets from nine countries (Renganathan et al. 2020; Upadhyaya et al. 2014). Other institutes include Vivekananda Parvatiya Krishi Anusandhan Sansthan (VPKAS) and Indian Institute of Millets Research (IIMR), India; National Institute of Agrobiological Sciences (NIAS), Japan, and Consultative Group on International Agricultural Research like

International Crop Research Institute for the Semi-Arid Tropics (ICRISAT), India (Table 3.5).

Little millet (*Panicum sumatrense*) is commonly known as Saamai or Kutki. In India, Sri Lanka, Pakistan, Myanmar, and other South East Asian nations, it is cultivated to a limited extent. It is thought to have been domesticated from a weedy species—*Panicum psilopodium*. Based on panicle morphology, little millet is classified into two races as nana and robusta. Race nana develops faster and produces less biomass than robusta (de Wet et al. 1983). In terms of fibre, fat, carbohydrates, and protein, little millet is equivalent to other cereals and it is rich in phytochemicals such as phenolic acids, flavonoids, tannins, and phytate (Pradeep and Guha 2011). It is insect resistant, drought, and salt tolerant, like many other small millets (Bhaskaran and Panneerselvam 2013; Ajithkumar and Panneerselvam 2014). For many of the traits studied, a distinct collection of 460 accessions of little millet maintained by ICRISAT indicated genetic variation (Upadhyaya et al. 2014). Other organizations involved in germplasm accessions include All India Coordinated Minor Millet Project (AICMMP) and USDA Agricultural Research Service (USDA-ARS) (Table 3.5).

3.3.2 Buckwheat

Buckwheat (*Fagopyrum esculentum*), belonging to the Polygonaceae family, is a dicotyledonous herb. It is known by many names—Ogala, Phaphra, Bhares, Daryun, Kathu, and Kuttu in India, Jare in Bhutan, Soba in Japan, Mithe Phaapar in Nepal, Tian'qiao in China, and Grecicha kul'furnaja in Russia (Zhou et al. 2018). *Fagopyrum* genus has three cultivated species—*F. esculentum, F. tataricum,* and *F. cymosum. F. Esculentum* (common buckwheat) is a widely cultivated crop that grows from Asia to Europe, North America and South Africa, whereas *F. Tataricum* (tartary buckwheat) and *F. cymosum* are mainly confined to uplands of Southwest China and Himalayas (Zhou et al. 2016). The main buckwheat growing countries include, India, Russia, China, Ukraine, Poland, Brazil, USA, France, Japan, Kazakhstan, Canada, Nepal, Hungary, Slovenia, Croatia, Lithuania, Estonia, and Latvia (Bavec 2014). Buckwheat is a pseudocereal crop that is gaining attention due to its high nutritional profile, shorter growing period than commercial crops, ability to adapt under marginal areas of Himalayas and requires low inputs for cultivation (Zhou et al. 2016). Despite the fact that it is cultivated in many areas of the world, it is still considered an underutilized crop due to its failure to realize its full potential in comparison to major cereals due to the lack of crop improvement methods. Its natural supplementation with critical nutrients accounts for its superiority over major cereal crops. It is a rich source of flavonoids like orientin, quercetin, vitexin, isovitexin, and isoorientin (that are found in sprouts). The plant parts are also reported to contain vitamins such as B_1, B_2, B_3, B_5, B_6, C, and E, inositol, protein, and iminosugars (Comino et al. 2013; Gimenez-Bastida et al. 2015). It is high in sulphur-containing aminoacids—lysine and methionine, making it a good source of protein and nutritious replacement for conventional food crops. Rutin, present in

Table 3.6 Germplasm resources of Buckwheat in major gene banks globally

Common name	Institution/organization	Headquarters	Number of accessions
Buckwheat	Institute of Crop science, Chinese Academy of Agricultural Sciences (CAAS)	China	2804
	N.I. Vavilov All-Russian Scientific Research Institute of Plant Industry	Russia	2116
	V Y Yuriev Plant Production Institute UAAS	Ukraine	1600
	Podillya State Agricultural University	Ukraine	900
	National Bureau of Plant Genetic Resources (NBPGR)	India	1050
	National Agriculture Genetic Resource Centre	Nepal	511
	University of Ljubljana	Slovenia	378
	National Institute of Agrobiological Sciences	Japan	226
	Crop Research Institute	Prague	170

seeds, has antioxidant and anti-inflammatory properties that can help against skin carcinogenesis (Nile and Park 2014; Raina and Gupta 2015). Thus, it is a kind of superfood that should be cultivated on a commercial basis in hilly regions by further improvement.

As buckwheat is a nutrient dense crop with a wide diversity, significant efforts are being made for conservation of germplasm resources. More than 10,000 diverse accessions of buckwheat genetic resources are currently being conserved across the world, with half of them being preserved in East and South Asian gene banks. The Institute of Crop Sciences, Chinese Academy of Agricultural Sciences (CAAS) maintains the largest repository worldwide, with more than 2800 germplasm accessions (Zhou et al. 2018) (Table 3.6). Other major genebanks maintaining germplasm resources are N I Vavilov All-Russian Institute of Plant Industry, Russia; V Y Yuriev Institute of Plant Production, Ukraine; National Bureau of Plant Genetic Resources, India (Rana et al. 2016) and National Agriculture Genetic Resource Centre, Nepal (Paudel et al. 2016).

3.3.3 Grain Amaranths

Amaranthus (*Amaranthus* sp.) species, also known as Amaranths, is one of the underutilized orphan crops grown in tropical and sub-tropical regions of the world. This genus, belonging to the family Amaranthaceae, is originated in South America (Janovská et al. 2012). The word 'Amaranthus' comes from the Greek word 'amarantos' which means 'one that does not wither' or 'never fading' (Anjali et al. 2013). About 400 species of Amaranthus species are found in temperate, sub-tropical, and tropical climate zones around the world, and are grown as leafy vegetables, cereals, and ornamental crops in many parts of the world. About 60 species of grain are native to South and Central America, whilst others are found in Europe, Asia, Africa, and Australia (Teutonico and Knorr 1985). There are three species for grain production in Amaranthus—*A. hypochondriacus*, *A. cruentus,* and *A. caudutus*, whilst the leafy types include *A. viridis*, *A. spinosus*, *A. reftroflexus,* and *A. hybridus* (Tony-Odigie et al. 2012). The tender plant of *A. cruentus* grain is also eaten as a leafy vegetable. The ornamental type (*A. tricolor*) is thought to have originated in India and then spread around the globe. Several ornamental varieties and vegetable types have been developed from this species (Nguyen et al. 2019).

Amaranth, classified as a C4 plant (Kauffman and Weber 1990), enables higher water-use efficiency and photosynthesis in high temperature and drought prone conditions, compared to C3 crops, thereby adding to the benefits achieved by expanding the adaptation and cultivation of amaranth to Eastern Africa (Stallknecht and Schulz-Schaeffer 1993; Barba de la Rosa et al. 2009). Seed amaranth require 53–58% less water than that required for wheat, 40–50% less than maize, and 21% less than cotton to grow (Kauffman and Weber 1990).

The underutilized amaranths have health benefits and nutritive value. As compared to wheat protein, the seeds of grain Amaranths have twice the amount of essential amino acids, particularly lysine, phenylalanine, and threonine and are rich in minerals such as calcium, iron, and zinc (Espitia 1992). Grain amaranths are gluten free and provide 30% more protein with a complete set of amino acids than wheat and rye. Amaranth grain can be flaked, popped, extruded and made into flour and can be used as a substitute in porridge or added into soup. Grain amaranths have a number of health benefits, including decreasing plasma cholesterol, protection of heart and improved hypertension, control of blood sugar, stimulation of immune system, anti-allergic, anti-cancer, and antioxidant properties due to the presence of important bioactive constituents (Girija et al. 2011; Das 2016). Vitamin E and squalene are abundant in amaranth seed and amaranth seed oil, which can benefit people suffering from hypertension and cardiovascular disease (Bressani et al. 1987). Amaranths, as a vegetable, provide energy, aid in mineral balance, lower bad cholesterol level, improve eyesight, and evade anaemia. With rising food demand and current levels of malnutrition, the development of amaranths as an alternative diet could be a life saviour for individuals suffering from malnutrition and hunger in underdeveloped countries (Brenner et al. 2010). Thus, amaranth is a nutritious and healthy food crop that could benefit people if it were grown commercially and consumed in greater quantities.

The proper collection, maintenance, and evaluation of germplasm are critical for genetic improvement of Amaranthus. A functional germplasm collection is maintained by a number of amaranth research centres and germplasm collections around the world. Amaranth germplasm is documented and maintained in germplasm banks in 11 countries (Toll and Van Sloten 1982). The germplasm collections are preserved in 61 collection centres (IPGRI 1999) from 12 different countries (Das 2016). With 3300 accessions from 40 nations, the USDA germplasm collection is currently the world's largest repository for amaranth (Trucco and Tranel 2011); National Bureau of Plant Genetic Resources (NBPGR) has 3081 (Das 2016); Univ. Nacional San Antonio Abad del Cusco (UNSAAC/CICA), Peru has 740 from 12 countries (Kalinowski et al. 1992); National Botanical Research Institute

Table 3.7 Amaranthus germplasm resources of the world

Common name	Institution/organization	Headquarters	Number of accessions
Amaranthus	Institute of Crop Germplasm Resources (CAAS) Beijing	China	438
	National Botanical Research Institute (NBRI)	India	2500
	Universidad Nacional del Altiplano Puno,	Peru	440
	Instituto Nacional de Investigaciones Forestales y Agropecuarias (INIFAP)	Mexico	495
	National Bureau of Plant Genetic Resources (NBPGR)	India	3081
	Univ. Nacional San Antonio Abad del Cusco (UNSAAC/CICA)	Peru	740
	North Central Regional Plant Introduction Station (NCRPIS)	USA	3380

(NBRI) has 2500 (Mathews 2001); Institute of Crop Germplasm Resources (CAAS), China has 438 accessions (IPGRI 1999) and Instituto Nacional de Investigaciones Forestales y Agropecuarias (INIFAP), Mexico has 495 from 5 countries (IPGRI 1999) (Table 3.7).

3.3.4 Quinoa

Quinoa (*Chenopodium quinoa*) is a significant crop that contributes to global food security. It is native to the Andean Plateau, where it was domesticated about 7000 years ago (Bazile et al. 2013). Till date, about 6000 varieties of quinoa are being cultivated worldwide (Rojas et al. 2015). The crop is well suited to a variety of agro-ecological conditions in different regions (Bazile et al. 2014) and is tolerant to frost, drought, and salinity stress (Ruiz et al. 2015). Bolivia and Peru are the world's top quinoa producers, followed by Ecuador, Argentina, and Chile. Quinoa cultivation has extended to countries as far as Tibet, Morocco, France, India, China, the United Kingdom, Netherlands, and Italy (Pulvento et al. 2010; Bazile 2015). Quinoa has been a staple diet for Andean people for thousands of years (Tagle and Planella 2002; Planella et al. 2011; Martinez et al. 2015). It has outstanding nutritional properties, with a higher protein content than cereals and a good balance of essential amino acids (Vega-Gálvez et al. 2010; Maureira and Martínez 2012; Miranda et al. 2012; Lutz et al. 2013).

The United Nations Food and Agriculture Organization's project 'American and European Test of Quinoa' was the first step toward increasing its cultivation in Europe. In 2013, the EU project 'Quinoa – a multiple crop for EC's Agriculture diversification, 1993–1997', organized by the University of Cambridge, UK, and the Food and Agriculture Organization of the United Nations (FAO) by the Ministry of Agriculture, Belgium, launched programmes in a number of countries to promote quinoa as a food for current and future generations.

Quinoa has the highest protein content (14%) amongst cereal grains, making it an excellent choice for vegetarians. It is gluten and cholesterol-free and contains all nine essential amino acids, making it a complete protein. The seeds are rich in vitamin B complex, vitamin E, and essential fatty acids like linoleic and alpha-linolenic acid. It is a good source of calcium, copper, fibre, iron, magnesium, manganese, and potassium as well. Cooking time is less than for other whole grains. Individual plant selection and/or mass selection, introduction of exotic germplasm, hybridization, backcrossing, and other breeding techniques have all been employed to improve it.

More than 88% of the accessions available worldwide are conserved in genebanks in the Andean area. Accessions of quinoa and its wild relatives are conserved in 59 genebanks distributed in 30 nations. Amongst the various countries in the world, Germany has 987 accessions, India has 294, the United States of America Department of Agriculture (USDA) maintains over 229, and Japan has 191 accessions of quinoa and its wild relatives (Rojas et al. 2015). In Italy, the Cereal Research Centre (CER-CRA) obtained and evaluated 100 accessions for cultivating and developing varieties adapted to Mediterranean conditions. The FAO Regional Office for Near East, based in Cairo, Egypt, aided in the distribution of quinoa accessions amongst National Research Institutions from eight countries (Algeria, Egypt, Iraq, Iran, Lebanon, Mauritania, Sudan, and Yemen) to conduct an evaluation of these genotypes in semi-arid and arid areas. Japan adopted the Nagoya Protocol on Access and Benefit-Sharing in 2010, which came into effect in 2014. It establishes an international agreement with the goal of fairly and equitably sharing the advantages of genetic resource usage whilst also contributing to the conservation of biological diversity. Despite wide diversity collected and preserved, the products produced in the market are drawn from a small set of land races. The crop's full potential is yet to be recognized and it is still underutilized. The interconnection of germplasm resources in genebanks amongst countries is inadequate, limiting its cultivation in underserved areas.

3.3.5 Ancient Wheat or Hulled Wheat

Wheat (*Triticum* sp.) is one of the most widely grown food crops on the planet. It is a complex polyploidy—diploid, tetraploid, and hexaploid, produced by multiple species of different ploidy levels, resulting from a combination of genomes from different species of the *Triticeae* tribe in the Poaceae family. The two primary varieties—durum or macaroni wheat (*Triticum turgidum* L. subsp. durum Desf.), which is a tetraploid wheat (AABB; $2n = 4\times = 28$) and bread wheat (*Triticum aestivum* L. em Thell; AABBDD; $2n = 6\times = 42$), are widely grown around the world (Costanzo et al. 2019).

Einkorn, Emmer, and Spelt wheat types named as ancient wheats, popularly called hulled wheats, are the earliest cultivated ones and have been neglected in modern agriculture. These wheat species originated in the Fertile Crescent of West Asia (Harlan and Zohary 1966). Any wild or cultivated *Triticum* population with non-threshable grains is referred to as Hulled wheat. In comparison to einkorn, emmer spread into more fruitful regions of the Old World during the diffusion of hulled wheats out of the fertile crescent (Zaharieva and Monneveux 2014). Both species, along with spelt wheat (*Triticum spelta* L.) collectively known as 'farro', were used to feed the Old world for ages before being mainly replaced by bread wheat (*Tritium aestivum* L.) 4000 years ago (Feldman and Kislev 2007). The most common Einkorn varieties are var. *tauricum* Drosd. and var. *monococcum*, with the sole difference between them being the awn colour. The most representative Emmer wheat cultivars were *dicoccon* Körn. and *tricoccum* (Schübl.) Körn, with the difference between the two being awn length. They are major food crops in Assyrian, Babylonian, and Egyptian civilizations (Jaradat 2018; Oliveira et al. 2011). Hulled wheats are produced in various distant mountainous regions and in low-input farming practices in areas of West Asia, North Africa, and Ethiopia, India and Southeast Europe, where crop genetic diversity has improved farmland biodiversity (Troccoli and Codianni 2005; Oliveira et al. 2011; Chateil et al. 2013). When compared to durum and bread wheats, traditional hulled wheats have tall plants, a large number of tillers, a huge biomass, and a relatively constant grain yield (Giraldo et al. 2016). In comparison to modern wheat cultivars, the grains are high in nutritional content (Shewry et al. 2016). Owing to its hulled grain, it is resistant to pests and diseases (Caballero et al. 2004). Furthermore, they outperform modern wheat in terms of resistance to abiotic and biotic stresses and soil nutrient shortage. Spelt wheat (*Triticum aestivum* ssp. *spelta* L.), the oldest known wheat species cultivated in ancient Egypt and Italy, was grown in the hilly and mountainous area of Central Europe and North America at the end of the twentieth century as a result of spontaneous crossings of wild grasses (Janković et al. 2013). From the Bronze Age to the turn of the twentieth century, it was a staple crop in Europe (Muller et al. 2018). Spelt wheat and its products are high in protein and have a great proportion of soluble fibre (Bonifácia et al. 2000; Sugár et al. 2019). Despite the fact that they contain a wealth of genetic diversity for improving agronomic and nutritional traits in durum and bread wheat, their potential remains unexplored.

Emmer or *dicoccum* wheat is a good source of protein, carbohydrate, minerals, and is low in fats, making it a very healthy cereal. Because of the complexity of the starch structure and the high amylose content, dicoccum starch is slow to digest (Galterio et al. 2003). Einkorn has been reported to have the highest lutein content, indicating a high antioxidant efficiency (Lachman et al. 2013; Ziegler et al. 2016). Gluten is an extremely important functional protein that gives wheat dough its viscoelastic property. Celiac Disease (CD) is a gluten intolerance-chronic gastrointestinal disorder in individuals that are sensitive to it. It has been reported that certain *dicoccum* wheat varieties are devoid of these noxious gluten proteins that cause CD (Vincentini et al. 2007; Zanini et al. 2015).

However, due to the advantages of hulled wheat species, such as their greater amenability to sustainable agriculture and superior health-related properties over bread wheat and durum wheat species, there has been a recent surge in interest in these underutilized crops (Arzani and Ashraf 2017).

3.3.6 Teff

Teff (*Eragrostis tef* (Zuccagni) Trotter) belongs to the grass family Poaceae, sub-family Chloridoideae (Eragrostoideae), tribe Eragrostidae, subtribe Eragrostae, and genus Eragrostis. There are more than 400 morphologically distinct species distributed throughout the world (Clayton et al. 2016), with *Eragrostis tef* (Zucc.) Trotter being the only species in the genus *Eragrostis* cultivated for human consumption. Teff is a C4 annual cereal and is self-pollinated. It is the major food crop in Ethiopia, located in East Africa (Clayton et al. 2016). Because of its beneficial agronomic and utilization characteristics, it is preferred by both farmers and consumers. Its flour is added as a thickening agent to soups and sauces, processed to make beer and ethnic drinks (Demeke and Di MArcantonio 2013) and is used to make porridge and puddings (Adebowale et al. 2011; de Oliveira do Nascimento et al. 2018). It is a multi-purpose crop where the grain is primarily used to make injera, an Ethiopian staple food and the crop residue is used to feed livestock (Heuzé et al. 2016; Habte et al. 2019). Its grains are rich in iron, calcium, and fibre (FAO 2015). It is an excellent source of essential amino acids, particularly lysine, which is the most commonly deficient aminoacid in grains (Ayalew et al. 2011). Teff's popularity has increased in the USA and Europe, owing to its appealing nutritional profile and gluten-free nature (Davison and Laca 2009; Gebremariam et al. 2012; Lee 2018). The Institute of Plant Sciences at the University of Bern in Switzerland, initiated Teff improvement project with the aim of improving the unused cereal crop teff and in collaboration with the Ethiopian Institute of Agricultural Research has taken the initiative to support the traditional breeding efforts in Ethiopia using modern molecular techniques. It also collaborated with National Teff Improvement Program in Ethiopia and employed key genetic and genomic methods to produce teff varieties with valuable characters (Cannarozzi et al. 2018; Girma et al. 2018; Kebede et al. 2018).

Diverse types of accessions are available in the organizations involved in the collection, assessment, and utilization of teff germplasm. Currently, more than 5000 *E. tef* accessions have been collected from different geographic regions of Ethiopia and are preserved in the seed gene bank of the Ethiopia Biodiversity Institute. With 5169 teff accessions, the Plant Genetic Resources Centre of Ethiopia (PGRCE), now called the Biodiversity Institute, is actively engaged in collecting, conservation, and characterization (Tesema 2013). Institute of Agricultural Research and the Debre Zeit Agricultural Research Centre are the other organizations that maintain teff germplasm accessions. Outside of Ethiopia, only a few accessions are found. The major ones are listed below, as per the FAO—World Information and Early Warning System on Plant Genetic Resources reports: Western Region Plant Introduction Station USDA-ARS (368 accessions), Washington State University; National Seed Storage Laboratory USDA-ARS (341 accessions), Colorado State University; Department of Genetic Resources (30 accessions), Japan, and Institute of Crop Science (30 accessions), Germany (Ketema 1997).

3.3.7 Bambara Groundnut

Bambara Groundnut (*Vigna subterranea* (L.) Verdc.) is a leguminous crop in the Fabaceae family that is one of the most underutilized crops. This crop originated in Africa and due to its nutritional properties and wide adaptability to climatic conditions, it is now largely cultivated in Central and West African regions. Its ability to fix nitrogen from the atmosphere in the soil offers it a competitive advantage over other crops. It is drought tolerant crop and resistant to abiotic stresses (Obidiebube et al. 2020).

Bambara groundnut is a low-cost, high-protein source that can enable rural households strengthen their food and nutrition security. Schaafsma (2012) reported that the amount of essential aminoacids is higher, with a protein score of 80 percent higher than other species such as soya bean (74%), groundnut (65%), and cowpea (64%). The biochemical profile of the carbohydrate, fat, protein, and mineral content reveals that it produces an almost balanced diet. However, it lacks sulphur-containing proteins like other legumes, so integrating it with a staple cereal like maize, which has higher levels of sulphur-containing amino acids, results in a complete diet (Akpapunam and Darbe 1994). It contains micronutrients, such as zinc, iron, calcium, and potassium. The iron content of red seeded varieties is nearly double that of cream seeded varieties. As a result, they are particularly useful in areas where iron deficiency is common (Muimba-Kankolongo and Muimba-Kankolongo 2018).

Despite its long history of cultivation, the crop is still underutilized for a variety of reasons. The research efforts are limited in the improvement of this crop, and lower crop yield and quality have resulted in underutilization of the crop. Its use has been replaced by the introduction of cash crops with greater export potential. Another factor for the bambara groundnut's underutilization is its hard-to-cook quality, as well as insufficient processing methods for this crop (Mubaiwa 2018). Its hard-to-cook characteristic is the prolonged cooking time (3–4 h) required to achieve suitable softening in cooking (Khan et al. 2016), and the hardness of the seed coat, which makes it extremely difficult to achieve better dehulling. The need for a short day photoperiod for pod filling also restricts its large-scale cultivation (Berchie et al. 2013; Kendabie et al. 2015). Suhairi et al. (2018) compared the suitability of this crop under various climatic conditions such as temperature, rainfall, and soil characteristics across Malaysia and observed that it was suitable on 59% of all soils compared to only 23% for maize.

Several countries have participated in Bambara Groundnut projects funded by the EU, FAO, McKnight Foundation, and local resources. Internal institute of Tropical Agriculture (IITA), Nigeria assembles germplasm accessions from countries across Sub-Saharan Africa. It characterized around 1910 Bambara groundnut accessions from different countries in Sub-Saharan Africa (FAO 2009). Crops for the Future (CFF), Seminyeh (Malaysia), places a strong emphasis on developing improved genotypes of bambara groundnut from landraces and wild relatives. Crop physiology is being explored in depth as part of a series of research grants led by the University of Nottingham in collaboration with international partners (BAMLINK-EU FP6

INCO-DEV 2006–2011). Other international organizations involved include Office De la Recherche Scientifique et Technique Outre-mer (ORSTOM), France; Department of Agricultural Research (DAR), Botswana; Plant Genetic Resources Research Institute (PGRRI), Ghana; National Plant Genetic Resource Centre (NPGRC), Tanzania.

3.3.8 Faba Bean

Faba bean (*Vicia faba*), belongs to the family Fabaceae, is an important legume crop cultivated all over the world. The area under cultivation of this crop is very less. Between 1980 and 2014, its global acreage under cultivation reduced from 3.7 to 2.1 million ha, and yields are highly variable within countries (FAO 2017a, b). The faba bean (*Vicia faba* L.) is popularly known as broad bean, horse bean, field bean in various languages, in Hindi it is popularly known Kala Matar and Bakala (Singh et al. 2013). It is recognized as a North African and Southwest Asian native crop. At present, they are major crop in many countries including China, Ethiopia and Egypt, Northern Europe, the Mediterranean, Ethiopia, Central Asia, East Asia, and Latin America.

The crop is tolerant towards acidic and alkaline soils, needs less water, and can withstand biotic and abiotic stresses than other crops. Since it is a leguminous crop, it can fix atmospheric nitrogen in the soil, resulting in nitrogen enrichment. It has proven to be effective in increasing the yield and seed protein content of successive cereal crops when rotated with cereal crops (Zou et al. 2015; Landry et al. 2016). It is also a rich source of essential amino acids—arginine, lysine, and leucine (Koivunen et al. 2016). Its seeds are rich source of nutrients like K, Ca, Mg, Fe, and Zn (Lizarazo et al. 2015; Longobardi et al. 2015; Neme et al. 2015). It has been proposed that faba bean could be used as a functional food as it contains macro-, micro-, and non-nutrient phytochemicals. Faba bean seeds contain L-3, 4-dihydroxyphenylalanine (L-DOPA), a precursor to the neurotransmitter catecholamine and a medication used to treat Parkinson's disease, according to Brauckmann and Latté (2010).

Vicia faba has a significant genetic diversity. More than 38,000 accessions of faba bean germplasm are conserved internationally in various gene banks, as well as at the International Center for Agricultural Research in Dry Areas (ICARDA), according to Duc et al. (2010, 2015). Several more genotypes are accessible locally in Europe, at farms and in breeder collections, according to studies undertaken by the EUROLEGUME group (Lepse et al. 2017; Karkanis et al. 2018).

3.3.9 Moringa

Moringa (*Moringa oleifera*), which belongs to the Moringaceae family, is a perennial vegetable that grows in tropical and sub-tropical climates around the world. It is also known as the Miracle tree or drumstick or horseradish tree or West Indian Ben.

It is of the shape of the pod, resembling the slender and curved stick used for beating the drum, thus deriving drumstick its name. The name 'horseradish tree' gets its name from the pendulous, slender and thin shape of the immature fruits of the tree resembling very much the siliqua of the radish (Sandeep et al. 2018). It is known as the miracle tree, because almost every part of this tree has been found to be very useful (Oyeyinka and Oyeyinka 2016). The plant is native to the Himalayan ranges of Northwestern India and is grown commercially in Pakistan, Singapore, West Indies, Cuba, Jamaica, Nigeria, South and Central America, Mexico, Philippines, Ceylon, Thailand, Malaysia, and Burma. There are more than 13 species of Moringa, with *M. oleifera* being the most widely used and cultivated species. Many species, including *M. arborea*, *M. borziana*, *M. longituba*, *M. rivae*, *M. ruspoliana*, and *M. stenopetala* are in danger of extinction (Stephenson and Fahey 2004). The genetic diversity of *M. oleifera* is poorly understood, and further research is required for employing different genetic enhancement programmes (Sarwar 2017).

The leaves, pods, seeds, flowers, fruits, and roots are the edible parts of moringa tree. Leaves are used as forage, tree trunks are used to make gums, flower nectar is used to make honey and powdered seeds for water purification (Fuglie 1999). *M. Oleifera* leaf has been used as an alternative food source to overcome malnutrition, especially amongst children and infants (Anwar et al. 2007). Vitamins A, C, and E are reported to be found in substantial amounts in *M. Oleifera* leaves (Hekmat et al. 2015). The leaves also contain significant amounts of total phenols, proteins, calcium, potassium, magnesium, iron, manganese, and copper (Hekmat et al. 2015; Issa 2012; Tahir et al. 2020). *M. Oleifera* leaves are rich in phytonutrients such as carotenoids, tocopherols, and ascorbic acid (Saini et al. 2014, 2016) and essential amino acids such as, Arg, His, Lys, Trp, Phe, Thr, Leu, Met, Ile, Val (Dahot 1988; Gopalakrishnan et al. 2016). These nutrients are known to scavenge free radicals and may have immunosuppressive effects (DanMalam et al. 2001). Moringa shoots may be used as a green manure to enrich agricultural soil. Over 300 ailments are treated or prevented using the leaves, roots, and bark of this plant (Ramachandran et al. 1980; National Institutes of Health (NIH) 2008). Every part of the plant (seeds, flowers, stems, leaves, and roots) is a rich source of nutrients and has therapeutic properties (Ganguly 2013). It has been used as a fertilizer and a natural biopesticide against a variety of pathogens (Abd El-Hack et al. 2018). Biodiesel and cosmetic oil may be extracted from the seeds for industrial use. Seeds of the plant have coagulant and bio sorbent property, allowing them to be used in water purification and microbial load reduction (Singh et al. 2019a). Various compounds present in *M. oleifera*, such as glucosinolates (Bennett et al. 2003), flavonoids (Makita et al. 2016), phenolic acids (Ramabulana et al. 2016) have been investigated. 1,1-diphenyl 2-picrylhydrazyl (DPPH), linoleic acid, Superoxide dismutase (SOD), Glutathione peroxidise, β-carotene and catalase are antioxidant extracts identified from various parts of Moringa (Singh et al. 2019a). A thiocarbamate, niaziminin, a compound derived from *M. oleifera* leaf, inhibits tumour progression induced by Epstein–Barr virus (EBV) (Jung 2014). *M. oleifera* seeds and leaves have a high nutritional value, making them ideal for livestock feed supplementation to either increase livestock growth efficiency or to replace conventional crops and provide an economically viable source of feed (Abd El-Hack et al. 2018). *M. oleifera* also contains potentially nerve-paralysing phytochemicals, such as alkaloids and other phytotoxins, when taken in high doses (Maizuwo et al. 2017). Despite the lack of safety information on *M. Oleifera*'s usefulness, especially in humans, no scientifically proven side effects are found to date.

Moringa is a multi-purpose tree. The conservation of the moringa species is necessary for biodiversity, dietary, and pharmacological potentials. AVDRC (Taiwan), Rural development initiative (Zambia), Moringa Philippines foundation (Philippines), and Moringa community (Zambia) are amongst the many research centres contributing to *M. oleifera* improvement around the world. The National Bureau of Plant Genetic Resources (NBPGR), New Delhi, which is the country's nodal body for managing plant genetic resources, has gathered over 200 accessions mostly of cultivated *M. oleifera* from various geographical areas (Plant Germplasm Reporter 2000–2010). Moreover, in 2004, the Indian Council of Agricultural Research funded a project on "*Improvement of Underutilized Vegetable Crops*" at several cooperative centres of Indian Institute for Horticultural Research (IIHR), Bangalore. A collaborative network amongst National and International Research Centre, NGOs, and farmers who already work on *M. oleifera* must be built. Policymakers, research centres, and extension offices must develop programmes aimed at raising awareness amongst local communities and farmers. In order to gain a true understanding of the genetic diversity and structure of *M. Oleifera*, it is necessary to prioritize the collection and characterisation of both cultivated and natural accessions. To grow the moringa industry and improve livelihood, nutrition, and poverty, further research is required.

3.3.10 Winged Bean

Winged bean (*Psophocarpus tetragonolobus* L.) is grown mainly in hot and humid countries, such as India, Southeast Asia, Indonesia, Africa, Sri Lanka, Bangladesh, Myanmar, Thailand, Laos, Vietnam, Cambodia, Malaysia, Indonesia, and Philippines (Verdcourt and Halliday 1978). It is also known as asparagus pea or Goa bean. Winged bean is a dicotyledonous plant included under the Fabaceae family. It is an underutilized species, but it has the potential to become a major multi-purpose food crop. It is the only cultivated member of the genus. Almost every parts of the plant are edible such as seeds, pods, flowers, leaves, and tuberous roots whilst the stems and leaves are used as fodder. The young tender pods can be stewed, boiled, fried or roasted. Its tubers are used in a variety of cuisines in Ghana, Burma, and Papua New Guinea. It is used to make traditional Indonesian dishes like tempeh kecipir and snacks in Thailand (Erskine 1979). The roots are used to treat vertigo and the leaves are used in the treatment of small pox (Massart 1895). The pod extract is used to treat boils and ulcers in New Guinea (Perry and Metzger 1980). The seeds, tubers, pods, foliage, and flowers of the winged bean are especially nutritious. The leaves and flowers are high in protein (10–15%), whilst the seeds are about 35% protein and 18% fat (Kumar et al. 2018). Essential amino acids and proteins are similar to those found in soyabeans (Misra et al. 1987; Adegboyega et al. 2019) and thus this crop can be consumed as a substitute to soyabean. The crude protein content of winged bean seed (33.82%) is higher than that of cowpea (22.5%), pigeon pea (22.4%), and lima beans (23.3%), and comparable to that of soybeans (35%) (Makeri et al. 2017). It is also rich in vitamins A and C as well as calcium, and iron (Yang et al. 2018). Winged bean oil contains 60–70% poly unsaturated fatty acids which is important for oil stability against autooxidation compared to soybean and corn oil (Khor et al. 1980). These findings indicate that winged bean seed can be considered a source of good oil quality (Mohantya et al. 2020).

Winged bean has sufficient amount of protein in seeds (32–37%) followed by tubers (17–19%) and leaves (5–8% by weight) (Kantha and Erdman 1984; Lepcha et al. 2017). These properties make it highly beneficial crop for alleviating malnutrition in the population, particularly amongst the rural poor, where important vegetable crops demand large inputs (Tanzi et al. 2019). As a result, this crop has the potential to make a substantial contribution to global food and nutritional security.

The majority of PGRFA-based agricultural research in developing countries has been connected to the Consultative Group on International Agriculture Research (CGIAR's) network of germplasm conservation, documentation, characterisation, and distribution services (Renkow and Byerlee 2010). Other institutions with large, well-documented, and easily accessed germplasm collections, that are important in providing the genetic basis for breeding and development projects, include the USDA-ARS (Agricultural Research Service of the US Department of Agriculture), European genebanks, AVRDC (The World Vegetable Center), and CATIE (Centro Agronómico Tropical de Investigación y Enseñanza).

3.4 Organizations Involved in Improving and Maintaining Genetic Resources

Various international organizations have been established for research, development, and maintenance of germplasm collections in their gene banks as mentioned below:

3.4.1 International Centre for Underutilized Crops (ICUC)

It is a scientific research centre that was founded in 1988 with the goal of increasing the usage of underutilized crops with greater potential for enhanced output and environmental safety. It provides training and support research by germplasm collections of underutilized species in their seed banks. The main aim is to initiate, strengthen, and integrate an international network of scientists and programmes such as scientific conferences and training events committed to the development of these crops. ICUC has built professional networks such as UTFANET (Underutilized Tropical Fruits in Asia Network), UTVAPNET (Underutilized Tropical Vegetables for Asia and the Pacific Network), SEANUC (Southern and East Africa Network for

Underutilized Crops and ACUC (Asian Centre for Underutilized Crops) in 21 countries. It has released a collection of monographs on ten priority species and has organized suitable training sessions.

3.4.2 Global Facilitation Unit (GFU)

The International organization—Global Facilitation Unit for Underutilized Species (GFU) is an international organization that promotes the use of neglected plant species by disseminating knowledge and increasing awareness about their importance in improving lifestyles. It also examines national and international policies that allow or prevent the usage of these species.

3.4.3 Convention on Biological Diversity (CBD)

The Convention on biological diversity (CBD) was signed in Rio de Janeiro in 1992 and has been ratified by over 180 countries. Its targets include in situ and ex situ biodiversity protection, research and training, public awareness and education, assessing project impacts on biological diversity, providing financial resources, and sharing the profits of commercial and a fair and equitable sharing of income from commercial and other usage of genetic resources. The CBD provides institutional frameworks to serve as a tool for evaluating success and facilitating future growth.

3.4.4 Crops for the Future (CFF)

It is an independent, worldwide organization based in Sem, Malaysia, dedicated to promoting and facilitating the increased use of neglected and underutilized plant species in collaboration with its partners. The International Centre for Underutilized Crops (ICUC), based in Sri Lanka, and the Global Facilitation Unit for Underutilized Species, based in Italy, merged in 2008 to form it. It promotes cutting-edge research and policies aimed at diversifying agricultural systems and diets that improve both human health and the environment by utilizing underutilized crops. The Malaysian government and the University of Nottingham are funding the research through contracts acquired from a variety of sources, which include the UK Newton Fund, the EU, FAO, and the YSM Foundation (Gregory et al. 2019).

3.4.5 The International Plant Genetic Resources Institute (IPGRI)

The International Plant Genetic Resources Institute (IPGRI) is centred in Rome, Italy, and employs people from eight different nations throughout the world. It is one of the 16 international agricultural research centres funded by the Consultative

Group on International Agricultural Research (CGIAR), a collaboration of 41 donor agencies located all over the world. Its goal is to encourage plant genetic resource conservation and usage. It has published a series of monographs on 25 neglected crops with productive capacity (Padulosi et al. 2013). It also conducted a study on behalf of the System-wide Genetic Resources Programme (SGRP) to investigate into the number of species and collections maintained in the genebanks of the CGIAR Centres, as well as their characterization. IPGRI will give recommendations to SGRP and the CGIAR system based on these findings on how to best strategize their work on these species for the benefit of the community.

3.4.6 The United Nation Food and Agriculture Organization (FAO)

The United Nations Food and Agriculture Organization (FAO) promotes the use of locally adapted crops, cultivars, and underutilized species. It involves educating, research, and training on development of plant resources. It includes policy initiatives, pilot actions, and publications that document the widespread utilization of plant resources for food, nutrition, and livelihoods. FAO also negotiates and hosts the International Treaty on Plant Genetic Resources, which is a global accord that aims to protect plant genetic resources.

3.5 Constraints in Development of Underutilized Crops

Underutilized crops encounter enormous hurdles in expanding their use outside their conventional niches. Furthermore, certain crops have unpleasant features that must be modified in order to improve yield and nutritional quality. The following are the limitations that are directly related to some of these crops:

Low productivity: These crops provide inadequate yields in terms of both quality and quantity due to limited information on production, consumption, and utilization of many underutilized plant products, limited germplasm availability, and a lack of superior quality material. Seed yields of African rice (*Oryza glaberrima*), teff, and millets, for example, are exceedingly low. Lodging, seed breakage, and other crop-specific barriers are the main causes of low yield in these crops (Assefa et al. 2011). There is no well-catalogued technical information on genetic diversity, developmental stages, and life cycle of these crops as well as agronomy and breeding methods, are all lacking in study. Furthermore, due to a lack of scientific resources for testing, evaluation, and post-harvest management of several species, production is lowered.

Toxic products: Cassava and grass pea, two extensively farmed underutilized crops, contain harmful compounds that harm human and cattle health. The toxic compound cyanogenic glycosides are produced by the edible parts of cassava and releases cyanide, resulting in the paralytic illness called konzo (Ceballos et al. 2004). Similarly, grass pea grains contain ODAP (-*N*-oxalyl-L-,

-diaminopropanoic acid), a neurotoxic chemical that causes irreversible paralysis of the lower body when taken over time (Yan et al. 2006). Some compounds, particularly tannins and phenolic compounds found in winged bean, block enzyme function in a non-specific manner and make a complex with dietary proteins, lowering their quality (Kantha et al. 1986).

3.6 Conclusion

Increasing food demand requires the exploitation of less known, i.e. underutilized crops as a source of functional food as these crops have multiple uses, such as food, medicinal, ecological, ornamental, etc. They also lead to a healthy ecosystem because of their ability to thrive in arid environments and resist a variety of abiotic stress factors. In comparison to the biofortification of major crops, underutilized crops are better options, as they are naturally fortified and have potential to fight malnutrition in developing countries. So the germplasm of these crops should be characterized for tracing important traits employed in breeding programme, identification of candidate genes responsible for nutritional traits and abiotic stress tolerance response in these crops. Various institutions have been working on these aspects and some initiatives are made to research on their improvement and marketing. Many efforts are urgently required in research for the improvement of these crops as well as formulation of strategic work plans to improve germplasm selection, supply, production, processing, and commercialization.

References

Abd El-Hack M, Alagawany M, Elrys A, Desoky ES, Tolba H, Elnahal A, Elnesr S, Swelum A (2018) Effect of Forage *Moringa oleifera* L. (moringa) on animal health and nutrition and its beneficial applications in soil, plants and water-purification. Agriculture 8:145

Abraham EM, Ganopoulos I, Madesis P, Mavromatis A, Mylona P, Nianiou-Obeidat I et al (2019) The use of Lupin as a source of protein in animal feeding: genomic tools and breeding approaches. Int J Mol Sci 20:851. https://doi.org/10.3390/20040851

Adebowale ARA, Emmambux MN, Beukes M, Taylor JR (2011) Fractionation and characterization of teff proteins. J Cereal Sci 54(3):380–386

Adegboyega TT, Abberton MT, AbdelGadir AH, Dianda M, Maziya-Dixon B, Oyatomi OA, Ofodile S, Babalola OO (2019) Nutrient and antinutrient composition of winged bean (*Psophocarpus tetragonolobus* (L.) DC.) seeds and tubers. J Food Qual 2019:3075208. https://doi.org/10.1155/2019/3075208

Aizat WM, Ahmad-Hashim FH, Syed Jaafar SN (2019a) Valorization of mangosteen, "the queen of fruits," and new advances in postharvest and in food and engineering applications: a review. J Adv Res 20:61–70

Aizat WM, Jamil IN, Ahmad-Hashim FH, Noor NM (2019b) Recent updates on metabolite composition and medicinal benefits of mangosteen plant. PeerJ 7:e6324

Ajithkumar IP, Panneerselvam R (2014) ROS scavenging system, osmotic maintenance, pigment and growth status of *Panicum sumatrense* roth. under drought stress. Cell Biochem Biophys 68:587–595. https://doi.org/10.1007/s12013-013-9746-x

Akande K (2016) Effects of anti-nutritive constituents of unprocessed jack bean (*Canavalia ensiformis*) (L) (DC) seed meal on poultry performance. Am J Exp Agric 10:1–10

Akpapunam M, Darbe J (1994) Chemical composition and functional properties of blends of maize and bambara groundnut flours for cookie production. Plant Foods Hum Nutr (Formerly Qualitas Plantarum) 46(2):147–155

Akubugwo IE, Obasi AN, Ginika SC (2007) Nutritional potential of the leaves and seeds of black nightshade-*solanum nigrum* l. Var virginicum from Afikpo-Nigeria. Pak J Nutr 6:323–326

Ali M, Kenganora M, Manjula SN (2016) Health benefits of *Morinda citrifolia* (Noni): a review. Pharm J 8(4):321–334

American Diabetes Association (2010) Diagnosis and classification of diabetes mellitus. Diabetes Care 33:S62–S69. https://doi.org/10.2337/dc10-S062

Anjali K, Joshi A, Maloo SR, Sharma R (2013) Assessment of the morphological and molecular diversity in *Amaranthus* spp. Afr J Agric Res 8:2307–2311

Anwar F, Latif S, Ashraf M, Gilani AH (2007) *Moringa oleifera*: a food plant with multiple medicinal uses. Phytother Res 21:17–25

AOCC (2018) African orphan crops consortium: healthy Africa through nutritious, diverse and local food crops. https://africanorphancrops.org/

Arzani A, Ashraf M (2017) Cultivated ancient wheats (*Triticum* spp.): a potential source of health-beneficial food products. Compr Rev Food Sci Food Saf 16(3):477–488. https://doi.org/10.1111/1541-4337.12262

Ashraf MY, Ashraf M, Ozturk M (2018) Underutilized vegetables: a tool to address nutritional issues, poverty reduction and food security. In: Ozturk M, Hakeem K, Ashraf M, Ahmad M (eds) Global perspectives on underutilized crops. Springer, Cham

Assefa K, Yu JK, Zeid M, Belay G, Tefera H, Sorrells ME (2011) Breeding tef [*Eragrostis tef* (Zucc.) trotter]:conventional and molecular approaches. Plant Breed 130(1):1–9. https://doi.org/10.1111/j.1439-0523.2010.01782.x

AVRDC (2019) World vegetable center. https://avrdc.org. Accessed 29 Jan 2019

Ayalew A, Kena K, Dejene T (2011) Application of NP Fertilizers for better production of Teff (*Eragrostis tef* (Zucc.) Trotter) on different types of soils in southern Ethiopia. J Nat Sci Res 1 (1):6–15

Ayenan MAT, Ezin VA (2016) Potential of Kersting's groundnut [*Macrotyloma geocarpum* (Harms) Maréchal & Baudet] and prospects for its promotion. Agric Food Secur 5:10 https://doi.org/10.1186/s40066-016-0058-4

Badrie N, Broomes J (2010) Beneficial uses of breadfruit (*Artocarpus altilis*): nutritional, medicinal and other uses. In: Bioactive foods in promoting health, pp 491–505. https://doi.org/10.1016/B978-0-12-374628-3.00033-5

Barba de la Rosa AP, Fomsgaard IS, Laursen B, Mortensen AG, Olvera-Martínez L, Silva-Sánchez C, Mendoza-Herrera A, González-Castañeda J, De León-Rodríguez A (2009) Amaranth (*Amaranthus hypochondriacus*) as an alternative crop for sustainable food production: phenolic acids and flavonoids with potential impact on its nutraceutical quality. J Cereal Sci 499 (1):117–121

Bavec F (2014) Organic production of cash cereals and pseudocereals. In: Pilipavicius V (ed) Organic agriculture towards sustainability. IntechOpen, Rijeka, pp 227–246. https://doi.org/10.5772/57475

Bazile D (2015) Le Quinoa, Les Enjeux D'une Conquête. Editions Quae, Versailles

Bazile D, Fuentes F, Mujica A (2013) Historical perspectives and domestication. In: Bhargava A, Srivastava S (eds) Quinoa: botany, production and uses. CABI, Wallingford, pp 16–35

Bazile D, Martinez EA, Fuentes F (2014) Diversity of quinoa in a biogeographical Island: a review of constraints and potential from arid to temperate regions of Chile. Nat Bot Horti Agrobot Cluj Napoca 42:289–298. https://doi.org/10.15835/nbha.42.2.9733

Bekkering CS, Tian L (2019) Thinking outside of the cereal box: breeding underutilized (pseudo) cereals for improved human nutrition. Front Genet 10:1289. https://doi.org/10.3389/fgene.2019.01289

Bennett RN, Mellon FA, Foidl N, Pratt JH, DuPont MS, Perkins L, Kroon PA (2003) Profiling glucosinolates and phenolics in vegetative and reproductive tissues of the multi-purpose trees *Moringa oleifera* L. (Horseradish tree) and *Moringa stenopetala* L. J Agric Food Chem 51:3546–3553

Berchie JN, Amelie G, McClymont S, Raizada M, Adu-Dapaah H, Sarkodie-Addo J (2013) Performance of 13 Bambara groundnut (*Vigna subterranea* (L.) Verdc.) landraces under 12 H and 14 H photoperiod. J Agron 12:20–28

Bermejo JEH, León J (eds) (1994) Neglected crops: 1492 from a different perspective. Plant Production and Protection Series no. 26. FAO, Rome. ISBN: 9251032173

Bhaskaran J, Panneerselvam R (2013) Accelerated reactive oxygen scavenging system and membrane integrity of two *Panicum* species varying in salt tolerance. Cell Biochem Biophys 67:885–892. https://doi.org/10.1007/s12013-013-9576-x

Bhatt D, Negi M, Sharma P, Saxena SC, Dobriyal AK, Arora S (2011) Responses to drought induced oxidative stress in five finger millet varieties differing in their geographical distribution. Physiol Mol Biol Plants 17:347–353. https://doi.org/10.1007/s12298-011-0084-4

Boakye AA, Wireko-Manu FD, Oduro I, Ellis WO, Gudjónsdóttir M, Chronakis IS (2018) Utilizing cocoyam (*Xanthosoma sagittifolium*) for food and nutrition security: a review. Food Sci Nutr 6 (4):703–713. https://doi.org/10.1002/fsn3.602

Bonifácia G, Galli V, Francisci R, Mair V, Skrabanja V, Kreft I (2000) Characteristics of spelt wheat products and nutritional value of spelt wheat-based bread. Food Chem 68:437–441

Brauckmann BM, Latté KP (2010) L-Dopa deriving from the beans of *Vicia faba* and *Mucuna pruriens* as a remedy for the treatment of Parkinson's disease. Schweiz Z Ganzheitsmed 22:292–300

Brenner DM, Baltensperger DD, Kulakow PA, Lehmann JW, Myers RL, Slabbert MM, Sleugh BB (2010) Genetic resources and breeding of *Amaranthus*. Plant Breed Rev 19:227–285

Bressani R, Gonzales JM, Zuniga J, Breuner M, Elias LG (1987) Yield, selected chemical composition and nutritive value of 14 selections of amaranth grain representing four species. J Sci Food Agric 38:347–356

Caballero L, Martin LM, Alvarez JB (2004) Genetic variability of the low-molecular-weight glutenin subunits in spelt wheat (*Triticum aestivum* ssp. spelta L. em Thell.). Theor Appl Genet 108:914–919

Caetano BFR, de Moura NA, Almeida APS, Dias MC, Sivieri K, Barbisan LF (2016) Yacon (*Smallanthus sonchifolius*) as a food supplement: health-promoting benefits of fructooloigosaccharides. Nutrients 8:436. https://doi.org/10.3390/nu8070436

Cannarozzi G, Chanyalew S, Assefa K, Bekele A, Blosch R, Weichert A, Klauser D, Plaza-Wuthrich S, Esfeld K, Jost M, Rindisbacher A, Jifar H, Johnson-Chadwick V, Abate E, Wang WY, Kamies R, Husein N, Kebede W, Tolosa K, Genet Y, Gebremeskel K, Ferede B, Mekbib F, Martinelli F, Pedersen HC, Rafudeen S, Hussein S, Tamiru M, Nakayama N, Robinson M, Barker I, Zeeman S, Tadele Z (2018) Technology generation to dissemination: lessons learned from the tef improvement project. Euphytica 214(2):31. https://doi.org/10.1007/s10681-018-2115-5

Cao Y, Ma ZF, Zhang H, Jin Y, Zhang Y, Hayford F (2018) Physiochemical properties and nutrigenomic implications of yacon as a potential source of prebiotic: Current evidence and future directions. Foods 7:59

Ceballos H, Iglesias CA, Pérez JC, Dixon AGO (2004) Cassava breeding: opportunities and challenges. Plant Mol Biol 56:503–515. https://doi.org/10.1007/s11103-004-5010-5

Chandel G, Meena R, Dubey M, Kumar M (2014) Nutritional properties of minor millets: neglected cereals with potentials to combat malnutrition. Curr Sci 107:1109–1111

Chateil C, Goldringer I, Tarall L, Kerbiriou C, Le Viol I, Ponge J-F, Salmon S, Gachet S, Porcher E (2013) Crop genetic diversity benefits farmland biodiversity in cultivated fields. Agric Ecosyst Environ 171:25–32

Cheng A (2018) Shaping a sustainable food future by rediscovering long-forgotten ancient grains. Plant Sci 269:136–142. https://doi.org/10.1016/j.plantsci.2018.01.018

Chingwaru W, Faria ML, Saravia C, Cencic A (2007) Indigenous knowledge of Health Benefits of morama plant among respondents in Ghantsi and Jwaneng of Botswana. Afr J Food Agric Nutr Dev 7(6)

Chingwaru W, Majinda RT, Yeboah SO, Jackson JC, Kapewangolo PT, Kandawa-Schulz M, Cencic A (2011) *Tylosema esculentum* (Marama) Tuber and bean extracts are strong antiviral agents against rotavirus infection. Evid Based Complement Alternat Med 2011(6):284795. https://doi.org/10.1155/2011/284795

Chivenge P, Mabhaudhi T, Modi AT, Mafongoya P (2015) The potential role of neglected and underutilised crop species as future crops under water scarce conditions in Sub-Saharan Africa. Int J Environ Res Public Health 12:5685–5711

Clayton WD, Vorontsova MS, Harman KT, Williamson H (2016) Grass base-the online world grass flora descriptions. http://www.kew.org/data/grassesdb/sppindex.htm. Accessed Sept 2019

Comino I, de Lourdes MM, Real A, Rodríguez-Herrera A, Barro F, Sousa C (2013) The gluten-free diet: testing alternative cereals tolerated by celiac patients. Nutrients 5:4250–4268. https://doi.org/10.3390/nu5104250

Coronel OADA, Leo-Garcia E, Vela-Gutierrez G, Medina JDC, Garcia-Varela R, Garcia H (2017) Chayote (*Secium edule* (Jacq.) Swartz). https://doi.org/10.1002/97811158042.ch47

Costanzo A, Amos DC, Dinelli G, Sferrazza RE, Accorsi G, Negri L, Bosi S (2019) Performance and nutritional properties of einkorn, emmer and rivet wheat in response to different rotational position and soil tillage. Sustainability 11:6304

Dahot MU (1988) Vitamin contents of the flowers and eeds of *Moringa oleifera*. Pak J Biochem 21:1–2

DanMalam HU, Abubakar Z, Katsayal UA (2001) Pharmacognosostic studies on the leaves of *Moringa oleifera*. Nig J Nat Prod Med 5:45–49

Das S (2016) Amaranthus: a promising crop of future. Springer, Singapore. https://doi.org/10.1007/978-981-10-1469-7_1

Davison J, Laca M (2009) Grain production of 15 teff varieties grown in Churchill County Nevada, during 2009, University of Nevada Cooperative Extension

de Oliveira do Nascimento K, do Nascimento Dias Paes S, Reis de Oliveira I, Pereira Reis I, Maria Augusta I (2018) Teff: suitability for different food applications and as a raw material of gluten-free, a literature review. J Food Nutr Res 6(2):74–81. https://doi.org/10.12691/jfnr-6-2-2

de Wet JMJ, Prasada Rao KE, Brink DE (1983) Systematics and domestication of *Panicum sumatrense* (Graminae). J d'agriculture Tradit Bot appliquée 30:159–168

Demeke M, Di Marcantonio F (2013) Analysis of incentives and disincentives for teff in Ethiopia. Technical notes series, MAFAP, FAO, Rome

Devi ND (2017) Medicinal values of *Trichosanthus cucumerina* L.(Snake Gourd)-A Review. Br J Pharm Res 16:1–10

Devi PB, Vijayabharathi R, Sathyabama S, Malleshi NG, Priyadarisini VB (2014) Health benefits of finger millet (*Eleusine coracana* L.) polyphenols and dietary fiber: a review. J Food Sci Technol 51:1021–1040. https://doi.org/10.1007/s13197-011-0584-9

Duc G, Bao S, Baum M, Redden B, Sadiki M, Suso MJ, Vishniakova M, Zong X (2010) Diversity maintenance and use of *Vicia faba* L. genetic resources. Field Crops Res 115:270–278

Duc G, Aleksić JM, Marget P, Mikić A, Paull J, Redden RJ, Sass O, Stoddard FL, Vandenberg A, Vishnyakova M, Torres AM (2015) Faba bean. In: De Ron AM (ed) Grain legumes. Springer, New York, pp 141–178. https://doi.org/10.1007/978-1-4939-2797-5_5

Dwivedi S, Upadhyaya H, Senthilvel S, Hash C, Fukunaga K, Diao X, Santra X, Baltensperge D, Prasad M (2012) Millets: genetic and genomic resources. In: Janick J (ed) Plant breeding reviews. Wiley, Hoboken, NJ, pp 247–374

Ebifa-Othieno E, Kabasa JD, Nyeko P, Nakimbugwe D, Mugisha A (2020) Nutritional potential of tamarind (*Tamarindus indica* L.) from semi-arid and subhumid zones of Uganda. Food Measure 14:1125–1134. https://doi.org/10.1007/s11694-019-00362-7

EcoBusiness (2015) What can nearly 12,000 'orphan crops' do to address the nutrition gap? https://www.ecobusiness.com/news/what-can-nearly-12000-orphan-crops-do-toaddress-the-nutrition-gap/. Accessed 7 Jan 2019

Erskine W (1979) The exploitation of genetic diversity in the winged bean, *Psophocarpus tetragonolobus* (L.) DC. for grain yield. Dissertation, Cambridge University

Espitia E (1992) Amaranth germplasm development and agronomic studies in Mexico. Food Rev Int 8:71–86

FAO (2009) FAO Stat. Food and Agriculture Organisation of the United Nations, Rome, Italy. http://faostat.fao.org/default.aspx. Accessed 1 Oct 2018

FAO (2015) Analysis of price incentives for Teff in Ethiopia Technical notes series, MAFAP, by Assefa B. Demeke M., Lanos B, Rome

FAO (2017a) Promoting neglected and underutilized crop species. Food and Agriculture Organization (FAO), Rome. http://www.fao.org/news/story/en/item/1032516/icode/. Accessed 7 Apr 2019

FAO (2017b) FAOSTAT database. Food and Agriculture Organization of the United Nations. www.fao.org/faostat/

FAOSTAT (2018) FAO (Food and Agricultural Organization) statistical data. FAO statistics division. http://www.fao.org/faostat/en/#data. Accessed 8 Jan 2019

Fayaz M, Bhati MH, Kumar A, Jain AK (2019) Phytochemical screening and nutritional analysis of some parts of *Celosia argentea* L. Chem Sci Trans 8(1):12–19

Feldman M, Kislev M (2007) Domestication of emmer wheat and evolution of free-threshing tetraploid wheat. Isr J Plant Sci 55:207–221

Fuglie L (1999) The miracle tree: *Moringa oleifera*: natural nutrition for the tropics. Church World Service, Moringa. Church World Service, Dakar, 172pp

Galterio G, Codianni P, Giusti B, Pezzarosa B, Cannella C (2003) Assessment of the agro- nomical and technological characteristics of *Triticum turgidum* ssp. *dicoccum* Shrank and spelta. Nahrung/Food 47:54–59

Ganguly S (2013) Indian ayurvedic and traditional medicinal implications of indigenously available plants, herbs and fruits: a review. Int J Res Ayurveda Pharm 4:623–625

Gebauer J, Adam YO, Sanchez AC, Muneer DD, Eltahir ES, Fadl EM, Fernsebner G, Frei M, Habte TY, Hammer K, Hunsche M, Johnson H, Kordofani M, Krawinkel M, Kugler F, Luedeling E, Mahmoud TE, Maina A, Mithofer D, Chimuleke RY, Georg Noga M, Willis RN, Owino O, Prinz K, Rimberia FK, Saied A, Schüring M, Sennhenn A, Spath MA, Taha MEN, Triebel A, Wichern F, Wiehle M, Wrage-Monnig N, Kehlenbeck K (2016) Africa's wooden elephant: the baobab tree (*Adansonia digitata* L.) in Sudan and Kenya: a review. Genet Resour Crop Evol 63:377–399

Gebremariam MM, Zarnkow M, Becker T (2012) Teff (*Eragrostis teff*) as a raw material for malting, brewing and manufacturing of gluten-free foods and beverages: a review. J Food Sci Tech 51:2881–2895

Geervani P, Eggum BO (1989) Nutrient composition and protein quality of minor millets. Plant Foods Hum Nutr 39:201–208. https://doi.org/10.1007/BF01091900

Gimenez-Bastida J, Piskuła M, Zielinski H (2015) Recent advances in processing and development of buckwheat derived bakery and non-bakery products-a review. Pol J Food Nutr Sci 65(1):9–20

Giraldo P, Royo C, Gonza'lez M, Carrillo J, Ruiz M (2016) Genetic diversity and association mapping for agro-morphological and grain Quality traits of a structured collection of durum wheat landraces including subsp. durum, turgidum and *dicoccon*. PLoS One 11:0166577. https://doi.org/10.1371/journal.pone.0166577

Girija K, Lakshman K, Chandrika U, Ghosh SS, Divya T (2011) Anti-diabetic and anti-cholesterolemic activity of methanol extracts of three species of Amaranthus. Asian Pac J Trop Biomed:133–138. ISSN 2221-1691

Girma D, Cannarozzi G, Weichert A, Tadele Z (2018) Genotyping by sequencing reasserts the close relationship between tef and its putative wild Eragrostis progenitors. Diversity 10(2):17

Gomashe SS (2016) Barnyard millet: present status and futurethrust areas. In: Millets and sorghum: biology and genetic improvement. Wiley, Hoboken, NJ, pp 184–198

Gopalakrishnan L, Doriya K, Kumar DS (2016) *Moringa oleifera*: a review on nutritive importance and its medicinal application. Food Sci Hum Wellness 5(2):49–56

Goron TL, Raizada MN (2015) Genetic diversity and genomic resources available for the small millet crops to accelerate a New Green Revolution. Front Plant Sci 6:157. https://doi.org/10.3389/fpls.2015.00157

Gregory PJ, Mayes S, Hui CH, Jahanshiri E, Julkifle A, Kuppusamy G, Kuan HW, Lin TX, Massawe F, Suhairi T, Azam-Ali SN (2019) Crops For the Future (CFF): an overview of research efforts in the adoption of underutilized species. Planta 250:979–988. https://doi.org/10.1007/s00425-019-03179-2

Gupta A, Mahajan V, Kumar M, Gupta H (2009) Biodiversity in the barnyard millet (*Echinochloa frumentacea* Link. Poaceae) germplasm in India. Genet Resour Crop Evol 56:883–889. https://doi.org/10.1007/s10722-009-9462-y

Gürbüz N, Uluişik S, Frary A, Frary A, Doğanlar S (2018) Health benefits and bioactive compounds of eggplant. Food Chem 268:602–610. https://doi.org/10.1016/j.foodchem.2018.06.093

Habiyaremye C, Matanguihan JB, D'Alpoim Guedes J, Ganjyal GM, Whiteman MR, Kidwell KK, Murphy KM (2017) Proso millet (*Panicum miliaceum* L.) and its potential for cultivation in the Pacific Northwest,U.S.: a review. Front Plant Sci 7:1961. https://doi.org/10.3389/fpls.2016.01961

Habte E, Muktar MS, Negawo AT, Lee S-H, Lee K-W, Jones CS (2019) An overview of Teff (*Eragrostis teff Zuccagni*) Trotter) as a potential summer forage crop in temperate systems. J Korean Soc Grassl Forage Sci 39(3):185–188. https://doi.org/10.5333/KGFS.2019.39.3.185

Harlan JR, Zohary D (1966) Distribution of wild wheats and barley. Science 153:1074–1080

Hegedus A, Papp N, Stefanovits-Banyai E (2013) A review of nutritional value and putative health-effects of quince (*Cydonia oblonga* Mill.) fruit. Int J Hortic Sci 19(3–4):29–32

Hekmat S, Morgan K, Soltani M, Gough R (2015) Sensory evaluation of locally grown fruit purees and inulin fibre on probiotic yogurt in Mwanza, Tanzania and the microbial analysis of probiotic yogurt fortified with *Moringa oleifera*. J Health Popul Nutr 33:60–67

Heuzé V, Thiollet H, Tran G, Bastianelli D, Lebas F (2016) Teff (*Eragrostis teff*) grain. Feedipedia. Food Agric Organ 13:39

Hiremath SC, Patil GNV, Salimath SS (1990) Genome homology and origin of *Panicum sumatrense* (Gramineae). Cytologia (Tokyo) 55:315–319

Ho P (1975) The cradle of the east. University of Chicago Press, Chicago, IL, p 440

Houndonougbo JSH, Kassa B, Mensah S, Salako VK, Glèlè Kakai R, Assogbadjo AE (2020) A global systematic review on conservation and domestication of *Parkia biglobosa* (Jacq.) R. Br. ex G. Don, an indigenous fruit tree species in Sub-Sahara African traditional parklands: Current knowledge and future directions. Genet Resour Crop Evol. https://doi.org/10.1007/s10722-020-00892-w

IIMR (2018) Annual report 2017–18. Indian Institute of Millets Research, Hyderabad, 164p

IPGRI (1999) Directory of germplasm collection. Online database, Rome. http://www.cgiar.org/ipgri/doc/dbintro.htm

Issa JY (2012) In vitro calcium bioaccessibility in *Moringa oleifera* vegetable leaves: potential plant food to increase dietary calcium intake in developing countries. A thesis submitted to the Graduate Faculty of North Carolina State University in partial fulfillment of the requirements for the degree of Master of Science Food Science. http://repository.lib.ncsu.edu/ir/bitstream/1840.16/7742/1/etd.pdf

Janković S, Ikanović J, Popović V, Rakić S, Kuzevski J (2013) Agro-ecological conditions and morpho-productive properties of spelt wheat. Biotechnol Anim Husb 29:547–554

Janovská D, Cepkova PH, Dzunkova M (2012) Characterisation of the amaranth genetic resources in the Czech Gene Bank. In: Caliskan M (ed) Genetic diversity in plants. Online: InTech, pp 457–478

Jaradat AA (2018) Comparative assessment of einkorn and emmer wheat phenomes: I. Plant architecture. Genet Resour Crop Evol 66:491–512. https://doi.org/10.1007/s10722-018-0729-z
Jayaraman A, Puranik S, Rai NK, Vidapu S, Sahu PP, Lata C, Prasad M (2008) cDNA-AFLP analysis reveals differential gene expression in response to salt stress in foxtail millet (*Setaria italica* L.). Mol Biotechnol 40:241–251. https://doi.org/10.1007/s12033-008-9081-4
Jimenez ME, Rossi A, Samman N (2015) Health properties of oca (*Oxalis tuberosa*) and yacon (*Smallanthus sonchifolius*). Food Funct 6:3266–3274
Jin X (2018) Jujuba—*Ziziphus jujube* exotic fruits. https://doi.org/10.1016/B978-0-12-803138-4.00034-4
Jung IL (2014) Soluble extract from *Moringa oleifera* leaves with a new anticancer activity. PLoS One 9:e95492
Kalinova J, Moudry J (2006) Content and quality of protein in proso millet (*Panicum miliaceum* L.) varieties. Plant Foods Hum Nutr 61(1):45–49
Kalinowski LS, Navarro JP, Concha AIR, Hermoza GC, Pacheco RA, Choquevilca YC, Jara EV (1992) Grain amaranth research in Peru. Food Rev Int 8:87–124
Kandoliya UK, Marviya GV, Bodar NP, Bhadja NV, Golakiya BA (2016) Nutritional and antioxidant components of ridge Gourd (*Luffa acutangula* L.Roxb) fruits of promising genotypes and varieties. Sch J Agric Vet Sci 3(5):397–401
Kantha SS, Erdman JW (1984) The winged bean as an oil and protein source: a review. J Am Oil Chem Soc 61(3):515–525
Kantha SS, Hettiarachchy NS, Erdman JJW (1986) Nutrient, antinutrient contents and solubility profiles of nitrogen, phytic acid and selected minerals in winged bean flour. Cereal Chem 63 (1):9–13
Karkanis A, Ntatsi G, Lepse L, Fernández JA, Vågen IM, Rewald B, Alsina I, Kronberga A, Balliu A, Olle M, Bodner G, Dubova L, Rosa E, Savvas D (2018) Faba bean cultivation–revealing novel managing practices for more sustainable and competitive European cropping systems. Front Plant Sci 9:1115. https://doi.org/10.3389/fpls.2018.01115
Katoch R (2013) Nutritional potential of rice bean (*Vigna Umbellata*): an underutilized legume. J Food Sci 78(1):C8–C16. https://doi.org/10.1111/j.1750-3841.2012.02989.x
Kauffman CS, Weber LE (1990) Grain amaranth. In: Janick J, Simon JE (eds) Advances in new crops. Timber Press, Portland, pp 127–139
Kebede W, Tolossa K, Hussein N, Fikre T, Genet Y, Bekele A, Gebremeskel K, Fentahun A, Daba H, Plaza-Wüthrich S, Blösch R, Chanyalew S, Asefa K, Tadele Z (2018) Tef (*Eragrostis tef*) Variety Tesfa. Ethiop J Agric Sci 28(2):107–112
Kendabie P, Massawe F, Mayes S (2015) Developing genetic mapping resources from landrace-derived genotypes that differ for photoperiod sensitivity in Bambara groundnut (*Vigna subterranea* L.). Aspects Appl Biol Breed Plants Future Clim Chan 124:49–56
Ketema S (1997) Tef. *Eragrostis tef* (Zucc.) Trotter. Promoting the conservation and use of underutilized and neglected crops, vol 12. Institute of Plant Genetics and Crop Plant Research, Gatersleben/International Plant Genetic Resources Institute, Rome
Khan F, Azman R, Chai HH, Mayes S, Lu C (2016) Genomic and transcriptomic approaches towards the genetic improvement of an underutilised crops: the case of Bambara groundnut. Afr Crop Sci J 24:429–458
Khor HT, Tan NH, Wong KC (1980) Winged bean seed: potential food source of the tropics. In: Proc 6th Conf Malays Biochem Soc, pp 157–62
Koivunen E, Partanen K, Perttilä S, Palander S, Tuunainen P, Valaja J (2016) Digestibility and energy value of pea (*Pisum sativum* L.), faba bean (*Vicia faba* L.) and blue lupin (narrow-leaf) (*Lupinus angustifolius*) seeds in broilers. Anim Feed Sci Technol 218:120–127. https://doi.org/10.1016/j.anifeedsci.2016.05.007
Kulkarni LR, Naik RK (2000) Nutritive value, protein quality and organoleptic quality of kodo millet (*Paspalum scrobiculatum*). Karnataka J Agric Sci 13:125–129

Kumar R, Rajasree V, Sagar L, Tripura U, Karthick K, Karthick R, Prasath G (2018) Underutilized vegetables as rich source of medicinal value: a boon for the pharmaceutical industries and farmer income. Int J Chem Stud 6(4):3320–3323

Kuru P (2014) *Tamarindus indica* and its health related effects. Asian Pac J Trop Biomed 4:676–681

Lachman J, Hejtmankova K, Kotikova Z (2013) Tocols and carotenoids of einkorn, emmer and sprint wheat varieties: selection for breeding and production. J Agric Food Chem 57:207–214

Lambein F, Travella S, Kuo Y-H, Montagu MV, Heijde M (2019) Grass pea (*Lathyrus sativus* L.): orphan crop, neutraceutical or just plain food? Planta 250:821–838. https://doi.org/10.1007/s00425-018-03084-0

Landry EJ, Coyne CJ, McGee RJ, Hu J (2016) Adaptation of autumn sown faba bean germplasm to south eastern Washington. Agron J 108:301–308

Langley P, Pergolizzi J, Taylor R, Ridgway C (2014) Antioxidant and associated capacities of camu-camu (*Myrciaria dubia*): a systematic review. J Altern Complement Med. https://doi.org/10.1089/acm.2014.0130

Lee H (2018) Teff, a rising global crop: current status of teff production and value chain. Open Agric J 12:194–206

Lepcha P, Egan AN, Doyle JJ, Sathyanarayana N (2017) A review on current status and future prospects of winged bean (*Psophocarpus tetragonolobus*) in tropical agriculture. Plant Foods Hum Nutr 72:225–235. https://doi.org/10.1007/s11130-017-0627-0

Lepse L, Dane S, Zeipiņa S, Domínguez-Perles R, Rosa EA (2017) Evaluation of vegetable-faba bean (*Vicia faba* L.) intercropping under Latvian agro-ecological conditions. J Sci Food Agric 97:4334–4342. https://doi.org/10.1002/jsfa.8239

Li X, Siddique KHM (2020) Future smart food: harnessing the potential of neglected and underutilized species for Zero Hunger. Matern Child Nutr 16(S3):e13008

Li J, Chen L, Li J, Duan Z, Zhu S, Fan L (2017) The composition analysis of maca (*Lepidium meyenii* Walp.) from Xinjiang and its antifatigue activity. J Food Qual:1–7. https://doi.org/10.1155/2017/8785121

Liu R, Zheng Y, Cai Z, Xu B (2017) Saponins and flavonoids from adzuki bean (*Vigna angularis* L.) ameliorate high-fat diet-induced obesity in ICR mice. Front Pharmacol 8:1–8

Lizarazo CI, Lampi AM, Sontag-Strohm T, Liu J, Piironen V, Stoddard FL (2015) Nutritive quality and protein production from grain legumes in a boreal climate. J Sci Food Agric 95:2053–2064. https://doi.org/10.1002/jsfa.6920

Longobardi F, Sacco D, Casiello G, Ventrella A, Sacco A (2015) Chemical profile of the carpino broad bean by conventional and innovative physicochemical analyses. J Food Qual 38:273–284. https://doi.org/10.1111/jfq.12143

Lutz M, Martínez EA, Martínez A (2013) Daidzein and genistein contents in seeds of quinoa (*Chenopodium quinoa* Willd) from local ecotypes grown in arid Chile. Ind Crop Prod 49:117–121. https://doi.org/10.1016/j.indcrop.2013.04.023

Maizuwo AI, Hassan AS, Momoh H, Muhammad JA (2017) Phytochemical constituents, biological activities, therapeutic potentials and nutritional values of *Moringa oleifera* (Zogale): a review. J Drug Des Med 3(4):60–66

Makeri MU, Abdulmannan F, Ilowefah MA, Chiemela C, Bala SM, Muhammad K (2017) Comparative physio-chemical, functional and structural characteristics of winged bean [*Psophocarpus tetragonolobus* DC] and soybean [*Glycine max*] protein isolates. J Food Meas Charact 11:835–846. https://doi.org/10.1007/s11694-016-9455-4

Makita C, Chimuka L, Steenkamp P, Cukrowska E, Madala E (2016) Comparative analyses of flavonoid content in *Moringa oleifera* and *Moringa ovalifolia* with the aid of UHPLC-qTOF-MS fingerprinting. S Afr J Bot 105:116–122

Malik AA, Chaudhary G (2018) Global food security: a truncated yield of underutilized and orphan crops. https://doi.org/10.1007/978-3-319-92399-4_11

Mariod AA, Abdelwahab SI (2012) *Sclerocarya birrea* (Marula), an African tree of nutritional and medicinal uses: a review. Food Rev Int 28(4):375–388

Martinez EA, Fuentes F, Bazile D (2015) History of Quinoa: its origin, domestication, diversification and cultivation with particular reference to the chilean context. In: Murphy K, Matanguihan J (eds) Quinoa: improvement and sustainable production. Wiley-Blackwell, Hoboken, pp 19–24. https://doi.org/10.1002/9781118628041.ch2

Massart J (1895) Un Botaniste in Malaise. C. Annoot-Braeckman, Brussells

Massawe FJ, Mayes S, Cheng A, Chai HH, Cleasby P, Symonds R, Ho WK, Siise A, Wong QN, Kendabie P, Yanusa Y, Jamalluddin N, Singh A, Azman R, Azam-Ali SN (2015) The potential for underutilised crops to improve food security in the face of climate change. Agric Clim Change 29:140–141. https://doi.org/10.1016/j.proenv.2015.07.228

Mathews MM (2001) India: facts and figures. Sterling Publishers Pvt. Ltd, New Delhi

Maureira H, Martínez EA (2012) Nutritional aspects of six quinoa (*Chenopodium quinoa* Willd.) ecotypes from three geographical areas of Chile. Chilean J Agric Res 72:175–181. https://doi.org/10.4067/S0718-58392012000200002

Mayes S, Massawe FJ, Alderson PG, Roberts JA, Azam-Ali SN, Hermann M (2012) The potential for underutilized crops to improve security of food production. J Exp Bot 63(3):1075–1079

Mir NA, Riar CS, Singh S (2018) Nutritional constituents of pseudo cereals and their potential use in food systems: a review. Trends Food Sci Tech 75:170–180. https://doi.org/10.1016/j.tifs.2018.03.016

Miranda M, Vega-Gálvez A, Martinez EA, López J, Rodríguez MJ, Henríquez K, Fuentes F (2012) Genetic diversity and comparison of physicochemical and nutritional characteristics of six quinoa (*Chenopodium quinoa* Willd.) genotypes cultivated in Chile. Food Sci Tech 32 (4):835–843. https://doi.org/10.1590/s0101-20612012005000114

Misra PS, Misra G, Prakash D, Tripathi RD, Chaudhary AR, Mishra PN (1987) Assay of some nutritional and anti-nutritional factors in different cultivars of winged bean *Psophocarpus tetragonolobus* (L.) DC. seeds. Plant Foods Hum Nutr 36:367–337

Mohantya CS, Singha V, Chapman MA (2020) Winged bean: an underutilized tropical legume on the path of improvement, to help mitigate food and nutrition security. Sci Hor 260:108789

Mubaiwa J (2018) Managing the Hard-to-Cook (HTC) phenomenon in bambara groundnut (*Vigna subterranea* (L.) Verdc.). In: Processing for resource limited communities in Zimbabwe. Ph.D. Thesis, Wageningen University, Wageningen, The Netherlands

Muimba-Kankolongo A, Muimba-Kankolongo A (2018) Chapter 10—Leguminous crops. In: Food crop production by smallholder farmers in Southern Africa. Academic Press, pp 173–203

Muller T, Scieriescher-Viret B, Fossati D, Brabant C, Schori A, Keller B, Krattinger SG (2018) Unlocking the diversity of genebanks: whole-genome marker analysis of Swiss bread wheat and spelt. Theor Appl Genet 131(2):407–416. https://doi.org/10.1007/s00122-017-3010-5

National Institutes of Health (NIH) (2008) NIH record, vol LX, No. 6. NIH, New York. www.nih.gov

National Research Council (1996) Lost crops of Africa: grains, vol I. The National Academies Press, Washington, DC

Neme K, Bultosa G, Bussa N (2015) Nutrient and functional properties of composite flours processed from pregelatinised barley, sprouted faba bean and carrot flours. Int J Food Sci Technol 50:2375–2382. https://doi.org/10.1111/ijfs.12903

Nguyen DC, Tran DS, Tran TTH, Ohsawa R, Yoshioka Y (2019) Genetic diversity of leafy amaranth (*Amaranthus tricolor* L.) resources in Vietnam. Breed Sci 69(4):640–650. https://doi.org/10.1270/jsbbs.19050

Nile SH, Park SW (2014) HPTLC analysis, antioxidant, anti-infammatory and antiproliferative activities of *Arisaema tortuosum* tuber extract. Pharm Biol 52:221–227

Obidiebube EA, Eruotor PG, Akparobi SO, Okolie H, Obasi CC (2020) Assessment of bambara groundnut (*Vigna subterranea* (L.) Verdc.) varieties for adaptation to rainforest agro-ecological zone of Anambra state of Nigeria. Can J Agric Crop 5:1–6

Olivares-Tenorio ML, Dekker M, Verkerk R, van Boekel MAJS (2016) Health-promoting compounds in cape gooseberry (*Physalis peruviana* L.): review from a supply chain perspective. Trends Food Sci Technol 57:83–92

Oliveira H, Jones H, Leigh F, Lister D, Jones M, Pena-Chocarro L (2011) Phylogeography of einkorn landraces in the Mediterranean basin and Central Europe: population structure and cultivation history. Archaeol Anthropol Sci 3:327–341. https://doi.org/10.1007/s12520-011-0076-x

Omondi EO, Engels C, Nambafu G, Schreiner M, Neugart S, Abukutsa-Onyango M et al (2017) Nutritional compound analysis and morphological characterization of spider plant (*Cleome gynandra*)—an African indigenous leafy vegetable. Food Res Int 100(1):284–295. https://doi.org/10.1016/j.foodres.2017.06.050

Omotayo AO, Ncama K, Aremu AO (2019) Exploring the diverse potential of underutilized Kei-Apple [*Dovyalis caffra* (Hook.f. & Harv.) Sim]: a multi-purpose fruit tree. Hum Ecol 47:613–618. https://doi.org/10.1007/s10745-019-00093

Oyeyinka AT, Oyeyinka SA (2016) *Moringa oleifera* as a food fortificant: recent trends and prospects. J Saudi Soc Agric Sci. https://doi.org/10.1016/j.jssas.2016.02.002

Padulosi S (2017) Bring NUS back on the table. Great Insights 6(4):21–22

Padulosi S, Hoeschle-Zeledon I (2004) Underutilized plant species: what are they? Leisa-Leusden 20:5–6

Padulosi S, Hoeschle-Zeledon I (2008) Crops for the future: paths out of poverty. Strategic plan 2009-2013, Bioversity International Regional Office for Asia, the Pacific and Oceania, Selangor, Malaysia. 16 p

Padulosi S, Thompson J, Rudebjer P (2013) Fighting poverty, hunger and malnutrition with neglected and underutilized species (NUS): needs, challenges and the way forward. Bioversity International, Rome

Paudel MN, Joshi BK, Ghimire KH (2016) Management status of agriculture plant genetic resources in Nepal. Agron JN 4:74–90

Perry LM, Metzger J (1980) Medicinal plants of east and southeast Asia: attributed properties and uses. MIT Press, Cambridge

Planella MT, Scherson R, McRostie V (2011) Sitio El Plomo y nuevos registros de cultígenos iniciales en cazadores del Arcaico IV en alto Maipo, Chile central. Chungara Rev Antropol Chil 43:189–202. https://doi.org/10.4067/S0717-73562011000200003

Plucknett DL (1976) Giant Swamp Taro, a little known Asian Pacific food crop. In: Proc 4th Int Symp Trop Root Tuber Crops, Cali, California

Polo S, Tardio J, Velez-del-Burgo A, Molina M, Pardo-de-Santayana M (2009) Knowledge use and ecology of golden thistle (*Scolymus hispanicus* L.) in Central Spain. J Ethnobiol Ethnomed 5:42

Pradeep SR, Guha M (2011) Effect of processing methods on the nutraceutical and antioxidant properties of little millet (*Panicum sumatrense*) extracts. Food Chem 126:1643–1647. https://doi.org/10.1016/j.foodchem.2010.12.047

Pradhan A, Thakur A, Patel S, Mishra N (2011) Effect of different nitrogen levels on kodo and finger millet under rainfed conditions. Res J Agric Sci 2:136–138

Prasada Rao KE, de Wet JMJ, Brink DE, Mengesha MH (1987) Intraspecific variation and systematics of cultivated *Setaria italica*, foxtail millet (Poaceae). Econ Bot 41:108–116

Pulvento C, Riccardi M, Lavini A, D'Andria R, Iafelice G, Marconi E (2010) Field trial evaluation of two *Chenopodium quinoa* genotypes grown under rainfed conditions in a typical Mediterranean environment in south Italy. J Agron Crop Sci 196:407–411. https://doi.org/10.1111/j.1439-037X.2010.00431.x

Rahul J, Jain MK, Singh SP (2015) *Adansonia digitata* L. (baobab): a review of traditional information and taxonomic description. Asian Pac J Trop Biomed 5(1):79–84

Rai M (2002) Nutritive cereals. In: Survey of Indian agriculture. The Hindu, Chennai, pp 59–62

Raina A, Gupta V (2015) Evaluation of buckwheat (*Fagopyrum* species) germplasm for rutin content in seeds. Indian J Plant Physiol 20:167–171

Rajput SG, Plyler-harveson T, Santra DK (2014) Development and characterization of SSR markers in proso millet based on switchgrass genomics. Am J Plant Sci 5:175–186

Ramabulana T, Mavunda RD, Steenkamp PA, Piater LA, Dubery IA, Madala NE (2016) Perturbation of pharmacologically relevant polyphenolic compounds in *Moringa oleifera* against photo-oxidative damages imposed by gamma radiation. J Photochem Photobiol B Biol 156:79–86

Ramachandran C, Peter KVG, Opalakrishnan PK (1980) Drumstick (*Moringa oleifera*) a multipurpose Indian vegetable. Econ Bot 34(3):276–283

Ramprasad V (2005) Foxtail millet enjoys revival in India. Bioversity International Geneflow

Rana JC, Singh M, Chauhan RS, Chahota RK, Sharma TR, Yadav R, Archak S (2016) Genetic resources of buckwheat in India. In: Zhou M, Kreft I, Woo S-H, Chrungoo N, Wieslander G (eds) Molecular breeding and nutritional aspects of buckwheat. Academic Press, Cambridge, pp 109–135

Ranasinghe RASN, Maduwanthi SDT, Marapana RAUJ (2019) A nutritional and health benefits of jackfruit (*Artocarpus heterophyllus* Lam.): a review. Int J Food Sci 6:1–12

Rashmi DR, Anitha B, Anjum Sahair R, Raghu N, Gopenath TS, Chandrashekrappa GK, Basalingappa KM (2018) An overview of Taro (*Colocasia esculenta*): a review. Acad J Agric Res 6(10):346–353

Ravi SB, Hrideek TK, Kumar ATK, Prabhakaran TR, Mal B, Padulosi S (2010) Mobilizing neglected and underutilized crops to strengthen food security and alleviate poverty in India. Indian J Plant Genet Resour 1(23):110–116

Reddy VG, Upadhyaya HD, Gowda CLL (2007) Morphological characterization of world's proso millet germplasm. SAT J 3:1–4

Renganathan VG, Vanniarajan C, Karthikeyan A, Ramalingam J (2020) Barnyard millet for food and nutritional security: Current status and future research direction. Front Genet 11:500. https://doi.org/10.3389/fgene.2020.00500

Renkow M, Byerlee D (2010) The impacts of CGIAR research: a review of recent evidence. Food Policy 35:391–402

Rojas W, Pinto M, Alanoca C, Gómez-Pando L, León-Lobos P, Alercia A, Diulgheroff S, Padulosi S, Bazile D (2015) Quinoa genetic resources and ex situ conservation. In: Bazile D, Bertero D, Nieto C (eds) State of the art report on quinoa around the world 2013. FAO; CIRAD, Rome, pp 56–82

Ruiz KB, Biondi S, Martínez EA, Orsini F, Antognoni F, Jacobsen S-E (2015) Quinoa—a model crop for understanding salt tolerance mechanisms in halophytes. Plant Biosyst 150:357–371. https://doi.org/10.1080/11263504.2015.1027317

Saini RK, Shetty NP, Giridhar P (2014) Carotenoid content in vegetative and reproductive parts of commercially grown *Moringa oleifera* Lam. cultivars from India by LC–APCI–MS. Eur Food Res Technol 238:971–978

Saini RK, Sivanesan I, Keum Y (2016) Phytochemicals of *Moringa oleifera*: a review of their nutritional, therapetic and industrial significance. 3. Biotech 6:203. https://doi.org/10.1007/s13205-016-0526-3

Saleh A, Zhang Q, Chen J, Shen Q (2013) Millet grains: nutritional quality, processing, and potential health benefits. Compr Rev Food Sci Food Saf 12:281–295. https://doi.org/10.1111/1541-4337.12012

Salgotra RK, Gupta BB (2015). Plant genetic resources and traditional knowledge for food security. https://doi.org/10.1007/978-981-10-0060-7_1

Salvi J, Katewa SS (2016) A review: underutilized wild edible plants as a potential source of alternative nutrition. Int J Bot Stud 1(4):32–36

Sanchez AC (2018) Baobabs are Africa's oldest and most beautiful trees, but they are under threat. https://www.independent.co.uk/environment/baobab-trees-africa-dying-oldest-under-threat-a8404136.html

Sandeep G, Anitha T, Vijayalatha KR, Sadasakth A (2018) Moringa for nutritional security (*Moringa oleifera* Lam.). Int J Bot Stud 4(1):21–24

Sarwar M (2017) Biodiversity and conservation of Moringa species. Biodivers Conserv BC-APRF-1(1): 5-5

Schaafsma G (2012) Advantages and limitations of the protein digestibility-corrected amino acid score (PDCAAS) as a method for evaluating protein quality in human diets. Br J Nutr 108(S2): S333–S336. https://doi.org/10.1017/S0007114512002541

Sharma S, Agarwal N, Verma P (2011a) Pigeon pea (*Cajanus cajan* L.): a hidden treasure of regime nutrition. J Funct Environ Bot 1(2):91–101

Sharma K, Joshi VK, Rana JC (2011b) Nutritional composition and processed products of Quince (*Cydonia oblonga* Mill.). Indian J Nat Prod Resour 2(3):354–357

Shendge PN, Belemkar S (2018) Therapeutic potential of *Luffa acutangula*: a review on its traditional uses, phytochemistry, pharmacology and toxicological aspescts. Front Pharmacol 9:1177

Shewry PR, Pellny TK, Lovegrove A (2016) Is modern wheat bad for health? Nat Plant 2:16097. https://doi.org/10.1038/nplants.2016.97

Shukla K, Srivastava S (2014) Evaluation of finger millet incorporated noodles for nutritive value and glycemic index. J Food Sci Technol 51:527–534. https://doi.org/10.1007/s13197-011-0530-x

Singh A, Sharma S (2017) Bioactive components and functional properties of biologically activated cereal grains: a bibliographic review. Crit Rev Food Sci Nutr 57:3051–3071. https://doi.org/10.1080/10408398.2015.1085828

Singh AK, Bhat BP, Sundaram PK, Gupta AK, Singh D (2013) Planting geometry to optimize growth and productivity faba bean (*Vicia faba* L.) and soil fertility. J Environ Biol 34 (1):117–122

Singh AK, Rana HK, Tshabalala T, Kumar R, Gupta A, Ndhlala AR, Pandey AK (2019a) Phytochemical, nutraceutical and pharmacological attributes of a functional crop *Moringa oleifera* Lam: an overview. S Afr J Bot. https://doi.org/10.1016/j.sajb.2019.06.017

Singh N, Singh S, Maurya P, Arya M, Khan F, Dwivedi DH, Saraf SA (2019b) An updated review on *Physalis peruviana* fruit: Cultivational, nutraceutical and pharmaceutical aspects. Indian J Nat Prod Resour 10(2):97–110

Singh S, Singh DR, Banu VS, Singh LB (2019c) Determination of functional constituents (Phytochemicals, micronutrients), anti-nutrients and antioxidant activity in commonly grown genotypes of *Basella* L. Int J Chem Stud 7(3):403–409

Sood S, Khulbe R, Kumar RA, Agrawal PK, Upadhyaya H (2015) Barnyard millet global core collection evaluation in the sub mountain Himalayan region of India using multivariate analysis. Crop J 3:517–525. https://doi.org/10.1016/j.cj.2015.07.005

Sood S, Joshi DC, Chandra AK, Kumar A (2019) Phenomics and genomics of finger millet: current status and future. Planta 250:731–751. https://doi.org/10.1007/s00425-019-03159-6

Spaenij-Dekking L, Kooy-Winkelaar Y, Koning F (2005) The Ethiopian cereal tef in celiac disease. N Engl J Med 353:1748–1749. https://doi.org/10.1056/NEJMc051492

Stallknecht GF, Schulz-Schaeffer JR (1993) Amaranth rediscovered. In: Janick J, Simon JE (eds) New crops. Wiley, New York, pp 211–218

Stephenson KK, Fahey JW (2004) Development of tissue culture methods for the rescue and propagation of endangered *Moringa* spp. germplasm. Econ Bot 58(1):116–124. https://doi.org/10.1663/0013-0001(2004)58[S116:DOTCMF]2.0.CO;2

Subiria-Cueto R, Larqué-Saavedra A, Reyes-Vega ML, de la Rosa LA, Santana-Contreras LE, Gaytán-Martínez M, Vázquez-Flores AA, Rodrigo-García J, Corral-Avitia AY, Núñez-Gastélum JA, Martínez-Ruiz NR (2019) *Brosimum alicastrum* Sw. (Ramón): an alternative to improve the nutritional properties and functional potential of the wheat flour tortilla. Foods 8:613. https://doi.org/10.3390/foods8120613

Sugár E, Fodor N, Sándor R, Bónis P, Vida G, Árendás T (2019) Spelt wheat: an alternative for sustainable plant production at low N-levels. Sustainability 11:6726. https://doi.org/10.3390/su11236726

Suhairi TASTM, Jahanshiri E, Nizar NMM (2018) Multicriteria land suitability assessment for growing underutilized crop, bambara groundnut in Peninsular Malaysia. IOP conference series: earth and environmental science, vol 169. IOP, Brstol, p 012044

Tagle MB, Planella MT (2002) La Quínoa en la Zona Central de Chile, Supervivencia de una Tradición Prehispánica. Editorial IKU, Santiago

Tahir NA, Majeed HO, Azeez HA, Omer DA, Faraj JM, Palani WRM (2020) Allelopathic plants: Moringa species. Allelopath J 50(1):35–48. https://doi.org/10.26651/allelo.j/2020-50-1-1272

Tang Y, Tsao R (2017) Phytochemicals in quinoa and amaranth grains and their antioxidant, anti-inflammatory, and potential health beneficial effects: a review. Mol Nutr Food Res 61:1600767. https://doi.org/10.1002/mnfr.201600767

Tanzi AS, Eagleton GE, Ho WK, Wong QN, Mayes S, Massawe F (2019) Winged bean (*Psophocarpus tetragonolobus* (L.) DC.) for food and nutritional security: synthesis of past research and future direction. Planta 250:911–931. https://doi.org/10.1007/s00425-019-03141-2

Tesema A (2013) Genetic diversity of tef in Ethiopia. In: Assefa A, Chanyalew S, Tadele A (eds) Achievements and prospects of tef improvement. EIAR University of Bern, Bern, pp 15–20

Teutonico RA, Knorr D (1985) Non destructive method for oxalate determination of cultured Amaranthus tricolor cells. J Agric Food Chem 33:60–64

Thrupp LA (2000) Linking agricultural biodiversity and food security: the valuable role of agrobiodiversity for sustainable agriculture. Int Aff 76:283–297

Toledo Á, Burlingame B (2006) Biodiversity and nutrition: a common path toward global food security and sustainable development. J Food Compos Anal 19:477–483

Toll J, Van Sloten DH (1982) Directory of germplasm collections. International Board for Plant Genetic Resources, Rome

Tony-Odigie AE, Adekoya KO, Makinde SCO, Oboh BO, Ogunkanmi LA, Fowora MA (2012) Assessment of genetic interspecies relationships among five selected Amaranthus species using phenotypic and RAPD markers. Int J Bot 8:145–152

Troccoli A, Codianni P (2005) Appropriate seeding rate of einkorn, emmer, and spelt grown under rainfed conditions in southern Italy. Eur J Agron 22:293–300

Trucco F, Tranel PL (2011) Amaranthus. In: Kole C (ed) Wild crop relatives: genomics and breeding resources, vegetables. Springer, Berlin/Heidelberg, pp 11–21

Uarrota VG, Bairros Angela FM, Gindri D, Leolato, Lucieli LS, Andrade GC, Nerling D, Stefen D, Arijama M, Razao EH, Rocha M (2019) From neglected and underutilized crops to powerful sources of vitamin A: three case studies of mozambican cultivated *Tacca leontopetaloides*, Cowpea, and Cassava. https://doi.org/10.5772/intechopen.84829

Ugare R, Chimmad B, Naik R, Bharati P, Itagi S (2014) Glycemic index and significance of barnyard millet (*Echinochloa frumentacae*) in type II diabetics. J Food Sci Technol 51:392–395. https://doi.org/10.1007/s13197-011-0516-8

UNEP (1995) Global biodiversity assessment. United Nations Environment Programme University Press, Cambridge

Upadhyaya HD, Sharma S, Gowda CLL, Reddy VG, Singh S (2011) Developing proso millet (*Panicum miliaceum* L.) core collection using geographic and morpho-agronomic data. Crop Pasture Sci 62:383–389

Upadhyaya HD, Dwivedi SL, Singh SK, Singh S, Vetriventhan M, Sharma S (2014) Forming core collections in barnyard, kodo, and little millets using morphoagronomic descriptors. Crop Sci 54:1–10. https://doi.org/10.2135/cropsci2014.03.0221

Upadhyaya HD, Vetriventhan M, Dwivedi SL, Pattanashetti SK, Singh SK (2015) Proso, barnyard, little and kodo millets. In: Singh M, Upadhyaya HD (eds) Genetic and genomic resources for grain cereals improvement. Academic Press/Elsevier, pp 321–343

Valdivia-López MÁ, Tecante A (2015) Chia (*Salvia hispanica*): a review of native Mexican seed and its nutritional and functional properties. Adv Food Nutr Res 75:53–75. https://doi.org/10.1016/bs.afnr.2015.06.002

Vega-Gálvez A, Miranda M, Vergara J, Uribe E, Puente L, Martínez EA (2010) Nutrition facts and functional potential of quinoa (*Chenopodium quinoa* Willd.), an ancient Andean grain: a review. J Sci Food Agric 90:2541–2547. https://doi.org/10.1002/jsfa.4158

Verdcourt B, Halliday P (1978) A revision of *Psophocarpus* (Leguminosae: Papilionoideae: Phaseoleae). Kew Bull 33:191–227

Vetriventhan M, Upadhyaya H, Anandakumar C, Senthilvel S, Parzies H, Bharathi A, Varshney R, Gowda C (2012) Assessing genetic diversity, allelic richness and genetic relationship among races in ICRISAT foxtail millet core collection. Plant Genet Resour 10:214–223

Vetriventhan M, Upadhyaya HD, Dwivedi SL, Pattanashetti SK, Singh SK (2015) 7–Finger and foxtail millets. In: Genetic and genomic resources for grain cereals improvement. Academic Press, Elsevier, pp 291–319. https://doi.org/10.1016/B978-0-12-802000-5.00007-1

Vincentini O, Maialetti F, Gazza L, Silano M, Dessi M, De Vincenzi M, Pogna NE (2007) Environmental factors of celiac disease: cytotoxicity of hulled wheat species *Triticum monococcum. T. turgidum* ssp. *dicoccum* and *T. aestivum* ssp. spelta. J Gastroenter Hepatol 22:1816–1822

Wang C, Jia G, Zhi H, Niu Z, Chai Y, Li W et al (2012) Genetic diversity and population structure of Chinese foxtail millet [*Setaria italica* (L.) Beauv.] landraces. G3 2:769–777. https://doi.org/10.1534/g3.112.002907

WCMC (1992) Global biodiversity: status of the earth' living resources–World Conservation Monitoring Centre. Chapman and Hall, London, p 5494

Williams JT, Haq N (2002) Global research on underutilized crops. An assessment of current activities and proposals for enhanced cooperation. ICUC, Southampton, p 46

Wilson EO (1992) The Diversity Of Life. Penguin, London

Yan ZY, Spencer PS, Li ZX, Liang YM, Wang YF, Wang CY, Li FM (2006) *Lathyrus sativus* (grass pea) and its neurotoxin ODAP. Phytochemistry 67(2):107–121. https://doi.org/10.1016/j.phytochem.2005.10.022

Yang S, Grall A, Chapman MA (2018) Origin and diversification of winged bean (*Psophocarpus tetragonolobus* (L.) DC.), a multipurpose underutilized legume. Am J Bot 105(5):888–897. https://doi.org/10.1002/ajb2.1093

Zaharieva M, Monneveux P (2014) Cultivated einkorn wheat (*Triticum monococcum* L. subsp. *monococcum*): the long life of a founder crop of agriculture. Genet Resour Crop Evol. https://doi.org/10.1007/s10722-014-0084-7

Zanini B, Villanacci V, De Leo L, Lanzini A (2015) *Triticum monococcum* in patients with celiac disease: a phase II open study on safety of prolonged daily administration. Eur J Nutr 54:1027–1029

Zarnkow M, Mauch A, Burberg F, Back W, Arendt EK, Kreisz S, Gastl M (2009) Proso millet (*Panicum miliaceum* L.) a sustainable raw material for the malting and brewing process: a review. Brew Sci 62:119–140

Zhang L, Liu R, Niu W (2014) Phytochemical and antiproliferative activity of proso millet. PLoS One 9:e104058. https://doi.org/10.1371/journal.pone.0104058

Zhou M, Kreft I, Woo SH, Chrungoo N, Wieslander G (2016) Molecular breeding and nutritional aspects of buckwheat. Academic Press, Cambridge, p 482

Zhou M, Kreft I, Suvorova G, Yu T, Sun-Hee W (2018) Buckwheat germplasm in the world, 1st edn. Academic Press, Cambridge, p 382

Zhu F (2020) Underutilized and unconventional starches: why should we care? Trends Food Sci Technol. https://doi.org/10.1016/j.tifs.2020.04.018

Ziegler JU, Schweiggert RM, Wurschum T, Longin CFH, Carle R (2016) Lipophilic antioxidants in wheat (*Triticum* spp.): a target for breeding new varieties for future functional cereal products. J Funct Foods 20:594–605

Zou L, Yli-Halla M, Stoddard FL, Mäkel PSA (2015) Effects of break crops on yield and grain protein concentration of barley in a boreal climate. PLoS One 10:e0130765. https://doi.org/10.1371/journal.pone.013076

Underutilized Vegetable Species of Indian Himalayas

4

Geetika Malik, D. B. Singh, Ajaz Ahmed Malik, and Ariza Gulzar

Abstract

Phytomedicine has always been a source of cure and prevention of human and animal diseases. Since antiquity man has been exploiting wild plant species as food and to sustain health. While few of these species have gradually been domesticated, many continue to remain wild while playing pivotal role in tribal food, feed and folklore medicine. India being among the most flora rich countries of the world is host to hundreds of wild plant species with established health benefits well known to the indigenous tribes and locals but remain largely untapped. However, growing awareness about bioactive phyto-compounds and their role in prevention and cure of modern human diseases and disorders is steadily bringing such underutilized plant species to a significant place in modern pharmaceutical and medical research. Considering such a scenario, the aim of writing this chapter is to highlight indigenous species of Indian North-Western Himalayan states and territories with respect to their health promoting attributes and potential application in the practice of alternative as well as mainstream medicine.

Keywords

Underutilized · Phytomedicine · Antioxidant · Indian Himalayan flora · Hidden hunger

G. Malik
Vegetable Science, ICAR-Central Institute of temperate Horticulture (CITH), Srinagar, Jammu and Kashmir, India

D. B. Singh
ICAR-Central Institute of temperate Horticulture (CITH), Srinagar, Jammu and Kashmir, India

A. A. Malik (✉) · A. Gulzar
Division of Vegetable Science, SKUAST-K, Srinagar, Jammu and Kashmir, India

S. M. Zargar et al. (eds.), *Neglected and Underutilized Crops - Towards Nutritional Security and Sustainability*, https://doi.org/10.1007/978-981-16-3876-3_4

4.1 Introduction

Underutilized plant species are highly valued as they are used as food medicine and in ethno-cultural traditions in their niches of adaptation. Their propagation and significance in the tribe/community is maintained by virtue of inherited knowledge and conserved by peasants and kitchen gardeners for consumption with little or no trade at all. Major portion of human diet and to some extent animal feed in technologically compromised and climatically challenged zones of the country are comprised of these untapped vegetables. From cold arid/dry cold Ladakh, Lahaul Spiti, and Kinnaur to snowy cool Jammu and Kashmir, Uttarakhand, Arunachal Pradesh, Mizoram, Nagaland, and Meghalaya these species play crucial role in nutrition and health of indigenous tribes.

Although the tribal people understood since long the importance of these species, it took plant genetic erosion and endangerment, hidden hunger, drug resistance, and medication side effects for the long after the start of monoculture era in agriculture, wherein, only few vegetables remained on the scene of world vegetable production and consumption, need was felt during the mid-twentieth century to not only preserve underutilized species from round the world but also popularize them among common people. Need to create income opportunities for tribal people, conserve endangered biodiversity and enhance germplasm richness for use by future generations were also behind the changed attitude.

Being endowed with prophylactic and curative properties against many human and animal diseases, leafy vegetable species are rightly touted as the only natural source of folic acid, which happens to be critical during early pregnancy in humans. The bioactive compounds in the form of phenolics, vitamins, and omega-3-fatty acids in these species are available in particularly ample amount, which makes them strong antioxidant sources helpful in countering human degenerative diseases like cancer and heart problems. Such effectiveness to prevent and cure modern human ailments is not generally found in common vegetables.

4.2 Status of Underutilized Vegetables in India

In India, in 1960s, the ICAR started working on identification and exploration of underutilized crops under the aegis of IARI, New Delhi. The work was further encouraged in 1980s by forming AICRP on underutilized crops by the NBPGR, New Delhi. This project has been instrumental in exploring, collecting, and preserving the exotic crops across India (Thakur 2014). As far as, temperate India is concerned, the state of Jammu and Kashmir, Himachal Pradesh, Uttarakhand, and some North East region represent their own share of underutilized vegetables, which are unique to rest of the country with respect to food, gastronomical, nutritional, and medicinal importance. Most of the underutilized crops exist as wild species and very few are cultivated by local inhabitants. An attempt has been made in this chapter to compile information on underutilized crop species from the perspective of their region of abundance, botany, and nutritional profile (as described in Table 4.1 and Fig. 4.1) in order to discuss their role as potential food crops and critical application in pharmaceutical industry.

4.3 Jammu and Kashmir

Embedded in Western Himalayas are three major regions of Jammu and Kashmir which include Jammu, Kashmir valley, and Ladakh. These regions cover the largest part of Indian Himalayas (40%). The union territory (UT) dwells between 300 and 7000 m height above mean sea level (amsl). Situated between 32° 17′, 37° 5′ N latitude, 72° 40′, and 80° 30′ E longitudes, the UT of Jammu and Kashmir is host to many underutilized vegetable species that are mostly of leafy, bulbous, and stem categories. They are adapted to extremes environmental conditions like freezing temperatures (up to −30 °C in Drass of Ladakh), drought conditions (102 mm precipitation in Leh of Ladakh) experienced by Ladakh valleys (Leh, Nubra, Changthang, Suru, and Zanskar) and temperate conditions witnessed by Kashmir and high altitude Jammu including Poonch, Rajouri, Banihal, Ramban, Bhadarwah, and Kishtwar districts. The local people routinely utilize these species in daily meals and are preferred over the common vegetables. Few of them are very expensive in local markets because of specialized use in popular Kashmiri cuisine "Wazwan" and gourmet dishes.

The Ladakh valleys represent different elevation, displaying different agro-climatic conditions and the region is host to many indigenous vegetables that are well adapted to extended sub-zero temperatures and semi-arid environment. Due to very short cropping season (March to August) and fewer vegetable crops in Ladakh, its inhabitants are dependent on vegetable imports from rest of the country. Their needs are further met by drying the excess vegetables for winter consumption as well as consumption of wild plant species adapted to such harsh climate. These species have ethno-botanical importance and are consumed as food and in *Amchi* system of medicine synonymous to Tibetan medicine practiced in Ladakh. The most popular indigenous and underutilized vegetables of Jammu and Kashmir include local forms of kale, other *Brassica* species, weedy plants, Alliaceae species, and some aquatic species. There are reports on tens of the underutilized vegetables of each of the many botanical families (Pal Murugan et al. 2010; Lamo et al. 2012; Namtak and Sharma 2018), however, only the most popular ones among local inhabitants have been discussed below.

4.3.1 Wand haaq (*Brassica oleracea* var. *acephala*)

One of the varieties of vegetables famously called as *Kashmiri haaq*. The *Wand haaq* is also known as *GM Dari* in the valley. Botanically, it is Kale (*Brassica oleracea* var. *acephala*) and best suited to winter season. The plant grows upright with alternate leaves on thick stem. The leaves are large, petiolated and ovate with

Table 4.1 Species, common names, and beneficial features of underutilized vegetable species of Himalayan origin

Family	Species	Common/ vernacular name	Major nutritional/health components	Beneficial in
Brassicaceae	*Brassica oleracea* var. *acephala*	Kale/Haaq	Folic acid, provitamin-A, vitamin K, chlorophyll	Pregnancy, vision development
	Lepidium latifolium	Perennial pepperweed/ Shangsho	Proteins, beneficial fatty acids, anti-carcinogens, and antioxidants	Oxidative stress, carcinogenesis
	Capsella bursa-pastoris	Shepherds purse/ Shamsho	Provitamin-A, ascorbic acid, omega-3-fatty acid	Infections, bleeding, oxidative stress
	Nasturtium officinale	Watercress/ Chala	Lutein, zeaxanthin, iodine, vitamin E, and glucosinolates	Inflammation, oxidative stress, infection
Alliaceae	*Allium cepa* var. *proliferum*	Pran	Fructan, vitamin B complex, selenium, manganese	Infection, pain, flatulence, diabetes, hypertension, carcinogenesis, improving gut microbiome
	Allium przewalskianum	Skotche		
	Allium tuberosum	Garlic chives/Zimu		
	Allium consanguineum and *Allium wallichii*	Jambu/Faran		
	Allium humile	Small Alpine onion/Duna		
Amaranthaceae	*Amaranthus viridis*	Green amaranth/ Jangli chulai	Amino acids, phenols, protein, beneficial fatty acids, ascorbic acid	Parasitic infections, gonorrhea, leprosy
	Amaranthus spinosus	Prickly amaranth/ Kantachulai		
Others	*Rheum emodi*	Himalayan rhubarb, Indian rhubarb/ pamb haaq	Anthraquinone, stilbene	Infection, carcinogenesis, oxidative stress, platelet aggregation, ulcers, liver and kidney problems

(continued)

Table 4.1 (continued)

Family	Species	Common/ vernacular name	Major nutritional/health components	Beneficial in
	Chenopodium album	Fat hen/bathua/ bathu	Leucin, isoleucin, lysine, ascorbic acid	Inducing contraception, immobilizing sperm, infections, parasitic infections, inflammation
	Asphodelus tenuifolius	Onion weed/ piyazi	Polyphenols, anthraquinone (aloe-emodin, chrysophanol), flavonoids	Hypertension, diabetes, infection, hemorrhoids, rheumatic pain, oxidative stress, insulin resistance
	Diplazium esculentum	Vegetable fern/Lungru, Kasrod	Flavonoids, vitamin E, folic acid, calcium, iron	Diabetes, oxidative stress, dysentery, indigestion, skin infections
	Portulaca oleracea	Common purslane/ Nunnar	Potassium, magnesium, calcium, ascorbic acid, omega-3-fatty acid	Osteoporosis, psoriasis, arthritis, burns,inflammation, cardiovascular diseases
	Taraxacum officinale	Common dandelion/ Hannd	Luteolin, caffeic acid, cholorgenic acid, provitamin-A, ascorbic acid, linolenic acid	Inflammation, carcinogenesis, oxidative stress, hypertension, diabetes
	Malva sylvestris	Common mallow/ Sonchal	Malvin (anthocyanin glycoside), calcium, potassium, omega-3-fatty acid	Burns, asthma, bronchitis, emphysema, inflammation, gall/ kidney stones,
	Rumex obtusifolius	Broad leaved dock/ Abuj	Phenolics, flavonoids, flavonols	Skin ailments, jaundice, whooping cough, viral infection, oxidativestress
	Nelumbo nucifera	Kamal kakdi/Nadru	Polyphenolics, alkaloids, potassium, calcium, protein	Diabetes, inflammation, fever, epistaxis, hematuria, diarrhea, cholera, hyperdipsia

(continued)

Table 4.1 (continued)

Family	Species	Common/ vernacular name	Major nutritional/health components	Beneficial in
	Urtica dioica	Stinging nettle/ Bichhubooti/ Soye	Provitamin-A, calcium, iron, protein, palmitic acid, polyphenolics, isolectins, iron	Benign prostatic hyperplasia, prostrate cancer, UTIs, bladder infections, osteoarthritis

dentate margins, mid-ribs and veins are creamish white. At flowering, flower heads are borne at the tip of inflorescence stalk and bears numerous yellow florets. The fruit is siliqua with brown seeds. Health benefits of kale are well known, because of which it is also called super-food. Kale is an excellent antioxidant food and is rich source of vitamin K, ascorbic acid, provitamin-A, vitamin B6, carotenoids, ascorbic acid, phenolics, chlorophyll, folate, iron, manganese, copper, and omega-3 fatty acids (Armesto et al. 2015).

In Kashmir valley, *Wand haaq* is consumed in the form of boiled and fried leaves in combination with other vegetables or meat. It is routine part of the Kashmiri meals as well as part of delicious pickles. Kale in all its popular forms is probably the most locally traded underutilized vegetable crop and is also used as popular food in Jammu region of UT.

4.3.2 Khanyari Haaq (*Brassica oleracea* var. *acephala*)

Just like *Wand haaq, Khanyari haaq* is also a form of kale and highly relished by Kashmiri in habitants. Unlike *Wand haaq*, which is winter-loving, it is more suited to summer season and liked for the meatiness of its thick and highly puckered leaves. The plant grows upright as single stem with alternate leaves, that are large, rounded with entire margins, highly puckered, dark green with creamish mid-rib and veins. The flowering characteristics are similar to *Wand haaq*. The nutritional profile, bioactivity, and local usage are also akin to it.

4.3.3 Hanz haaq (*Brassica oleracea* var. *acephala*)

Similar to *Wand haaq* and *Khanyari, Hanz haaq* is a kale but with considerably smaller leaves, which are numerous in number and facilitate multiple harvesting unlike 3–5 pickings in the above two types. *Hanz haaq* also grows better in winter. The plant is single stemmed with large number of small light green alternate leaves. The leaves are ovate, dentate, and pedicillate with whitish mid-rib and veins. The usage, flowering characteristics, nutritional, and bioactivity profile are similar to the above mentioned kales.

Nunnar
(www.worldofsucculents.com)

Wasthaaq
(www.allthatgrows.com)

Hannd
(www.keyserver.lucidcentral.org)

Sonchal
(www.gobotany.newenglandwild.org)

Abuj
(www.notjustgreenfingers.wordpress.com)

Pambhaaq
(www.indianmedicinalplants.info)

Shamsho
(www.rejoiff.info)

Shangsho
(www.flowermedia.com)

Fig. 4.1 Visual presentation of different untapped vegetables of Indian Himalayas

Skotche
(www.growingdelights.co.uk)

Pran
(ICAR-CITH, Srinagar)

Nadru
(www.colourbox.com)

Zimu
(pfaf.org)

Janglichulai
(en.hortipedia.com)

Kantachulai
(gobotany.newenglandwild.org)

Bathua
(keyserver.lucidcentral.org)

Fig. 4.1 (continued)

Chala
(pondplantsonline.com)

Piyaji
(purdue.edu)

Lungru/Kasrod
(tropical.theferns.info)

Duna
(flowersofindia.net)

Jambu/ Faran
(commons.wikimedia.org)

Bichhubooti
(Wikipedia)

Fig. 4.1 (continued)

4.3.4 Wast haaq (*Amaranthus* spp.)

It is a species of Amaranthus, belonging to family *Amaranthaceae*. The entire plant is usually purple in color, even though green leaved variants are also cultivated Young and tender leaves are cooked and consumed as *haaq/saag*. It has diverse nutritional and phytochemical benefits and is known for its antioxidant properties, richness of β-carotene, ascorbic acid, provitamin-A, iron, and calcium (Achigan-

Dako et al. 2014). The benefits of eating red/purple are even more, as the pigment anthocyanin responsible for the color acts as antioxidant (Amin et al. 2006).

4.3.5 Nunnar (*Portulaca oleracea*)

It belongs to family Portulacaceae and is commonly known as common purslane. It is a cosmopolitan and noxious terrestrial weed, in agricultural terms, that grows in fields, gardens, and fallow lands thorough out the year. It is annual and the plant growth habit is prostrate with succulent, glabrous, oblong to egg shaped green leaves with or without purple margins. It bears yellow pentamerous flowers that open until early afternoon in sunny weather. It is considered as powerhouse store of omega-3-fatty acid. Nunnar is a rich source of ascorbic acid, vitamin A, and minerals like potassium, magnesium, and calcium In addition, it has great antioxidant and medicinal potential and seen as prospective herb/vegetable source of health benefits to humans and animals (Simopoulos et al. 1992; Kamal Uddin et al. 2014). In the valley, it is cooked for its savory leaves and believed to be one of the most wholesome yet easily available foods.

4.3.6 Hannd/Han (*Taraxacum officinale*)

Popularly known as common dandelion, it is a terrestrial, intermediate to prostrate growing perennial weed belonging to family Asteraceae. The leaves are oblong, oblanceolate or obovate with shallow or deep lobes and are borne in rosette at the top of root just above the ground. The leaf margins are dentate and petioles may be winged or un-winged. Each flowering stem leads to single flower head with 50–100 bisexual yellow florets. The fruit is an achene, having several seeds bearing white/gray hair tuft for dissemination by wind.

Hannd is antioxidant, diuretic, antitumor, and anti-inflammatory by virtue of the presence of luteolin, caffeic acid, and cholorgenic acid (Colle et al. 2012). It has been found to contain high quantity of provitamin-A, ascorbic acid, linolenic acid, and good calcium/phosphorus ratio of 1:1 (Escudero et al. 2003). In Kashmir and Ladakh, it is cooked as a leafy vegetable to supplement the food with iron, especially for pregnant and lactating women.

4.3.7 Sonchal/Swatsal (*Malva sylvestris*)

Belonging to family Malvaceae, sonchal or globally known as Mallow, is an annual/biennial/perennial plant that grows in cultivated as well as fallow fields, gardens, and grounds. It grows upright and is spreading in nature. The main stem and branches may be pubescent or glabrous. The leaves are dark green, soft, round shaped, and shallow lobed. Flowers are borne in clusters, are pentamerous and purple/mauve in color. Sonchal has been found to be a good source of calcium and potassium and *at par* with spinach for overall mineral content (Tabaraki et al. 2012). The antioxidant activity of sonchal is also comparable to that of spinach. It is a good source of omega fatty acids and is having antibacterial as well as antifungal properties (Razavi et al. 2011). Sonchal is usually applied directly on wounds and its seeds are boiled as concoction to cure diuretic conditions. In Kashmir, it is cooked and relished due to its numerous health benefits.

4.3.8 Abuj (*Rumex obtusifolius*)

Terrestrial weedy plant with rosette appearance, due to appearance of the leaves that are borne at single point just above the ground belongs to family Polygonaceae and is commonly known as dock. The leaves are oval with obtuse tips, dark green with or without purplish tinge and are glabrous. The flowers are borne in clusters, are inconspicuous and green in color, which turn reddish at maturity. Abuj has been documented as medicinal herbs to treat many animal as well as human ailments most commonly skin eruptions, boils, burns, blisters, and scalds. The root is also used to treat diseases like jaundice and whooping cough. Many studies have demonstrated its antibacterial (Koochak et al. 2010), antiviral (Orhan et al. 2009), and free radical scavenging activity (Harshaw et al. 2010). This herb is especially abundant in oxalic acid, which gives its leaves acidic taste. Excess and undercooked consumption of this herb may aggravate conditions like gout, kidney stones, arthritis, and rheumatism and may fix some minerals to render them unusable in the body. The older leaves are avoided due to higher bitterness and younger leaves are cooked as leafy vegetables or haaq (saag) in Kashmir valley.

4.3.9 Pamb haaq (*Rheum emodi*)

Rhubarb, locally known as *Pamb haaq* in Kashmir, belongs to family Polygonaceae. It is endemic to Indian Himalayan region, is perennial, grows erect, and bears large triangular leaves. Both the rhizomes and roots are consumed as medicine and food. Its roots are known to have strong laxative property, used as expectorant, diuretic, appetizer, anti-dysentery (Rehman et al. 2014), anti-carcinogen(Cai et al. 2008) and anti-cholesterol (Goel et al. 1997). Kashmiris consume its leaves, which are steamed and fried like common *haaq* preparation or cooked with fish.

4.3.10 Shamsho (*Capsella bursa-pastoris*)

It belongs to family Brassicaceae and commonly called Shephard's purse. The plant is annual, ever flowering, frost hardy and has weed potential. The plant growth looks like a rosette of leaves just above ground. Leaves are oblong, dentate or lobed. Flowers are white and borne on umbellate inflorescence that forms at the end of single flowering stock appearing from within the leaf rosette. The fruit shape, which gives this plant its vernacular name of Shephard's purse is flat heart shaped or triangular. This species is particularly known for its medicinal, antimicrobial, antioxidant, antibleeding, anti-carcinogenic, and nutritional (high provitamin-A, ascorbic acid, omega-3 fatty acids, proteins) properties (Grosso et al. 2011). Its leaves are cooked as vegetable or consumed as salad in Ladakh.

4.3.11 Shangsho (*Lepidium latifolium*)

It also belongs to family Brassicaceae and plant architecture is rosette of leaves at ground level. The leaves are long and petiolated. The leaves that are borne on stalk later in lifecycle are sessile and smaller in size. The panicle inflorescence is borne at the tip of stalk and flowers are white in color. The fruits are silique with two chambers and one seed per chamber. This species is known for its antioxidant and anti-carcinogenic properties and has great nutritional profile that consists of proteins, beneficial fatty acids (Kaur et al. 2013). The people of Ladakh consume it as salad or cooked form in popular dish *Shangshochonma*.

4.3.12 Khala (*Lactuca dolichophylla*)

Also called long leaved lettuce, it belongs to the same family as lettuce; Asteraceae. It is narrow linear leaved tall plant, perennial with solitary stem. It bears minute blue to purple flower heads in branched clusters. Each flower head is numerous florets. Both stem as well as leaves are consumed as vegetable in Ladakh.

4.3.13 Pran (*Allium cepa* var. *proliferum*)

Pranis a non-weedy, bulbous vegetable of Alliaceae family. It bears resemblance with multiplier onion, however, the bulb shape is oval and aroma is stronger and different. It does not bear flowers and thus does not set true seeds. It propagates vegetatively through multiplication of bulbs and to some extent bulblets borne aerially on scape. The plant is a monocot with linear and cylindrical leaves borne alternatively on pseudostem little above the ground. Proliferum onion is known to be antiseptic, antispasmodic, carminative, expectorant, diuretic, hypo-glycemic, and hypotensive. It is highly priced commodity in local Kashmiri market because of its specific aroma that is relished more than onion and made use of in Kashmiri *wazwan* during wedding season. It is indispensible part of "spice cake" indigenous to Kashmiri cuisine.

4.3.14 Skotche (*Allium przewalskianum*)

Also called *jimbu* in Nepal and NE India, *skotche* is a wild species of Alliaceae. It is a frost hardy bulb species, non-weedy, shorter than onion plant. It has narrower leaves than common onion and bears purple umbels. The flowers are hermaphrodite and bear true seeds, which act as propagating material. Ladakhi people consume it as flavoring agent in place of onion and its leaves and not bulbs are used for the purpose. The juice is consumed to heal stomach pain.

4.3.15 Nadru (*Nelumbo nucifera*)

Botanically, *Nadru* is stem of lotus plant, which is an aquatic species found in abundance in various lakes of Jammu and Kashmir, most popularly the Dal lake of Srinagar. It is rich in iodine and recommended for treatment of goiter. The plant has anti-cancerous, antioxidant, antiviral, antibacterial, antifungal, antipyretic, and diuretic properties (Mukherjee et al. 2009). In Kashmir valley, it is popularly cooked with fish and *haaq*, extensively used in *yakhni* preparation and also pickled with mixed vegetables.

4.4 Himachal Pradesh, Uttarakhand and Temperate NE Region

These Western Himalayan states represent temperate underutilized flora of India, apart from Jammu and Kashmir and comprise about 60% of Western Himalayas. This region lies lower than J&K but hosts the highest point of India at 8000 m amsl (The Kanchenjunga massif). The region fosters hundreds of identified valuable wild plant species, among which are those consumed as vegetables by local tribes and communities. These species hold promise for nutritional security and health promoting pharmaceutical industry of the nation. They belong mainly to Amaranthaceae, Alliaceae, and Brassicaceae families. Most of them grow wild in valleys and mountains experiencing extremes of temperature and precipitation means on the minus scale. The semi-arid cold deserts of Lahaul Spiti, the cold mountains of Kumaon and Garhwal and temperate areas of Sikkim and Arunachal Pradesh host many frost hardy species that are locally consumed as food and ethno-medicinal ingredients. Partial accessibility to these areas for centuries have made the locals adept enough to recognize, exploit, minimally process and to some extent domesticate and trade these species to meet their demand for food, feed, and medicine, especially in times of extreme temperature falls and complete shutdown of roads due to snow and landslides. Some of the most commonly known and exploited underutilized vegetables of the region are discussed below.

4.4.1 Zimu (*Allium tuberosum*)

A species of Alliaceae, zimu is a bulbous crop with common onion like pungency and flavor. Commonly known as garlic chives, it is perennial upright growing plant with strap leaves and white star like flowers. The bulb is elongated in shape and much smaller in size than common onion. Propagation occurs through bulbs as well as seeds. It grows in the wild and also cultivated in private gardens. Apart from consumption as food, it also serves as ornamental and medicinal herb.

Both leaves and bulbs are used as condiment in curries and soups and in addition, leaves are also dried for use in slack period. Apart from Himachal Pradesh it also grows and is consumed in NE states and Uttarakhand.

4.4.2 Jangli chulai (*Amaranthus viridis*) and Kantachulai (*Amaranthus spinosus*)

Also known as slender Amaranthus or green Amaranthus, *Jangli chulai* grows as a weed in cultivated fields and fallow lands. The plant grows upright and bears many branches. Leaves are ovate and flowers are borne as panicles at the top ends of flower stalks. *Kantachulai* has similar plant structure except the leaf axils bear a pair of spines. The leaves of both the species may be green or green with v-shaped watermark like marking. These are not frost hardy species although they grow well in temperate zones when temperature is well above freezing point. They are subtropical as well as temperate species. The flowers are small, green, monoecious, and wind pollinated. Seeds are very small and hydrophilic.

Both are consumed for their leaves and to some extent seeds. The leaves are cooked like spinach or any other *saag* like leafy vegetable. All species of Amaranthus are considered to be underutilized vegetables as well as underutilized pseudocereal owing to the edibility of both leaves and seeds. Because of the health and nutritional benefits of their seeds, Amaranthus seeds have been categorized as miracle/super grain of future. Its seeds have been valued by many civilizations of the ancient world and used in preparation of health restoring and health promoting concoctions, potions, and confections. Amaranthus species are great antioxidants, anti-helmintic, anti-gonorrheal, anti-leprotic, immunomodulatory agents (Amabye 2015), and excellent sources of nutrients like proteins, beneficial fatty acids, ascorbic acid, and various elements (Alegbejo 2013).

4.4.3 Bathua (*Chenopodium album* Linn.)

It is a short stature weedy species that grows in various crop fields including wheat, potato, and gram. On reaching flowering, it can grow several feet tall and bear inflorescences at the apices of main stem and primary branches. It is annual, soft stemmed, erect growing. The stem may have purple stripes. Leaves are small, broadly lanceolate to ovate in shape and irregularly dentate. The youngest leaves are remarkably silvery due to the presence of mealy granules. The inconspicuous flowers are densely borne on spikes, are greenish, without petals and wind pollinated.

Bathu contains high amounts of proteins and balanced amino acids, especially high amounts of lysine and methionine. It has high amounts of fiber, iron, ascorbic acid, carotenoids, and phenolics to act as highly nutritious and functional food. It has also been found as sperm immobilizer and contraceptive in nature (Poonia and Upadhayay 2015).Its leaves are consumed in cooked form more often than seeds.

4.4.4 Chala (*Nasturtium officinale*)

Chala commonly known as watercress is an aquatic *Brassica* species. It is perennial and frost hardy, grows naturally near streams and rivers, in crevices and ditches near water bodies. The upright growing plant bears leaves that are small, may be simple or compound. The leaf margins may be entire or dentate. The flowers are typical of Brassicaceae family; white in color with four petals and radially symmetrical. The fruit is dry siliqua and splits open to shatter seeds on maturity.The plant is valued for its strong mustardy flavor and cooked as *saag*. It is characterized by the presence of glucosinolates, carotenoids, chlorophyll, ascorbic acid, provitamin-A, and vitamin E that are antimicrobial, antioxidant, anti-inflammatory and richness of lutein, zeaxanthin, iron, calcium, iodine, and folic acid, which make it excellent source of nutrition (Chaudhary et al. 2018).

4.4.5 Piyaji (*Asphodelus tenuifolius*)

It is a species of Asphodelaceae and grows as weed in cultivated fields and in fallow lands. It looks like onion and thus also commonly called wild onion. However, the plant is smaller and narrower than onion, bears larger flowers in panicles at the ends of stalks. The petals are white with purplish brown stripes. The root is fibrous and not bulbous like onion. The species has condiment as well as medicinal importance. It has antioxidant, diuretic, antipyretic, and anti-inflammatory properties (Malmir et al. 2018).

4.4.6 Lungru/Kasrod (*Diplazium esculentum* Retz.)

It is a moisture and shade loving Pteridophyte species belonging to family Athyriaceae and commonly called vegetable fern. It is a perennial fern with main stalk arising from erect rhizome. It is eaten at immature frond stage, when it is in curled and unopened lamina stage.

It has been found to be laxative, anti-inflammatory, antioxidant, anti-helmintic, and antimicrobial (Akter et al. 2014) and contains nutrients like folic acid, calcium, iron, and phosphorus in high amounts (Archana et al. 2012). In Himachal Pradesh, people consume unopened curled fronds by boiling, frying, and pickling. Apart from immature fronds, rhizomes and opened laminas are also consumed as medicine in various parts of the world.

4.4.7 Duna (*Allium humile* Kunth.)

Duna is a wild species of Alliaceae that grows in temperate and cold semi-arid valleys and mountains of Indian Himalayas. To some extent it is grown in kitchen gardens for domestic consumption, especially Ladakh, Lahaul Spiti, Kumaon, and Sikkim. The plant is onion like but shorter in height, leaves are broader and compressed. The scape arises from pseudostem and ends at hemispherical umbels bearing numerous flowers that are hermaphrodite, white in color and pollinated by bees. The bulbs are small and cylindrical.

It is consumed for its onion like flavor and pungency of leaves that are eaten green in chopped or paste form or dried for use in curries and soups for garnishing and flavor enhancement. It is also used as alternative medicine for infectious diseases and as analgesic. Both these properties have been scientifically proven (Agrawal et al. 2010; Singh et al. 2014).

4.4.8 Jambu or Faran (*Allium consanguineum* and *Allium wallichii* Kunth.)

They grow wild in Ladakh and Uttrakhand and collectively called *Jambu* as well as *Faran* in local languages of Garhwal and Kumaon. The aerial parts; stems and leaves are used as condiment and spice and as alternative medicine in Tibetan, *Amchi* and Unani systems of medicine. It is sold and consumed as fresh as well as dehydrated. All the health and nutritional benefits of Alliaceae family can be found in these species too.

4.4.9 Bichhubooti (*Urtica dioica*)

Commonly known as stinging nettle, *bichhubooti* grows profusely in all Himalayan states. In Kashmir it is known as *Soye*. Its leaves are cooked as *saag* after thorough boiling and straining and used extensively as folk medicine. Its antioxidant, antidiabetic, anti-hyperlipidemic, antiviral, anticancer, analgesic, and anti-helmintic properties are also well documented (Joshi et al. 2014).

The nutritional/health benefits and remarkable features of the above discussed untapped vegetables are summarized in Table 4.1.

4.5 Anti-nutritional Factors

Many of the herbs in addition to having beneficial bioactivity have anti-nutritional metabolites that need to be removed before consumption. Nitrates, tannins, phytates, and oxalates are by far the most commonly known anti-nutritional factors and unfortunately found to be most prevalent in leafy vegetables. The reason for having higher nitrates compared to other vegetables is heavy feeding on soil nitrogen. High nitrates and nitrites in human body lead to fatal Blue Baby Syndrome in infants. Tannins are normal plant metabolites and tend to form complexes with protein to block them from assimilating in body. Phytates are the complexes of phytic acid with different minerals and found in leafy vegetables. They bind with essential minerals in the body to make them unavailable for absorption. Oxalates are produced by combination of oxalic acid with mineral salts and are found in plenty in the leaves of rhubarb, amaranths, and spinach. The oxalates bind the calcium and render it unassimilated in human body. Fortunately, blanching and cooking have been found to be effective in reducing or eliminating these anti-nutritional factors from food and their consumption in moderation has also been reported to avoid any harm. However, phytates being heat stable need to be removed through fermentation. Hence, the advantages of these underutilized species outweigh any harmful properties and must be preferred for nutrition and general well-being.

4.6 Scope of Underutilized Vegetables in India

The above mentioned vegetable species along with many other unmentioned and untapped species are consumed as local vegetables and phytomedicines. There is almost no trade at national or international level per se. However, with increasing revelations about their potential benefits and awareness among common people about the possibility of including them in their diets and medicines, the herbal nutraceutical, pharmaceutical, and cosmeceutical industries are burgeoning fast. They have started to relegate the chemical regime of nutrition and medicine to a lower position. In current global scenario of herbal food additives, diet supplements and health promoting/anti-aging/beauty enhancing/anti-carcinogenic formulations, China, India, and Europe are the biggest market for alternative medicine and thus the hub for exploitation of underutilized wild species of Indian, Chinese, and Nepalese Himalayan regions. Undoubtedly, India is also the biggest players in supplying the herbal inputs, as the Indian herbal companies are right now in the forefront of global herbal suppliers owing to rich flora biodiversity of Himalayas and trans-Himalayas. The current leaders in this arena are Himalaya Drug Company, Zandu Pharmaceutical Works Ltd., Hamdard Laboratories, Dabur and Patanjali Ayurved Ltd. (https://www.reuters.com/brandfeatures/venture-capital/article?id=32992).

Unaffordable chemical medicines and their aggressive side effects are increasingly driving the people towards cheaper herbal medicines and supplements that have least or no side effects at all. However, currently there is a lack of sufficient research into the biochemical properties and bioactivity of the herbs and no proper regulatory and legal framework at international level, which has created a lapse in standardization of effective medicinal formulations. This leads to formulation of sub-standard products that may be dubious in their effectiveness. Nevertheless, research and development work in identification of bioactives, their quantification and isolation for industrial use is growing at steady state. Many drugs and products are available on the market in over the counter module. The most consumed products include diet supplements like lycopene, chlorophyll, beta-carotene, folic acid, multivitamins, minerals, antioxidants, and dyes in powder, pill, and liquid as well as dried plant part form. Once every step starting from identification of source species, its harvesting at appropriate stage, mode of harvesting, storage and carriage, quantification and isolation of usable phyto-compounds, determination of its optimal dosage and target disease/condition, regulation of manufacturing process are standardized. India is anticipated to be the leader in supplier of health products extracted from underutilized temperate as well as other vegetables. Furthermore, India being an agriculture based country will be most benefitted by utilizing its natural wealth, the most part of which is comprised by flora of Himalayas. The tribes and communities of habitable Himalayas should be accessed and provided with minimal technical help, incentives and facilities to cultivate and supply underutilized vegetable species for use by stakeholders. This will not only help food and pharmaceutical industry, but provide efficient means of livelihood to these underprivileged communities. Moreover, the gift of nature to human and animal life on the Earth will be properly utilized and benefitted from.

References

Achigan-Dako EG, Sogbohossou OED, Maundu P (2014) Current knowledge on *Amaranthus* spp.: research avenues for improved nutritional value and yield in leafy amaranths in sub-Saharan Africa. Euphytica 197:303–317. https://doi.org/10.1007/s10681-014-1081-9

Agrawal SR, Garima KJP, Bhatt SAJP (2010) In-vitro antibacterial activity of *Allium humile*. Acad Aren 2(6):234–239

Akter S, Hossain MM, Ara I, Akhtar P (2014) Investigation of *In vitro* antioxidant, antimicrobial and cytotoxic activity of *Diplazium esculentum* (Retz.). SW. Int J Adv Pharm Biol Chem 3 (3):723–733

Alegbejo JO (2013) Nutritional value and utilization of amaranthus(*Amaranthus* spp.)—a review. Bayero J Pure Appl Sci 6(1):136–143

Amabye TG (2015) Evaluation of physiochemical, phytochemical, antioxidant and antimicrobial screening parameters of *Amaranthus spinosus* leaves. Nat Prod Chem Res 4:199. https://doi.org/10.4172/2329-6836.1000199

Amin Y, Norazaidah KI, Hainida E (2006) Antioxidant activity and phenolic content of raw and blanched *Amaranthus* species. Food Chem 94(1):47–52

Archana GN, Pradeesh S, Chinmayee MD, Mini I, Swapna TS (2012) *Diplazium esculentum:* a wild nutrient-rich leafy vegetable from Western Ghats. In: Sabu A, Augustine A (eds) Prospects in bioscience: addressing the issues. Springer, India

Armesto J, Carballo J, Martínez S (2015) Physicochemical and phytochemical properties of two phenotypes of *Galega*lale (*Brassica oleracea* L. var. *Acephala* cv. Galega). J Food Biochem 39 (4):439–448

Cai J, Razzak A, Hering J, Syed A, Babcock TA, Helton S, Espat J (2008) Feasibility evaluation of Emodin (rhubarb extract) as an inhibitor of pancreatic cancer cell proliferation *in vitro*. J Parenter Enter Nutr 32(2):190–196

Chaudhary S, Hisham H, Mohamed D (2018) A review on phytochemical and pharmacological potential of watercress plant. Asian J Pharm Clin Res 11(12):102–107

Colle D, Arantes LP, Rauber R, Campos de Mattos SE, Teixeira da Rocha JB, Nogueira CW, Soares FAA (2012) Antioxidant properties of *Taraxacum officinale* fruit extract are involved in the protective effect against cellular death induced by sodium nitroprusside in brain of rats. Pharm Biol 50(7):883–891

Escudero NL, De Arellano ML, Fernández S, Albarracínand G, Mucciarelli S (2003) *Taraxacum officinale* as a food source. Plant Foods Hum Nutr 58:1–10

Goel V, Ooraikul B, Basu TK (1997) Cholesterol lowering effects of rhubarb stalk fiber in hypercholesterolemic men. J Am Coll Nutr 16(6):600–604. https://doi.org/10.1080/07315724.1997.10718725

Grosso C, Vinholes J, Silva LR, Guedes de Pinho P, Gonçalves RF, Valentão P, Jäger AK, Andrade PB (2011) Chemical composition and biological screening of *Capsella bursa-pastoris*. Rev Bras Farmacogn 21(4). https://doi.org/10.1590/S0102-695X2011005000107

Harshaw D, Nahar L, Vadla B, Saif-E-Naser GM, Sarker SD (2010) Bioactivity of *Rumexobtusifolius* (Polygonaceae). Arch Biol Sci 62(2):387–392. https://doi.org/10.2298/ABS1002387H

Joshi BC, Mukhija M, Kalia AN (2014) Pharmacognostical review of Urticadioica L. Int J Green Pharm 2014:201–209

Lamo K, Akbar PI, Mir MS (2012) Underexplored and underutilized traditional vegetables of cold arid Ladakh region of India. Vegetos 25(1):271–273

Kamal Uddin M, Juraimi AS, Hossain MS, Altaf Un Nahar M, Ali ME, Rahman MM (2014) Purslane weed (*Portulaca oleracea*): a prospective plant source of nutrition, Omega-3 fatty acid and antioxidant attributes. Sci World J 2014: 6 pages. https://doi.org/10.1155/2014/951019

Kaur T, Hussain K, Koul S, Vishwakarma R, Vyas D (2013) Evaluation of nutritional and antioxidant status of *Lepidium latifolium* Linn.: a novel phytofood from Ladakh. PLoS One 8 (8):e69112. https://doi.org/10.1371/journal.pone.0069112

Koochak H, Mansour S, Nejad S, Motamedi H (2010) Preliminary study on the antibacterial activity of some medicinal plants of Khuzestan (Iran). Asian Pac J Trop Med 3(3):180–184

Malmir M, Serrano R, Caniça M, Silva-Lima B, Silva O (2018) A comprehensive review on the medicinal plants from the genus *Asphodelus*. Plants (Basel) 7(1):20

Mukherjee PK, Mukherjee D, Maji AK, Rai S, Heinrich M (2009) The sacred lotus (*Nelumbo nucifera*)—phytochemical and therapeutic profile. J Pharm Pharmacol 61(4):407–422. https://doi.org/10.1211/jpp/61.04.0001

Namtak S, Sharma RC (2018) Medicinal plant resources in Skuru watershed of Karakoram wildlife sanctuary and their uses in traditional medicines system of Ladakh, India. Int J Complement Alternat Med 11(5):294–302

Orhan I, Deliorman-Orhan D, Özçelik B (2009) Antiviral activity and cytotoxicity of the lipophilic extracts of various edible plants and their fatty acids. Food Chem 115(2):701–705

Pal Murugan M, Janifer Raj X, Phani Kumar G, Gupta S, Singh SB (2010) Phytofoods of Nubra valley, Ladakh—the cold desert. Indian J Tradit Knowl 9(2):303–308

Poonia A, Upadhayay A (2015) *Chenopodium album* Linn: review of nutritive value and biological properties. J Food Sci Technol 52(7):3977–3985

Razavi SM, Zarrini G, Molavi G, Ghasemi G (2011) Bioactivity of *Malva Sylvestris* L., a medicinal plant from Iran. Iran J Basic Med Sci 14(6):574–579

Rehman H, Begum W, Anjum F, Tabasum H (2014) *Rheumemodi* (rhubarb): a fascinating herb. J Pharmacogn Phytochem 3(2):89–94

Simopoulos AP, Norman HA, Gillaspy JE, Duke JA (1992) Common purslane: a source of omega-3 fatty acids and antioxidants. J Am Coll Nutr 11(4):374–382

Singh K, Gangwar RK, Singh G, Jadon VS, Ranjan S (2014) Studies on the analgesic potential of leaf extracts of *Allium humile*on Swiss albino mice. Eur Res 83(9-2):1677–1681. https://doi.org/10.13187/er.2014.83.1677

Tabaraki R, Yosefi Z, Gharneh HAA (2012) Chemical composition and antioxidant properties of Malva sylvestris L. J Res Agric Sci 8(1):59–68

Thakur M (2014) Underutilized food crops: treasure for the future India. Food Sci Res J 5 (2):174–183. https://doi.org/10.15740/HAS/FSRJ/5.2/174-183

Nutritional and Medicinal Values of Neglected Spice Crops of Indian Himalayas

5

Pavan Kumar Kancharla, Aejaz Ahmad Dar, Ankit Banik, Srishti Dev, and Susheel Sharma

Abstract

The Indian Himalayas are hub for more than 8000 species of vascular plants. Among these, only 1748 are known to man for their medicinal and nutritional properties. The diversity of crop plants is seen majorly on the southern regions of the Himalayas, which is influenced by climatic variations created by altitudinal gradients. The decline in crop diversity is grossly accelerated by different factors like abiotic and biotic stresses and thus many species of crops are going endangered. Among majority of these crops, researchers and policymakers initiate least conservation strategies to neglected and underutilized species (NUS). NUS crop species are essential in economically sustainable farming, human nutrition, and food security to local farmers. There are many neglected and underutilized species of spices in Himalayan region, rich in vitamins, aromatic compounds, antiseptics and other nutritional as well as medicinal compounds. There is enough scope for the production of value added products from the neglected spices. This chapter highlights importance of these crops in nutrition and plant-based medicinal applications to human health and food security. In addition, the status of cultivation and utilization of NUS is discussed to create public awareness to instigate their timely conservation strategies.

Keywords

Spices · Himalayas · Neglected crops · Nutritional value · Antimicrobial · Anti-inflammatory · Anticancer

P. K. Kancharla · A. Banik · S. Dev
Department of Biotechnology, School of Life Sciences, Pondicherry University, Puducherry, India

A. A. Dar (✉) · S. Sharma
School of Biotechnology, Sher-e-Kashmir University of Agricultural Sciences and Technology of Jammu (SKUAST-J), Jammu, Jammu and Kashmir, India

S. M. Zargar et al. (eds.), *Neglected and Underutilized Crops - Towards Nutritional Security and Sustainability*, https://doi.org/10.1007/978-981-16-3876-3_5

5.1 Introduction

The Himalayan regions encompass a vast diversity of flora due to variation in the mountain ranges, as the climatic conditions differ from eastern Himalayas to western Himalayas. The floral diversity of the Himalayan region is mainly due to monsoonal rains accumulated in the summer season and altitudinal zonation of tropical lowland rainforests and alpine meadows (Joshi et al. 2016). Neglected and underutilized spices have high nutritional and medicinal importance, but their job in achieving nutritional security is not well understood and recognized and is not featuring in food and nutrition policies and programs of Hindu-Kush Himalayan (HKH) region (Adhikari et al. 2017). Spices are the herbs, which contain aromatic compounds such as flavonoids, carotenoids, etc., used to add flavor, color and aroma, stimulate appetite, and create visual attraction to dishes consumed on daily basis. Humans have consumed all types of spices since ancient times to fulfill body requirements on routine basis. The natural aromatic compounds present in such crops have curative activity and can be involved in pharmaceutical drug discovery and drug design. These aromatic compounds are present in rhizomes, roots, stems, bark, leaves, flowers, fruits, and seeds of the plants (Sachan et al. 2018).

The Indian Himalayan Range (IHR) has enormous treasury of spice crops found in altitudinal gradients and exhibit the highest collection of herbs and aromatic plants. In Ayurveda and many different mythological books, the great range of floral biodiversity has been explained in IHR and various unrecognized medical herbs, which can cure the basic diseases to the pandemic ones. Most of these plants are neglected and underutilized due to large focus of public and scientific communities on high yielding varieties and cash crops. Few important spice crops categorized under neglected category include *Dicliptera bupleuroides* and *Zanthoxylum armatum*. These plants have anti-helminthic, antifungal, and hepatoprotective activities and are also effective against stomach disorders and diabetes. These crops are at high risk of extinction, because of overexploitation (Sharma et al. 2010). Broad use of underutilized spice crops can provide more alternatives to construct spatial and temporal heterogeneity into uniform cropping systems and will improve flexibility to both abiotic and biotic stress. Thus significant research based on breeding, genetic engineering, and development efforts is desired for these promising crops to convert existing local cultivars into competitive varieties with broad range of adaptation and commercial potential (Ebert 2014).

5.2 Addressing Underutilization of NUS

The poor update and maintenance of databases pertaining to medicinal and economic importance of these underutilized crops is major barrier to their cultivation and utilization. At stakeholder level, the knowledge of these crops is limited to indigenous communities with respect to their language and culture practices. As a result, farmers from other regions are unaware of the nutritional and medicinal values of underutilized crops and impassive to induct these plants in their agricultural practices as major crop. Lack of demand from public sectors and farming communities of underutilized crops results in least cultivation, production, and commercialization (Thompson et al. 2007; Baldermann et al. 2016).

Unavailability of authenticated documents of indigenous knowledge on management of wild species results in unacceptance from the researcher to explore nutritive value of these herbs. Unrevealed and hidden knowledge from the non-modern science scripture of Ayurveda, Siddha and Unani and many other mythological books and Vedas, which was neglected for many decades by the world, became one of the reasons for diminishing crops. As a result, local knowledge on indigenous plants is being lost and sources of indigenous plants and animals are disappearing. Traditional knowledge about these herbal plants' transferred from older to younger generations is diminishing. Moreover, lack of interest in agriculture by younger generation's due to modernization and urbanization is also a serious concern (Ravishankar and Shukla 2007; Vidyarthi et al. 2013).

Overcultivation of mainstream crops by deforestation leads to the limitation or extinction of these neglected crops. Jhum cultivation and invasion by chir pine is one of the reasons for the extinction and negligence of many Himalayan medicinal crops (Singh 2019). The unknown loss of genetic diversity due to habitat loss, gap between male–female ratio, declination in types of natural habitat, increase in population density, overexploitation, industrialization, and urbanization results in obstruction of devising conservation and genetic improvement strategies of underutilized plants (Shankhwara et al. 2019). Due to lack of knowledge, regarding the environmental conditions and nutritional values, underutilized crops are much ignored by scientific as well as farmer communities resulting in their least production. In addition, unavailability of planting and seeding material poses additional hindrances to the production of these crops (Noun 2006).

At the level of post-harvest/processing, different crops have different characteristics and require different techniques for processing. Usually for indigenous plants, they do not meet the existing standards of safety and certification concerns due to presence of various toxic or hormone stimulating compounds, which may cause side effects. In the marketing level, the foremost challenge is low perception of crops as they are indigenous, which is not familiar to everyone. If the nutritional attribute is known and not easily available in popular food sources, the advertisement of the product may enhance the demand, if not, then limited demand leads to a limited supply (Koolman 2014).

From the marketing sector point-of-view including both the public and private sectors, there is poor response towards these crops. In government sector, limited support from research and development (R&D), lack of political support for promoting underutilized foods, whereas, in private sectors, supermarkets are difficult to convince regarding their successful commercialization. Market size is too small for commercial activities and there is a lack of consumer awareness on the value of indigenous foods. There is a limited and inadequate marketing support and infrastructure facility for transportation, storage, and processing (Durst and Bayasgalanbat 2014). In research laboratories, underutilized medicinal plants should be thoroughly investigated for phytochemicals, in order to inductee large-scale commercialization. However, there are no proper institutional facilities and agro-industrial based research laboratories due to financial limitations in this area. Researchers face many barriers for certification and traceability of these suppressed crops.

5.3 Scenario of Production

The successful production involves systemic approach to bring the underutilized, minor, orphan, or neglected crops in a better frequency of utilization and investigating its potential to withstand wide range of biotic and abiotic stress factors (Baldermann et al. 2016). Following are the major steps of production to be initiated for NUS cultivation:

5.3.1 Identification, Selection, and Cultivation

The underutilized crops are maintained by traditional practices and cultural preferences, thus remain poorly identified and neglected. Lack of gross attention has underestimated and underexploited their overall potential value in nutrition, medicine, and food security. It poses a threat of continued genetic erosion and disappearance, thus restricting their further developmental options and hinders the widening of their cultivation (Padulosi et al. 1999; Ghane et al. 2010; Dansi et al. 2012). The proper knowledge regarding the methods for identification and cultivation of underutilized/NUS can be retained from tribal and local inhabitants easily to study screening, chemical investigation, nutritive and medicinal values. Neglected and underutilized crop species are known globally based on their importance, domestication, adaptation, and rejection by mainstream agricultural scientists, breeders, technology providers, policy and decision makers, and consumers (Popoola et al. 2019).

The composite interactions of water dearth linked with climate change and population pressure need novel strategies to tackle food insecurity and undernourishment. Current research has identified neglected crops having promising potential to reduce food and nutritional insecurity. Cultivation techniques include proper seed selection, growing crops in its native or non-native environmental conditions. As neglected plants possess adaptability to less agricultural input systems and nutritional composition, and adaptation in its native environment without proper irrigation and care. Neglected crops have thus inbuilt tolerance to broad range of biotic and abiotic stress factors for enhancing nutritional food security. It is also necessary to promote neglected crops worldwide through scientific research such as breeding, agronomy, post-harvest handling, and linking farmers to mainstream markets (Chivenge et al. 2015).

5.3.2 Resources and Production

Very limited knowledge is available regarding neglected plants, due to least subscription to study their conservation and commercialization. Disseminating the facts about the introduction and importance of these crops to local farmers is of dire importance. Maintaining biodiversity in agriculture is an important parameter for proper functioning of ecosystems, sustainable agricultural production, and the achievement of food and nutritional security. However, till now only a few neglected crop species are utilized for food production globally (Padulosi et al. 2002; Toledo and Burlingame 2006; Chappell and LaValle 2011). The conservation of genetic resources is much needed, in order to enhance the cultivation of neglected plants (NPs) and protect the biodiversity by in situ methods of cultivation in their region of origin and at the same time ex situ methods via germplasm exchange program among cultivators and researchers (Durst and Bayasgalanbat 2014).

5.3.3 Storage and Processing

Stringent preservation techniques are prerequisite to maximize the nutrition percentage from these herbs even after long time storage. Reduction of nutritive quality can occur due to negligence of proper storage. For prolonged shelf-life and quality maintenance, slowing down the metabolism and aging of these herbs at low temperatures is done. Loss in percentage of food derived from neglected plants is about half of the production in rural areas (Kitinoja et al. 2011) and is generally due to lack of proper knowledge and adequate storage conditions. Another way for the preservation of processing is usually done by fermentation, for enhancing the validity and reducing contamination after harvesting (Sreeramulu et al. 2013; Zerio-Egli et al. 2014; Mishra et al. 2016). Food processing is responsible for the reduction of anti-nutritional ingredients, such as oxalic acid from leafy vegetables and cyanogens from roots of cassava (Betsche and Fretzdorff 2005; Chergui et al. 2015). In addition, one must also keep in mind that processing techniques like cooking and thermal treatments result in loss of the essential nutrients, such as provitamin A and C (Berni et al. 2015; Rodríguez-Roque et al. 2015).

5.3.4 Trade and Retail

Trade and retail involve both inherent and extrinsic properties. Inherent properties comprise nutritional and aromatic compounds present in the spices, which is a major factor responsible for efficient trading and retailing. Apart from inherent properties, extrinsic properties are major factors of gaining customer's trust for the product (Schreiner et al. 2013). Appearance and cost (Akdeniz et al. 2013) or, tagline of the product, communication policies, importation, and advertising (Richardson et al. 1994) play a major role in determining the complete marketing progress. Successful advertisement, recipe videos of the neglected plants promoted by top nutritionist and fitness models can enhance demand of neglected crop plants.

5.3.5 Consumer

At consumer level, NPs should be promoted as innovative products and information about them should be spread through advertisement and digital media. Due to negative perception with respect to yield, medicinal and nutritional values, NPs are widely rejected as potentially economically important crops. There is a stereotype towards the local and traditional foods, they being perceived as food for poor people (Fanzo et al. 2013). Introducing NPs as a source of innovative product will be helpful in seeking the customer's attention and acceptance (Robinson 2009).

5.4 Nutritional and Medicinal Scope of the Neglected Spice Crops

Several crop plants, which are not cultivated on large scale, have higher wealth of nutritional as well as medicinal properties. Crop plants categorized as underutilized horticultural crops (UUHC) are rarely planted commercially on large scale. The details about these neglected spice crops are presented in Table 5.1.

5.4.1 *Capsicum minimum*

Crop plants from North Eastern Himalayan region contribute 3.14% of the total geographical area of the region. One such spice crop is *Capsicum minimum* Syn. *Capsicum fastigiatum*, also called as Bird-eye-chili (Rai et al. 2004). Bird-eye-chili is widely consumed in the area of Northeastern India, specifically in Mizoram and Manipur (Vaishnavi et al. 2018). Commercially most of these crops are utilized to retrieve the oleoresin for use as spicing agents in foods and drugs (Blum et al. 2002). It is widely used to cure many diseases, disorders, and to control pest infestation. The bird-eye-chili is applied for the cure of dyspepsia, arthritis, rheumatism, toothache, flatulence and helps to stimulate appetite (Nkansah et al. 2011). The red color in its fruits is due to presence of capsanthin, carotenoids, and capsorubin, whereas, beta-carotene and violaxanthin impart yellowish orange color to the fruits. Majority of the carotenoids in ripe fruits is capsanthin, which is present in more than half of the total carotenoid concentration, whereas capsorubin and capsanthin increase proportionally during the advanced stages of ripeness (Bosland 1993).

Table 5.1 Medicinal values of different underutilized spice crop species of Indian Himalayas

S. No.	Underutilized spice crop	Common name	Medicinal importance	References
1.	*Capsicum minimum*	Bird-eye-chili	Cures dyspepsia, arthritis, rheumatism, toothache, and flatulence	Vaishnavi et al. (2018)
2.	*Zanthoxylum armatum*	Nepal pepper or toothache tree	It has stomachic, anti-helmintic, anti-larvicidal, antifungal, hepatoprotective, and allelopathic activities	Singh and Singh (2011)
3.	*Hedychium spicatum*	Spiked ginger lily	Rhizomes have analgesic, anti-inflammatory, antimicrobial, antioxidant, hypotensive, antispasmodic, pediculicidal, and cytotoxic properties	Rasool and Maqbool (2019)
4.	*Ajuga bracteosa*	Yellow bugle	Acts against various disorders, including diabetes, hypertension, fever, malaria stomach pain, swollen wounds, insect bites, bladder diseases, and tumors	Kayani et al. (2016)
5.	*Euphorbia hirta*	Asthma plant	Remedy for diarrhea, intestinal parasitosis, dysentery, asthma, bronchitis, and hay fever	Sharma et al. (2007)
6.	*Dicliptera bupleuroides*	Thorowax, Kaalu or Kirch	Effective against eczema, ear drop, eye diseases, stomach disorders, fever, and diabetes	Rehman et al. (2017)
7.	*Myrica esculenta*	Katphala	Cures asthma, bronchitis, inflammation, ulcers, diarrhea, anemia, respiratory and throat infections	Kabra et al. (2019)
8.	*Berberis arista*	Tree turmeric	Used in treatment of piles, pimples, boils, ulcers, sores, eye disorders, anorexia, dysentery, hepatitis, cough, jaundice, obesity, fever, diabetes, inflammations, and cancers	Semwal et al. (2018)
9.	*Flemingia vestita*	Sohphlang	It has wormicide, antioxidant, and antibacterial activities	Hrckova and Velebny (2012)
10.	*Perilla frutescens*	Perilla or Korean perilla	Protects from heart diseases, cholesterol problems, obesity, diabetes, bone related problems, and tumors	Asif (2011)
11.	*Thymus linearis*	Wild carom seed or Jungali ajwain	It has antibacterial, antifungal, antiviral, antiparasitic, anticancer, antioxidant, spasmolytic, antitussive, expectorant, antispasmodic, and anesthetic properties	Hussain et al. (2013)

(continued)

Table 5.1 (continued)

S. No.	Underutilized spice crop	Common name	Medicinal importance	References
12.	*Allium stracheyi*	Jambu faran or Himalayan seasoning Allium	Cures cold, stomachache, jaundice, diabetes, and reduces blood cholesterol level	Payal et al. (2020)
13.	*Angelica glauca*	Smooth Angelica or Chora	Remedy for common cold, hepatitis, indigestion, coughs, chronic bronchitis, typhoid, diarrhea, fever, rheumatism, fungal and bacterial infections	Sarker and Nahar (2004)
14.	Piper longum	Indian long pepper or pipli	Cures cough, asthma, bronchitis, muscular pain, and cardiovascular diseases	Manoj et al. (2004)

5.4.2 *Zanthoxylum armatum*

According to Prasad and Tomar (2020), spice crop *Zanthoxylum armatum* is one of the high value medicinal crops. *Z. armatum* is popularly known as toothache tree. It acts as carminative, stomachic, anti-helminthic, and in the treatment of toothache. This herb produces volatile oil with active constituents, such as linalool, limonene, and lignin. Various reports indicate that it exhibits anti-larvicidal, antifungal, hepatoprotective, and allelopathic activities (Singh and Singh 2011).

In a chemical investigation, Gilani et al. (2010) treated rabbit jejunum with crude extract of *Z. armatum* which caused relaxation of spontaneous and K^+ induced contractions, thus being more efficient against K^+ and Ca^{2+} antagonist effect. The *Z. armatum* inhibited diarrhea in mice and relaxed the carbachol and K^+ induced contractions in rabbit tracheal preparations. *Z. armatum* also induce vasodilator effects against phenylephrine and K^+ induced contractions in rabbit aortic rings. These results conclude that *Z. armatum* relieve spasm of smooth muscles, mediated by Ca^{++} antagonist mechanism, which act as a pharmacological base for its medicinal application in the respiratory, cardiovascular, and gastrointestinal diseases (Gilani et al. 2010).

5.4.3 *Hedychium spicatum*

Hedychium spicatum is popularly known as spiked ginger lily. It is distributed almost in every part of Himalayan region. Estimation of 4% essential oil is present in rhizome and phytochemical analysis reported different types of terpenoids like monoterpenoids, diterpenoids, and sesquiterpenoids in this herb. Rhizomes cure respiratory complications and have anti-inflammatory, analgesic, antimicrobial, antioxidant, hypotensive, antispasmodic, antifungal, pediculicidal, and cytotoxic properties. This spice plant has immense medicinal value and is considered to be an essential natural resource that dispenses many beneficial products for food, medicines, spices, dyes, perfumes, and esthetics, specifically in Southeast Asia (Rasool and Maqbool 2019).

5.4.4 *Ajuga bracteosa*

Ajuga bracteosa is commonly known as yellow bugle found mainly in grassy slopes in Sichuan and Yunnan. Decoction of the leaves is involved in the synthesis of various traditional medicines against various disorders, including hypertension, diabetes, malaria, fever, and stomachache (Chopra et al. 1986). The native folk uses leaf extracts as blood purifier and root extracts help in the recovery of normal body metabolism after diarrhea and dysentery (Manandhar 2002). In addition, leaves act as stimulants and enhance the amount of water and salt excretion from the body (Pal and Pawar 2011). The leaf extract has a potential healing activity towards swollen wounds, insect bites, eye troubles, bladder diseases, and tumors (Perry and Metzger 1980; Chang 1980; Kayani et al. 2016). It is reported that *A. bracteosa* produce compounds of the class withasteroids and diterpenoids which belong to naturally occurring C_{28} steroidal lactones. Withasteroids exhibit a broad range of biological activities, such as anti-inflammatory, immunomodulatory, antitumor, hepatoprotective, cytotoxic, antimicrobial, and insecticidal activities (Glotter 1991; Ahmed and Chaudhary 2009; Jan et al. 2014).

5.4.5 Euphorbia hirta

Euphorbia hirta is popularly known as asthma plant. It is reported that analgesic, anxiolytic, antipyretic, and anti-inflammatory activities are exhibited by the aqueous extracts of the root (Xia et al. 2018). The stem sap is remedy for eyelid styes and a leaf paste is useful in curing of boils and swelling (Kumar et al. 2010). *E. hirta* is a remedy for gastrointestinal problems like diarrhea, intestinal parasitosis and dysentery, and breathing disorders, such as asthma, bronchitis, and hay fever (Salles et al. 1999; Sharma et al. 2007). Hypotensive, tonic, and anticancer effects are also shown by extracts of E. hirta (Chopra and Chopra 1994; Joshi et al. 2005).

Broth prepared from dry herbs cures skin diseases, whereas broth prepared from fresh herbs is used as gargle for the thrush treatment (Kumar et al. 2010). Root broth is also beneficial for milk production in nursing mother. Root extracts are also used in neutralizing snake venom (Williamson 2002). The phenolic extract of *E. hirta* shows anti-amoebic (Tona et al. 2000) and antispasmodic properties (Gnecco et al. 1996). Antidiarrheal activity is also shown by this plant due to the presence of a flavonoid glycoside known as quercitrin (Galvez et al. 1993). It also shows a tranquillizing effect on the genitourinary tract (Kirtikar and Basu 2001). Methanolic extracts of leaves exhibit antifungal and antibacterial activities. For the cure of eyesores, latex of the plant is applied on lower eyelids, like kajal. The root exudates show nematocidal activity against juveniles of *Meloidogyne incognita*. The liquid extracts of the plant reduce the production of prostaglandins I_2, E_2, and D_2 and restrict aflatoxin contamination in wheat, rice, maize, and mustard crops (Singh and Sinha 1986; Joshi et al. 2005).

5.4.6 *Dicliptera bupleuroides*

Dicliptera bupleuroides is also known as thorowax, Kaalu, or Kirch by local people and belongs to family Acanthaceae. The plant is about 90 cm long, branched herb with hairy twigs. Flowers are mainly pink with purplish tinge. Seeds or whole plant decoction is useful in dysentery (Bhatt et al. 2013), leaf pastes are used for treatment of eczema and leaf extract for ear drop (Shah et al. 2013). *D. bupleuroides* juice reduces the body temperature of children during fever and helps in healing wounds of snake bites (Ajaib et al. 2015). The plant is also used to cure the eye diseases, its fresh leaves are used to cure diabetes, and juice of this plant is used to cure stomach disorders (Rehman et al. 2017). The aqueous fraction of the plant reported good antioxidant potential by 2, 2-diphenyl-1-picrylhydrazyl (DPPH) scavenging activity and fluorescence recovery after photo bleaching (FRAP) assay (Riaz et al. 2012). Many plants of this family have various ethno-medicinal importances like antimicrobial, anticancer, anti-inflammatory, and hepatoprotective properties (Komalavalli et al. 2014).

5.4.7 *Myrica esculenta*

Myrica esculenta is popularized with the name Katphala in Ayurveda. It is one of the underutilized species, but one of the socio-economically important species found in Meghalaya and sub-Himalayan region (Shankhwara et al. 2019). Almost all part of this species has a high nutritive and therapeutic value. The bark, roots, and leaves show a great impact to treat asthma, chronic bronchitis, inflammation, ulcers, diarrhea, anemia, and ear, nose, and throat disorders (Kabra et al. 2019).

Various Ayurvedic formulations of plant by different manufacturers used this crop as one of the main ingredient. For example, "Chyavanprash" by Dabur, Patanjali, Nature, and Nurture Healthcare to improve digestive disorders and body power, "Katphaladi Churna" by VHCA Ayurveda for the cure of fever, throat infections, breathing problem, and abdominal pain, "Pushyanuga Churna" for the treatment of bleeding disorders and candidiasis, "Katphala Taila" by VHCA Ayurveda for the treatment of joint pain, "Mahavisagarbha Taila" by VHCA Ayurveda for neuromuscular conditions, "Bala Taila" for cure of respiratory infections and weakness (Bao et al. 2005). The bioactive compounds like glycosides, flavonoids, terpenoids, tannins, saponins, and volatile oils provide the reason for the presence of all the pharmacological effects (Srivastava et al. 2016).

5.4.8 *Berberis aristata*

Another example of an underutilized crop is *Berberis aristata,* which has a high medicinal value. It is commonly known as tree turmeric, grown in the temperate and sub-tropical regions of Asia. It is a rare vulnerable species, which results in the underutilization of its benefits (Ray et al. 2011). Ayurveda explains this plant as a cure for diabetes, wound healing, swelling, piles, pimples, boils, ulcers, sores, eye disorders, leucorrhoea, anorexia, dysentery, hepatitis, liver disorders, cough, blood purification, jaundice, obesity, and fever (Semwal et al. 2018). Ayurvedic formulation of this plant known as "Rausat" cures eye diseases, skin disorders, and ulcers (Srivastava and Rawat 2013). A quaternary ammonium alkaloid of benzylisoquinoline class known as berberine is found mainly in *B.* aristata, is responsible for antimicrobial, antidiabetic, anti-obesity, hypotensive, hypolipidemic, anti-inflammatory, and anti-carcinogenic activities (Singh and Kakkar 2009; Tabeshpour et al. 2017).

It is found that berberine from methanolic stem extract of *B. aristata* has the potential to inhibit *H. pylori* growth isolated from gastroesophageal reflux disease (GERD) patients (Das et al. 2020). Along with berberine, the plant also produces alkaloids having major constituents of the plant, which include palmatine, berbamine, oxyberberine, oxycanthine, karachine, aromoline, epiberberine, dihyrokarachine, dehydrocaroline, jatrorrhizine, columbamine, taximaline, pakistanine, 1-*O*-methylpakistanine, pseudopalmatine, and pseudoberberine (Mokhber-Dezfuli et al. 2014).

5.4.9 *Flemingia vestita*

Flemingia vestita, commonly known as Sohphlang, is a very much neglected crop, rich in vitamins and micronutrients (Yang and Keding 2009). Exploration of literature evidently presents the property of tubers as wormicides (Singh and Arora 1973). Stomachache, dysentery, and intestinal worms can be prevented by including this plant in daily diet. Anti-helminthic use of tubers has been proven through various studies (Tandon and Das 2007; Hynniewta and Kumar 2010; Hrckova and Velebny 2012). The alcoholic crude root-peel extract of tubers reported vermifugal as well as vermicidal effect against the fowl tapeworm (*Raillietina echinobothrida*) (Tandon and Das 2007). Tuber extracts have potential antioxidant and antibacterial activity. Smaller tubers are generally more preferred over big sized fleshy tubers due to the presence of an isoflavone known as genistein, which is more efficacious (Hrckova and Velebny 2012).

5.4.10 *Perilla frutescens*

Perilla frutescens, commonly known as Perilla, is used as sedatives in Chinese traditional medicines. Various studies have reported the isolation of certain compounds from *P. frutescens* that confers anti-inflammatory, anti-allergic, antioxidant, and antitumor activities (del Carmen Recio et al. 1995; Safayhi and Sailer 1997; Hsu et al. 1997). Some of these compounds include different derivatives of cinnamic acid, anthocyanin, and flavonoids (Meng et al. 2006). The plant seeds are a good source of unsaturated fatty acids beneficial to our body system, protecting against heart diseases, cholesterol problems, obesity, diabetes, bone related problems, etc. (Asif 2011). A variety of volatile compounds constituted in the leaves of *P. frutescens,* such as linalool, limonene, caryophyllene, and perillaldehyde have antimicrobial, anti-dermatophytic, and antifungal activity (Kang et al. 1992; Duke 2002).

5.4.11 *Thymus linearis*

Thymus linearis is one of the indigenous and most threatened medicinal spice plant, which is also the reason for its underutilization (Kala 2005). It is commonly known as wild carom seed (Jungali ajwain). It is distributed in western Himalayan region and is an aroma producing flowering herb. This genus is characterized by unique morphological and chemical variability with a treasure of essential oils (Akcin 2006).

Recent studies report about their strong antifungal, antibacterial, antiviral, antiparasitic, anticancer, and antioxidant activities (Parajuli et al. 2005; Hafidh et al. 2009; Hussain et al. 2013). This spice herb has been found to exhibit antiseptic, antipyretic, analgesic, spasmolytic, antitussive, expectorant, and antispasmodic activities. They also possess anesthetic properties, because of the presence of geraniol, thymol, carvacrol, and other volatile compounds in the plant (Van Den Broucke and Lemli 1981; Guillen and Manzanos 1998; Keefover-Ring et al. 2009; Haq et al. 2011).

5.4.12 *Allium stracheyi*

Allium stracheyi is popularized by the name Jambu Faran in the native Himalayan folk. It is an endemic, endangered herb used in cuisine as spice that helps in the cure of cold, stomachache, and jaundice. It is used as an economic indicator for determining the plant-based economy and tradition of Bhotiya ethnic groups in the Mana and Niti Valleys of Uttarakhand, India (Payal et al. 2020). A. stracheyi has a high concentration of fiber and protein, but low in fat and carbohydrates concentration. Phyto-constituent investigation of *A. stracheyi* reported many compounds like hydrocarbons, terpenoids, terpenes, and sulfur-containing compounds (Mohan et al. 2019). Sulfur rich compounds have anti-inflammatory, antioxidant, and antimicrobial properties and have the property of reducing blood cholesterol (Abuajah et al. 2015). *Allium* species also control diabetes by consuming extracts, juice, and essential oil (Eidi et al. 2006; Mohamed et al. 2016).

5.4.13 *Angelica glauca*

Angelica glauca is one of the underutilized horizontal spice crops lying in the category of critically endangered plants (Peter 2008). Chora is the local name of *A. glauca*, distributed in temperate to alpine zone of Himachal Pradesh and Uttarakhand (Samant et al. 1998; Aldam 2002). Roots contain angelic acid, valeric acid, and angelisine resin and various essential oils (Chopra et al. 1986; Blake 2004). According to Sharma (1992), the roots are used for the preparation of several Ayurvedic medicines, such as Kumarpistikritghrita, Mahapaishachikghrit, and Rannadi cura.

Active compounds extracted from this spice crop mostly comprise different coumarins, sesquiterpenes, acetylenic compounds, and polysaccharides. This plant has expectorant, anti-inflammatory, and diaphoretic property for curing common cold, influenza, flu, indigestion, hepatitis, coughs, chronic bronchitis, fever, typhoid, rheumatism, fungal and bacterial infections, and urinary diseases (Sarker and Nahar 2004; Butola and Samant 2006). Root powder dissolved in cold water helps to cure diarrhea. The curry or dal (local dishes) flavored with root powder gives strength and vitality to women after delivery (Asolkar and Chopra 1992).

5.4.14 *Piper longum*

It is a species, which has not yet been explored much apart from its native region (Singh et al. 2012). *Piper longum* is known for the remedy of various disorders particularly of the respiratory tract like cough, asthma, bronchitis, etc. It is applied locally for muscular pain, i.e. having analgesic property and is a general tonic and hematinic. They are carminative and known to increase the bioavailability of food and drugs (Manoj et al. 2004). Roots and fruits are also considered for antidote of snake and scorpion venom (Chopra et al. 1956). One by one ratio of powdered seeds of *Embelia ribes* with fruit of *P. longum* and borax powder is a formulation for Ayurvedic contraceptive (Chaudhury et al. 2001). Broth prepared from roots is used for healing of joint swellings in cattle in the Northwestern Himalayan regions. The native folks used this plant in cardiovascular and anti-inflammatory activities. Alkaloids, lignans, and vola isolated from this herb have been found to exhibit immunomodulatory effect and hepatoprotective activity (Khushbu et al. 2011).

5.5 Conclusion and a Way Forward from Ecological Niche to Mainstream Purpose

The nutritive and medicinal values of neglected spice herbs from Himalayan region are suppressed, due to gap between male–female ratio, declination of types of natural habitat, increase in population density, overexploitation, industrialization, and urbanization. The production, harvesting, and marketing of these spice crops can be accelerated by the intervention of government and public sector agencies through strong financial support and dissemination of important data pertaining to their economic values. In addition, these spices crops can be potentially popularized for their medicinal properties and utilized as low cost food ingredient for all stakeholders. The development of these crops due to high stress tolerance can also be utilized to develop resilient plants.

Today world is suffering from malnutrition, poor health, hunger and main sufferers are usually lower middle class and poor people, who cannot even afford the major crops consisting of all the basic requirements. Malnutrition has a close relation with disease since malnutrition leads to disorders like iron and iodine deficiency, anemia, and neurological disorders and due to lack of retinol it can lead to loss of eye sight (Ransom and Elder 2003). These underutilized crops can provide nutrients and immunity to the low income urban people (Bhatt et al. 2019). These crops have the capability to eliminate the unemployment through job and income generation and improvise the small entrepreneurs by enhancing the profitability in urban and rural areas.

Underutilized herbs have ability to get adapted to harsh environmental conditions and ability to withstand in stress environment. So due to this property, NPs can be grown particularly in the xeric and semi-xeric land, mountains, and evergreen forest to meet the required nutritional demands (Thompson et al. 2007). The development of the neglected crops could lead to decrease over dependency on limited and selected major crops. These crops have ability to play a major role in global food security to contribute livelihoods not only by widening the option of edibles, but also lead to a path to enhance sustainability of farming by deduction in the investment, such as expensive and harmful fertilizers for agriculture.

Training courses pertaining to proper management of crops, proper selection of varieties, and producing new varieties (Padulosi et al. 2013) should be inculcated to local farmers, extension workers, and household members. Using these knowledge and skills they can get involved in production and household preparation of NPs. Making songs, slogans, etc., about these neglected spices can promote these crops at folk as well as at urban level. For example, the "Cassava Song," authored by Flora Nwapa in the 1980s (Nwapa 1986), is a protest against the import of the rice crop and a petition to stay with the traditional African cassava crop.

Acknowledgments AAD is thankful to DST SERB (NPDF Sanction order No. PDF/2017/001488) for financial support.

Authors Contribution PKK, AB and SD have presented the idea of contents. PKK, AAD, AB, SD and SS have written the chapter. AAD has finally edited the chapter.

References

Abuajah CI, Ogbonna AC, Osuji CM (2015) Functional components and medicinal properties of food: a review. J Food Sci Technol 52:2522–2529

Adhikari L, Hussain A, Rasul G (2017) Tapping the potential of neglected and underutilized food crops for sustainable nutrition security in the mountains of Pakistan and Nepal. Sustainability 9:291. https://doi.org/10.3390/su9020291

Ahmed D, Chaudhary MA (2009) Medicinal and nutritional aspects of various trace metals determined in *Ajuga bracteosa*. J Appl Sci Res 5:864–869

Ajaib M, Anjum M, Malik NZ, Sidiqui MF (2015) Ethnobotanical study of some plants of Darguti, tehsil Khuiratta, Azad Jammu and Kashmir. Int J Biol Res 3:101–107

Akcin TA (2006) Numerical taxonomic studies on some species of the genus Thymus L. (Labiatae) in Turkey. Asian J Plant Sci 5:782–788

Akdeniz B, Calantone RJ, Voorhees CM (2013) Effectiveness of marketing cues on consumer perceptions of quality: the moderating roles of brand reputation and third-party information. Psychol Mark 30:76–89

Aldam C (2002) Endangered medicinal plant species in Himachal Pradesh. Curr Sci 83:797

Asif M (2011) Health effects of omega-3, 6, 9 fatty acids: *Perilla frutescens* is a good example of plant oils. Orient Pharm Exp Med 11:51–59

Asolkar LV, Chopra RN (1992) Second supplement to glossary of Indian medicinal plants with active principles. Publications & Information Directorate

Baldermann S, Blagojević L, Frede K, Klopsch R, Neugart S et al (2016) Are neglected plants the food for the future? Crit Rev Plant Sci 35:106–119

Bao J, Cai Y, Sun M, Wang G, Corke H (2005) Anthocyanins, flavonols, and free radical scavenging activity of Chinese bayberry (*Myrica rubra*) extracts and their color properties and stability. J Agric Food Chem 53:2327–2332

Berni P, Chitchumroonchokchai C, Canniatti-Brazaca SG, De Moura FF, Failla ML (2015) Comparison of content and in vitro bioaccessibility of provitamin A carotenoids in home cooked and commercially processed orange fleshed sweet potato (*Ipomoea batatas* Lam). Plant Foods Hum Nutr 70:1–8

Betsche T, Fretzdorff B (2005) Biodegradation of oxalic acid from spinach using cereal radicles. J Agric Food Chem 53:9751–9758

Bhatt D, Kumar R, Joshi GC, Tewari LM (2013) Indigenous uses of medicinal plants by the Vanraji tribes of Kumaun Himalaya, India. J Med Plants Res 7:2747–2754

Bhatt L, Samota MK, Nautiyal MK (2019) Potential of underutilized, neglected or untrapped vegetables. J Pharmacogn Phytochem 8:1650–1653

Blake S (2004) Medicinal plant constituents. Medicinal plants from A to B 542:1–93

Blum E, Liu K, Mazourek M, Yoo EY, Jahn M, Paran I (2002) Molecular mapping of the C locus for presence of pungency in Capsicum. Genome 45:702–705

Bosland PW (1993) An effective plant field cage to increase the production of genetically pure chile (Capsicum spp.) seed. HortScience 28:1053–1053

Butola JS, Samant SS (2006) Physiological studies on seed germination of *Angelica glauca*. J Trop Med Plants 7:205

Chang BH (1980) Saengyak Hakhoe Chi 11:15

Chappell MJ, Lavalle (2011) Food security and biodiversity: can we have both? An agroecological analysis. Agric Hum Values 28:3–26

Chaudhury MR, Chandrasekaran R, Mishra S (2001) Embryotoxicity and teratogenicity studies of an ayurvedic contraceptive—pippaliyadi vati. J Ethnopharmacol 74:189–193

Chergui S, Yeddou AR, Chergui A, Halet F, Amaouche H, Nadjemi B, Ould-Dris A (2015) Removal of cyanide in aqueous solution by oxidation with hydrogen peroxide in presence of activated alumina. Toxicol Environ Chem 97:1289–1295

Chivenge P, Mabhaudhi T, Modi AT, Mafongoya P (2015) The potential role of neglected and underutilised crop species as future crops under water scarce conditions in sub-Saharan Africa. Int J Environ Res Public Health 12:5685–5711

Chopra RN, Chopra IC (1994) Indigenous drugs of India. Academic Publishers

Chopra RN, Nayar SL, Chopra IC, Asolkar L, Kakkar K (1956) Glossary of Indian medicinal plants. Council of Scientific & Industrial Research, New Delhi

Chopra RN, Nayar SL, Chopra IC (1986) Glossary of Indian medicinal plants, reprinted edition. Publications and Information Directorate, CSIR, New Delhi, p 241

Dansi A, Vodouhè R, Azokpota P, Yedomonhan H, Assogba P et al (2012) Diversity of the neglected and underutilized crop species of importance in Benin. Sci World J 2012:932947

Das S, Das MK, Das R, Gehlot V, Mahant S et al (2020) Isolation, characterization of Berberine from *Berberis aristata* DC for eradication of resistant *Helicobacter pylori*. Biocatal Agric Biotechnol 26:101622

del Carmen Recio M, Giner RM, Máñez S, Ríos JL (1995) Structural requirements for the anti-inflammatory activity of natural triterpenoids. Planta Med 61:182–185

Duke JA (2002) Handbook of medicinal herbs. CRC Press

Durst PB, Bayasgalanbat N (2014) Promotion of underutilized indigenous food resources for food security and nutrition in Asia and the Pacific. RAP Publication (FAO) eng no 2014/07

Ebert AW (2014) Potential of underutilized traditional vegetables and legume crops to contribute to food and nutritional security, income and more sustainable production systems. Sustainability 6:319–335

Eidi A, Eidi M, Esmaeili E (2006) Antidiabetic effect of garlic (*Allium sativum* L.) in normal and streptozotocin-induced diabetic rats. Phytomedicine 13:624–629

Fanzo J, Hunter D, Borelli T, Mattei F (eds) (2013) Diversifying food and diets: using agricultural biodiversity to improve nutrition and health. Routledge, New York. ISBN: 9781849714570

Galvez J, Zarzuelo A, Crespo ME, Lorente MD, Ocete MA, Jimenez J (1993) Antidiarrhoeic activity of *Euphorbia hirta* extract and isolation of an active flavonoid constituent. Planta Med 59:333–336

Ghane SG, Lokhande VH, Ahire ML, Nikam TD (2010) Indigofera glandulosa Wendl. (Barbada) a potential source of nutritious food: underutilized and neglected legume in India. Genet Resour Crop Evol 57:147–153

Gilani SN, Khan AU, Gilani AH (2010) Pharmacological basis for the medicinal use of *Zanthoxylum armatum* in gut, airways and cardiovascular disorders. Phytother Res 24:553–558

Glotter E (1991) Withanolides and related ergostane-type steroids. Nat Prod Rep 8:415–440

Gnecco S, Perez C, Bittner M, Silva YM (1996) Distribution pattern of n-alkanes in Chilean species from the Euphorbiaceae family. Boletin de la Sociedad Chilena de Quimica 41:229–233

Guillen MD, Manzanos MJ (1998) Study of the composition of the different parts of a Spanish *Thymus vulgaris* L. plant. Food Chem 63:373–383

Hafidh RR, Abdulamir AS, Jahanshiri F, Abas F, Abu Bakar F, Sekawi Z (2009) Asia is the mine of natural antiviral products for public health. Open Complement Med J 1:58–68

Haq F, Ahmad H, Alam M (2011) Traditional uses of medicinal plants of Nandiar Khuwarr catchment (district Battagram), Pakistan. J Med Plants Res 5:39–48

Hrckova G, Velebny S (2012) Pharmacological potential of selected natural compounds in the control of parasitic diseases. Springer, Wien

Hsu HY, Yang JJ, Lin CC (1997) Effects of oleanolic acid and ursolic acid on inhibiting tumor growth and enhancing the recovery of hematopoietic system postirradiation in mice. Cancer Lett 111:7–13

Hussain AI, Anwar F, Chatha SA, Latif S, Sherazi ST et al (2013) Chemical composition and bioactivity studies of the essential oils from two Thymus species from the Pakistani flora. LWT-Food Sci Technol 50:185–192

Hynniewta SR, Kumar Y (2010) The lesser-known medicine Ka Dawai Ñiangsohpet of the Khasis in Meghalaya, Northeast India. Indian J Tradit Knowl 9:475–479

Jan M, Singh S, Kaloo ZA, Maqbool F (2014) Medicinal importance of *Ajuga bracteosa* Wall ex Benth.—a review. Int J Adv Res 2:389–394

Joshi SK, Sharma BD, Bhatia CR, Singh RV, Thakur RS (2005) The wealth of India raw materials, vol 3. Council of Scientific and Industrial Research Publication, New Delhi, pp 270–271

Joshi RK, Satyal P, Setze WN (2016) Himalayan aromatic medicinal plants: a review of their ethnopharmacology, volatile phytochemistry, and biological activities. Medicines (Basel) 3:6. https://doi.org/10.3390/medicines3010006

Kabra A, Sharma R, Hano C, Kabra R, Martins N, Baghel US (2019) Phytochemical composition, antioxidant, and antimicrobial attributes of different solvent extracts from *Myrica esculenta* Buch-Ham. ex. D. Don leaves. Biomolecules 9:357

Kala CP (2005) Indigenous uses, population density, and conservation of threatened medicinal plants in protected areas of the Indian Himalayas. Conserv Biol 19:368–378

Kang R, Helms R, Stout MJ, Jaber H, Chen Z, Nakatsu T (1992) Antimicrobial activity of the volatile constituents of *Perilla frutescens* and its synergistic effects with polygodial. J Agric Food Chem 40:2328–2330

Kayani WK, Dilshad E, Ahmed T, Ismail H, Mirza B (2016) Evaluation of *Ajuga bracteosa* for antioxidant, anti-inflammatory, analgesic, antidepressant and anticoagulant activities. BMC Complement Altern Med 16:375

Keefover-Ring K, Thompson JD, Linhart YB (2009) Beyond six scents: defining a seventh *Thymus vulgaris* chemotype new to southern France by ethanol extraction. Flavour Fragr J 24:117–122

Khushbu C, Roshni S, Anar P, Carol M, Mayuree P (2011) Phytochemical and therapeutic potential of *Piper longum* Linn a review. Int J Res Ayurveda Pharm 2:157–161

Kirtikar KR, Basu BD (2001) Indian medicinal plants, vol 8, 2nd edn. Oriental Enterprises, Uttaranchal, p 2604

Kitinoja L, Saran S, Roy SK, Kader AA (2011) Postharvest technology for developing countries: challenges and opportunities in research, outreach and advocacy. J Sci Food Agric 91:597–603

Komalavalli T, Packia LM, Muthukumarasamy S, Mohan VR (2014) Anticancer activity of ethanol extract of *Asystasiatra vancorica* Bedd. (Acanthaceae) whole plant against Dalton Ascites Lymphoma. J Adv Pharm Educ Res 4:240–246

Koolman K (2014) Marketing neglected and underutilised species. Success factors of a market and consumer oriented approach. Germany

Kumar S, Malhotra R, Kumar D (2010) *Euphorbia hirta*: its chemistry, traditional and medicinal uses, and pharmacological activities. Pharmacogn Rev 4:58–61

Manandhar NP (2002) Plants and people of Nepal. Timber Press, Portland

Manoj P, Soniya EV, Banerjee NS, Ravichandran P (2004) Recent studies on well-known spice, *Piper longum* Linn. Nat Prod Radiance 3:222–227

Meng L, Lozano Y, Bombarda I, Gaydou E, Li B (2006) Anthocyanin and flavonoid production from *Perilla frutescens*: pilot plant scale processing including cross-flow microfiltration and reverse osmosis. J Agric Food Chem 54:4297–4303

Mishra A, Buchanan R, Schaffner D, Pradhan AK (2016) Cost, quality, and safety: a nonlinear programming approach to optimize the temperature during supply chain of leafy greens. Lebensmittel-Wissenschaft und-Technologie 73:412–418

Mohamed SM, Jaleel GAA, Abdallah HMI, Bashandy SAE, Salama AB, Mahmoud AH (2016) Hypoglycemic, hypolipidemic and antioxidant activities of *Allium porrum* leaves extract in streptozotocin-induced diabetic rats. Int J Pharm Tech Res 9:187–200

Mohan M, Haider SZ, Gautam SS, Singh P (2019) *Allium stracheyi* (Baker), the endangered and traditional medicinally important herb of Uttarakhand Himalaya, India: a review. Res Pharm 9:01–05

Mokhber-Dezfuli N, Saeidnia S, Gohari AR, Kurepaz-Mahmoodabadi M (2014) Phytochemistry and pharmacology of berberis species. Pharmacogn Rev 8:8–15

Nkansah GO, Ofosu-Budu KG, Ayarna AW (2011) Growth and yield performance of bird eye pepper in the forest ecological zone of Ghana. J Appl Biosci 47:3235–3241

Noun J (Producer) (2006) Neglected and underutilized plant species in the shouf-aley area of Lebanon: an ethnobotanical study. Region. Workshop Biodiver. Food. [Regional workshop on Biodiversity as Food, Beyrouth]

Nwapa F (1986) Cassava song & rice song. Tana Press, Enugu

Padulosi S, Eyzaquirre P, Hodgkin T (1999) Challenges and strategies in promoting conservation and use of neglected and underutilized crop species. In: Janick J (ed) Perspectives on new crops and new uses. ASHS Press, Alexandria

Padulosi S, Hodgkin T, Williams JT, Haq N (2002) Underutilized crops: trends, challenges and opportunities in the 21st century. In: Engels JMM, Ramanatha Rao V, Brown AHD, Jackson MT (eds) Managing plant genetic resources. CAB International, Wallingford, pp 323–338

Padulosi S, Thompson J, Rudebjer P (2013) Fighting poverty, hunger and malnutrition with neglected and underutilized species: needs, challenges and the way forward. Biodiversity International, Rome. ISBN 13:978-92-9043-941-7

Pal A, Pawar RS (2011) A study on *Ajuga bracteosa* wall ex. Benth for analgesic activity. Int J Curr Biol Med Sci 1:12–14

Parajuli RR, Tiwari RD, Chaudhary RP, Gupta VN (2005) Fungitoxicity of the essential oils of some aromatic plants of Manang against *Alternaria brassicicola*. Sci World 3:39–43

Payal K, Rao KS, Maikhuri RK, Kandari LS (2020) Ethnobotanical assessment of endangered *Allium stracheyi* (Baker): a lesser known species of the Nanda Devi Biosphere Reserve, Central Himalaya, India. In: Ethnomedicinal plant use and practice in traditional medicine, pp 1–16

Perry LM, Metzger J (1980) Medicinal plants of east and Southeast Asia: attributed properties and uses. MIT Press

Peter KV (ed) (2008) Underutilized and underexploited horticultural crops, vol 4. New India Publishing

Popoola JO, Ojuederie OB, Omonhinmin CA, Adegbite AE (2019) Neglected and underutilized legume crops: improvement and future prospects. In: Recent advances in grain crops research. Intechopen. https://doi.org/10.5772/intechopen.87069

Prasad S, Tomar JMS (2020) Distribution and utilization pattern of herbal medicinal plants in Uttarakhand Himalaya: a case study. J Med Plants 8:107–111

Rai N, Asati BS, Patel RK, Patel KK, Yadav DS (2004) Underutilized horticultural crops in north eastern region. ENVIS Bull Himalayan Ecol 13:46–52

Ransom E, Elder L (2003) Nutrition of women and adolescent girls: why it matters. Population Reference Bureau, Washington, DC

Rasool S, Maqbool M (2019) An overview about *Hedychium spicatum*: a review. J Drug Deliv Ther 9:476–480

Ravishankar B, Shukla VJ (2007) Indian systems of medicine: a brief profile. Afr J Tradit Complement Altern Med 4:319–337

Ray R, Gururaja KV, Ramchandra TV (2011) Predictive distribution modeling for rare Himalayan medicinal plant *Berberis aristata* DC. J Environ Biol 32:725–730

Rehman P, Ajaib M, Anjum M, Kamran SH, Ishtiaq S (2017) Phytochemical, antimicrobial and antidiabetic activity of *Dicliptera bupleuroides*: an ethnomedicinal non-toxic plant. Asian Pac J Trop Biomed. https://doi.org/10.1016/j.apjtb.2017.06.012

Riaz T, Abbasi MA, Rehman A, Shahzadi U, Qureshi MZ, Ajaib M (2012) *Dicliptera bupleuroides*: an imperative source for protection from oxidative stress. J Chem Soc Pak 34:326–332

Richardson PS, Dick AS, Jain AK (1994) Extrinsic and intrinsic cue effects on perceptions of store brand quality. J Mark 58:28–36

Robinson L (2009) A summary of diffusion of innovations. Enabling Change 5:10

Rodríguez-Roque MJ, de Ancos B, Sánchez-Moreno C, Cano MP, Elez-Martínez P, Martín-Belloso O (2015) Impact of food matrix and processing on the in vitro bioaccessibility of vitamin C, phenolic compounds, and hydrophilic antioxidant activity from fruit juice-based beverages. J Funct Foods 14:33–43

Sachan AKR, Kumar S, Kumari K, Singh D (2018) Medicinal uses of spices used in our traditional culture: world wide. J Med Plants Stud 6:116–122

Safayhi H, Sailer ER (1997) Anti-inflammatory actions of pentacyclic triterpenes. Planta Med 63:487–493

Salles B, Sattler U, Bozzato C, Calsou P (1999) Repair of oxidative DNA damage in vitro: a tool for screening antioxidative compounds. Food Chem Toxicol 37:1009–1014

Samant SS, Dhar U, Palni LMS (1998) Medicinal plants of Indian Himalaya: diversity distribution potential values. Gyanodaya Prakashan, Nainital

Sarker SD, Nahar L (2004) Natural medicine: the genus Angelica. Curr Med Chem 11:1479–1500

Schreiner M, Korn M, Stenger M, Holzgreve L, Altmann M (2013) Current understanding and use of quality characteristics of horticulture products. Sci Hortic 163:63–69

Semwal RB, Semwal DK, Aswal S, Kumar A (2018) Spectral studies and pharmacological relevance of berberine isolated from *Berberis aristata* roots. Curr Med Drug Res 2:181

Shah A, Sarfaraz KM, Faryal G, Ameer K, Khizar HB, Muhammad A et al (2013) Ethnobotanical study of medicinal plants of semi-tribal area of Makerwal & Gulla Khel (lying between Khyber Pakhtunkhwa and Punjab provinces), Pakistan. Am J Plant Sci 4:98–116

Shankhwara R, Bhandari MS, Meena RK, Shekhar C, Pandey VV et al (2019) Potential eco-distribution mapping of *Myrica esculenta* in northwestern Himalayas. Ecol Eng 128:98–111

Sharma DV (1992) Dravyaguna Viyanam, vol 2. Chaukhamba Bharati Academy, Varanasi, India, pp 34–36

Sharma JN, Al-Omran A, Parvathy SS (2007) Role of nitric oxide in inflammatory diseases. Inflammopharmacology 15:252–259

Sharma PK, Chauhan NS, Lal B, Husaini A, Teixeira da Silva J (2010) Conservation of phyto-diversity of Parvati Valley in Northwestern Himalayas of Himachal Pradesh-India. Med Aromatic Plant Sci Biotechnol 4:47–63

Singh MK (2019) Potential of underutilized legume tree Parkia timoriana (DC.) Merr. In eco-restoration of Jhum fallows of Manipur. J Pharmacogn Phytochem 8:1685–1687

Singh HB, Arora RK (1973) Soh-phlong, *Moghania vestita*: a leguminous root crop of India. Econ Bot 27:333–338

Singh J, Kakkar P (2009) Antihyperglycemic and antioxidant effect of *Berberis aristata* root extract and its role in regulating carbohydrate metabolism in diabetic rats. J Ethnopharmacol 123:22–26

Singh TP, Singh OM (2011) Phytochemical and pharmacological profile of *Zanthoxylum armatum* DC.—an overview. Indian J Nat Prod Resour 2:275–285

Singh P, Sinha KK (1986) Inhibition of aflatoxin production on some agricultural commodities through aqueous plant extracts. J Indian Bot Soc 65:30–32

Singh SJ, Batra VK, Singh SK, Singh TJ (2012) Diversity of underutilized vegetable crops species in North-East India with special reference to Manipur: a review. IJ NeBIO 3:87–95

Sreeramulu D, Reddy CVK, Chauhan A, Balakrishna N, Raghunath M (2013) Natural antioxidant activity of commonly consumed plant foods in India: effect of domestic processing. Oxidative Med Cell Longev 2013:369479

Srivastava S, Rawat AKS (2013) Quality evaluation of ayurvedic crude drug daruharidra, its allied species, and commercial samples from herbal drug markets of India. Evid Based Complement Alternat Med 2013:472973

Srivastava B, Sharma VC, Pant P, Pandey NK, Jadhav AD (2016) Evaluation for substitution of stem bark with small branches of *Myrica esculenta* for medicinal use—a comparative phytochemical study. J Ayurveda Integr Med 7:218–223

Tabeshpour J, Imenshahidi M, Hosseinzadeh H (2017) A review of the effects of *Berberis vulgaris* and its major component, berberine, in metabolic syndrome. Iran J Basic Med Sci 20:557–568

Tandon V, Das B (2007) In vitro testing of anthelmintic efficacy of *Flemingia vestita* (Fabaceae) on carbohydrate metabolism in *Rallietinae chinobothrida*. Methods 42:330–338

Thompson J, Hodgkin T, Atta-Krah K, Jarvis D, Hoogendoorn C, Padulosi S (2007) Biodiversity in agroecosystems farming with nature: the science and practice of ecoagriculture. In: Scherr SJ, McNeely JA (eds) The science and practice of ecoagriculture. Island Press, Washington, pp 46–60

Toledo A, Burlingame (2006) Biodiversity and nutrition; a common path toward global food security and sustainable development. J Food Compos Anal 19:477–483

Tona L, Kambu K, Ngimbi N, Mesia K, Penge O et al (2000) Antiamoebic and spasmolytic activities of extracts from some antidiarrhoeal traditional preparations used in Kinshasa, Congo. Phytomedicine 7:31–38

Vaishnavi BA, Bhoomika HR, Shetty GR (2018) Genetic parameters study for growth, yield and quality traits in Bird's eye chilli (*Capsicum frutescens* L.). Int J Curr Microbiol App Sci 7:1813–1817

Van Den Broucke CO, Lemli JA (1981) Pharmacological and chemical investigation of thyme liquid extracts. Planta Med 41:129–135

Vidyarthi S, Samant SS, Sharma P (2013) Traditional and indigenous uses of medicinal plants by local residents in Himachal Pradesh, North Western Himalaya, India. Int J Biodivers Sci Ecosyst Serv Manag 9:185–200

Williamson EM (2002) Dabur Ayurvet Limited, & Dabur Research Foundation. Major herbs of ayurveda. Churchill Livingstone, Edinburgh

Xia M, Liu L, Qiu R, Li M, Huang W et al (2018) Anti-inflammatory and anxiolytic activities of *Euphorbia hirta* extract in neonatal asthmatic rats. AMB Express 8:179

Yang RY, Keding GB (2009) Nutritional contributions of important African indigenous vegetables. African indigenous vegetables in urban agriculture. Earthscan, London, pp 105–143

Zerio-Egli CN, Sirsat SA, Neal JA (2014) Development of a novel economical device to improve post-harvest processing practices on small farms. Food Control 44:152–158

Genetics and Genomics Resources of Millets: Availability, Advancements, and Applications

6

Sabreena, Muslima Nazir, Bashir Ahmad Ganai, Rakeeb Ahmad Mir, and Sajad Majeed Zargar

Abstract

In the contemporary times, agricultural and food entities have gathered a lot of scientific attention confessing their research focus mainly on major food crops and at the same time undermining the local but highly vital crops. One such food crop of notable worth that demands further research exploration is millet crop. Millets are the nutritious food crops traditionally grown and used in Asia and Africa. Millets fit into the grass family *Poaceae* that include species such as finger millet (*Eleusine coracana*), kodo millet (*Paspalum scrobiculatum*), foxtail millet (*Setaria italica*), proso millet (*Panicum miliaceum*), little millet (*Panicum sumatrense*), job's tears (*Coix lacrymajobi*), and barnyard millet (*Echinochloa*

Sabreena (✉)
Proteomics Laboratory, Division of Plant Biotechnology, Sher-e-Kashmir University of Agricultural Sciences and Technology of Kashmir (SKUAST-K), Kashmir, Jammu and Kashmir, India

Center of Research for Development, University of Kashmir, Hazratbal, Srinagar, Jammu and Kashmir, India

Department of Environmental Sciences, University of Kashmir, Hazratbal, Srinagar, Jammu and Kashmir, India

M. Nazir · B. A. Ganai
Center of Research for Development, University of Kashmir, Hazratbal, Srinagar, Jammu and Kashmir, India

R. A. Mir
Department of Biotechnology, School of Biosciences & Biotechnology, Baba Ghulam Shah Badshah University, Rajouri, Jammu and Kashmir, India

S. M. Zargar
Proteomics Laboratory, Division of Plant Biotechnology, Sher-e-Kashmir University of Agricultural Sciences and Technology of Kashmir, Srinagar, Jammu and Kashmir, India
e-mail: smzargar@skuastkashmir.ac.in

S. M. Zargar et al. (eds.), *Neglected and Underutilized Crops - Towards Nutritional Security and Sustainability*, https://doi.org/10.1007/978-981-16-3876-3_6

spp.). Besides being immensely nutritional rich crop, millets pose tremendous proficiency to withstand hard stress conditions allowing them to endure the severity of climate change. These resilient features underline the prominence of directing research and development towards these crops. Except for finger millet and foxtail millet, other small millets have received little research consideration in terms of the development of genetic and genomic resources and breeding for yield improvement. In addition to germplasm diversity, the recent developments in phenotyping and genomics technologies can be used for the improvement of millets. This assessment offers an inclusive vision into the significance of millets, the global standing of their germplasm, diversity, assuring germplasm resources, and breeding approaches in order to accelerate the inclusion of climate-resilient and nutrient-rich millets for sustainable agriculture, ecology, and healthy food systems.

Keywords

Genomic resources · Millets · Underutilized crop · Biotechnological tools

6.1 Introduction

Cereal crops between 2004 and 2008 contributed an average of 255.1 MT (million tonnes) to universal food yield with millets accounting for 41.0% of that total. India, while realizing the importance of such food crop celebrated 2018 as the National year of millets. On the global front, 2023 has been designated to be the international year of millets in order to highlight its health benefits. Millets known for small grain size are principally C_4 photosynthetic crops grown in Asia and Africa's semi-arid areas. These crops consist of forage grasses with a distinguishing feature of being exceedingly tolerant to climate change. Millets thrive in soils with lesser nutrient content and places that are characterized by higher temperature and low moisture (Lata et al. 2013; Bergamini et al. 2013). Millets also have a number of distinguishing characteristics, including high nutritive value, low pathogen susceptibility, and drought and salinity tolerance (Lata et al. 2013). As the world is facing a serious population increase, crops like millet pose shorter life cycle and therefore it can be an ideal food source to meet its demands. Millets are typically grasses that belong to the *Gramineae* family and are divided into ten genera and at least 14 species with some prominent species (*Pennisetum glaucum*, *Eleusine, coracana*, *Setaria italic*, *Paspalum miliaceum*, *Panicum sumatrense*, *Echinochloa crus-galli*, *Paspalum scrobiculatum*, *Eragrostis tef*, *Digitaria exilis*, *Brachiaria deflexa*, *Coix lacrymajobi*) (Basavaraj et al. 2010). Millets are all members of the *Poaceae* family, but they vary in species, genus, tribe, and subfamily hierarchy. *Pennisetum glaucum*, *Eleusine, coracana* are diploids, whereas *Setaria italic, Paspalum miliaceum*, *Eragrotis tef*, *Digitaria exilis*, and *Coix lacrymajobi* are tetraploids, and *Echinochloa crus-galli* is hexaploid. Among the millets tested for genome size variation, *Setaria italic* has the smallest genome, while *Pennisetum glaucum* with high outbreeding and *Eleusine* coracana with mixed mating outbreeding has the largest genome (Bennett et al. 2000). Millets like *foxtail millet*, *Setaria italic*, and its wild parent, *S. viridis*, are categorized by advanced inbreeding proficiencies with impending outcrossing (Li et al. 1945; Till-Bottraud et al. 1992). Moreover, the wild relatives of these millets have much more taxonomic diversity. Millets are produced in 90 countries around the world, with India, China, Pakistan, Myanmar, Cameroon, Ghana, and the USA being the top producers. India has gained few yards in millet production over the years owing to its increased generation of hybrid millet cultivation such as pearl millet that is exceedingly resistant to environmental stresses (Gowda et al. 2006). Millet grains often keep their viability for a long time even though stored in poor conditions. Millets, compared to other cereals, are thought to have more grain per unit of water. Millet grains have nutritional value that is superior to or equal to that of other cereal grains (Mengesha et al. 1965; Joint FAO/WHO Expert Committee on Food Additives 1972). Carbohydrates, lipids, vitamins, and minerals are abundant in the grains. Millets are considered as nutrient-rich sources containing proteins and other essential micronutrients (National Research Council (NRC) 1996; Malleshi and Klopfenstein 1998; Lumen et al. 1993; Fernandez et al. 2003; Khairwal et al. 2004; Alaunyte et al. 2010). Millet grains in particular are used for treatment of diabetes and are recommended for lactating mothers (Kumari and Sumathi 2002). Its grains possess high medicinal properties and are utilized to synthesize essential amino acids besides being used for treatment of various ailments across the global medicinal system (Shimanuki et al. 2006; Zhang et al. 2007; Dwivedi et al. 2010). Millets are under-searched crops relative to maize, but the previous millet cropping method is still in use. In order to develop genetic and genomic methods for future breeding programs, pearl millet has attracted more research focus over the years in comparison to other species of millet.

6.2 Genetics and Genomic Relationships

For ascertaining the genomic basis of nutritional characters the selection of various genetic contrivances becomes extremely essential. In the contemporary times a number of "omic" approaches introduced in genetic and genome technology has augmented the nutritional profile by utilizing genes/QTLs in millet production. This is evident from the breeding and research programs conducted by the Rockefeller Foundation in the USA that is considered as a first pioneer research that collected millet germplasm. Different studies have used morphological and molecular data sets to study primitive diverging lineages of basal grasses that indicated the role of genomic evolution encompassing genome replications accompanied by gene deletions (Kellogg 2003; Malconber et al. 2006). Finger millet and rice genome are believed to have abundant macro-collinearity (Gale and Devos 1998), while foxtail millet possesses disturbed or incomplete micro-collinearity when analyzed at genome sequence level. The genome of pearl millet has endured various reallocations in disparity to foxtail millet and rice (Devos et al. 2000). There has been different point of views regarding the progeny of pearl millet and the literature findings indicate that *Pennisetum glaucum* spp. *Monodii,* a wild species is thought to be considered its potential progenitor besides few investigation also report it as a major domesticated food subjected to introgression that has been assisted by putative supergene or gene (Harlan 1975; Brunken et al. 1977). However other research investigations report that it is a single domestication crop (Harlan 1975; Porteres 1976; Brunken et al. 1977; Ibrahima et al. 2005; Miura and Terauchi 2005). The morphological differences due to the cross linkage between pearl millet and *Pennisetum glaucum* spp. *monodii* result in spikelet growth and structural control linked to QTLs (Poncet et al. 2002). Similarly *S. viridis* is been regarding as a wild progenitor of *Setaria italic* (Foxtail millet) which itself is a diploid species. The wild as well as domesticated accessions from China reported sturdy population structure supporting foxtail millet's monophyletic origin. However owing to the literature availability, the application of markers like RAPD and RFLP has been unable to track down the foxtail millet domestication centers (Schontz and Rether 1998). The presence of waxy phenotype in foxtail millet has been contributed to QTL mapping and human assortment of individual genes (Doust et al. 2004). The monophyletic origin in finger millet is revealed by DNA sequencing; however, there is a biased point of view regarding the basal relationship in the genus. Furthermore, two putative ITS homologous loci (A and B loci) were found in finger millet, *E. coracana*, and its putative A genome donor, *E. indica*; however, sequence evidence disproves the proposition that *E. floccifolia* is its binary genetic donor. As a result, the B genome donor is either unknown or extinct. Microsatellite markers have been utilized to investigate the ancestry of the finger millet and it has been reported that their progeny belongs to the Africa and India (Dida et al. 2008). Porso millet (*Panicum miliaceum*) obtained from *Panicum miliaceum* var. ruderale being tetraploid species similar to little millet (*Panicum sumatrense*) poses extraordinary floral basis but there is a key dissimilarity in its anatomical, physiological, and cytological features. Although ndhf genome analysis have been used to investigate the phylogeny of *panicum*, but genetic sequences related to porso millet and little millet are still to be exploited (Aliscioni et al. 2003). Similarly the porso and its species in North America have been analyzed using various RAPD markers (Colosi and Schaal 1997). The hexaploid Japanese barnyard millet (*Echinochloa crus-galli*) and the Indian barnyard millet (*Echinochloa colona*) are produced due to the hybridization between tetraploid *Echinochloa oryzicola* and an unknown diploid species. Phylogenetic analysis of *Oryza* sh4 indicates a genomic association of Asian *Echinochloa* revealing the *E. crus-galli* bonds binary genomes with its ancestral genotypes. Relationship between tef millet with *E. pilosa* has been studied by Ingram and Doyle (2003) using the phylogenetic tools and plastid locus rps16 to investigate the origin of Tef millet, an allotetraploid crop. The said investigation has revealed the similarity in morphological features and karyotypes but dissimilarity in spikelet shattering (Phillips 1995). Similarly Job's tears (*Coix lacrymajobi*), an Asian herb is cultivated extensively as a cereal food (Arora 1977). The concept of phylogeny has been studied in detail in *Gramineae* family employing cpDNAs endonucleases. Similar results were observed by Tateoka. Further, while comparing the hereditary alterations amid the chloroplast genomes of *Sorghum bicolor*, *Zea mays,* and *Coix lacryma-jobi*, Tateoka found that it is lower in sorghum as compared to maize. Digitaria is a genus of 230 species that occurs in six tropical and subtropical regions (Clayton and Renvoze 1986). Outcome measures for the study done by (Porteres 1976) revealed that *D. exilis* and *D. iburua* are domesticated in West Africa. *Digitaria horizontalis* and *Digitaria longiflora* are likely wild relatives of cultivated fonio, with fonio having diverse fascinating agronomic natures and appears expedient for its cultivation (Dansi et al. 2010).

6.3 Genomic Resources

Genomic tools are critical for the significant improvement of different crop species because they act as the primary breeding input. Genomic tools facilitate the effective and expedient depiction of genetic resources, as well as their ensuing application in the documentation of valuable genes for crop expansion. Although adequate genetic resources are at hand in case of millets, but due to the unavailability of biomarkers and genetic maps, the exploration of such a wonderful crop becomes restricted (Goron and Raizada 2015; Saha et al. 2016); however, the genome sequence of foxtail millet has provided new grounds for generation of high throughput genome analysis that can be a starting point for its exploration at greater extent (Jia et al. 2013; Kumari et al. 2013; Pandey et al. 2013; Muthamilarasan et al. 2014; Yadav et al. 2014, 2015; Zhang et al. 2014) and integrated marker databases (Jia et al. 2013; Muthamilarasan and Prasad 2015). The onset of omic concepts along with NGS, genome analysis has attained new levels in scientific world. NGS has been commonly used for de novo sequencing, whole-genome sequencing (WGS), whole-genome resequencing (WGRS), genotyping by sequencing (GBS), transcriptome and epigenetic research, and therefore, NGS may be a useful tool in the future. PCR based genetic markers and biomarkers have been utilized for the investigation of millets with foxtail millet is the species of millet that has been first sequenced for genome characterization (Zhang et al. 2014; Bennetzen et al. 2012). Gupta et al. (2014) reported the use of PCR based modalities in genomics-assisted breeding (GAB). Further research testified 25% of phenotypic variation due to mapping association of 20 agronomic traits among the 200 foxtail accessions (Gupta et al. 2014). SSR markers in foxtail millet have been found quite suitable for the documentation of genes/QTLs for controlling both agronomic and nutritional traits. The foxtail millet has by far the most single nucleotide polymorphisms (SNPs) owing to 1000 SNP markers being uniformly disseminated across all nine chromosomes. Across the genome, convenience of a drought genome sequence has been employed to ascertain and characterize the genes that are active in stress response in foxtail millet (Puranik et al. 2013; Mishra et al. 2014; Muthamilarasan et al. 2014; Yadav et al. 2014; Lata et al. 2013; Kumar et al. 2015). Similarly, sequencing the genome of allotetraploid tef (*Eragrostis tef*) has resulted in creation of new options SSR markers (Cannarozzi et al. 2014). This has remarkably improved the nutritional trajectories in millet research. Polymorphism in millet crop at the molecular level and cloning of its resistant gene homologues may lead to newfangled discoveries. Numerous conjoint expressed sequence tag (EST) markers can be found in switch grass (*Panicum virgatum*) and proso millet (Tobias et al. 2008). Tef (*E. tef*), a gluten-free crop with high protein, nutrient, and mineral content, has recently had its genome sequenced (Cannarozzi et al. 2014). The genome of tef accession "DZ-Cr-37" was sequenced using the Ilumina HiSeq2000 and 454-FLX platforms, yielding a total of 40Gbp from single- and paired-end reads, yielding 44-fold coverage (80% genome coverage). Among all the millet species that have been used to study population and genetic structure, it has been evaluated that tef millet poses transposable factors (Dwivedi et al. 2010). With the advancement of PCR assisted markers, new contrivances have been developed to determine the genetic and population paradigm of millet germplasm.

6.4 Advancements

Downy mildew resistance, drought resilience, grain harvest, yield mechanisms, and stover productivity have all been studied using various QTLs in pearl millet and for agronomic characters in foxtail millet and tef. The bulk of these investigations used linkage exploration to discover genes/QTLs which may otherwise would not have been apposite for marker-facilitated rearing of candidate gene documentation/cloning. There is a limited literature available regarding the genes involved in agronomic features in millets. Like it has been found that tb and bat gene are important for basal and auxiliary characterization in foxtail millet, similarly QTL genes are mandatory for tolerating the harsh drought conditions in pearl millet (Sehgal et al. 2009). Studies conducted to investigate salt resistance using gene expression profiles have indicated that cDNA-AFLP method in foxtail millet can be a viable option. Study reported by Lata et al. (2013) has discovered a 2.5-fold modification in nine up-mediated transcripts tangled in dehydration tolerance control. This has generated a lot of research experimentation to find out the candidate genes that regulate drought resistance in foxtail millet. Genetic resemblance between the closely related genomic areas is done by using techniques like gene collinearity and syntenic mapping to retrieve the information of genomic translation (Cannarozzi et al. 2014; Muthamilarasan et al. 2014). According to these sequence-based synteny maps, the utmost genomic collinearity was detected between foxtail millet and sorghum, followed by foxtail millet and maize, foxtail millet and rice. The maize ortholog of the teosinte branched1 (tb1) gene, which controls vegetative branching or tillering in foxtail millet, was discovered using comparative genomics. Recognizing orthologous sequences would be possible employing comparative genomics techniques. In order to perform genomic analysis for the documentation and confirmation of concerned genes allied with imperative features, more exertion should be put into generating enormous amounts of genic and genomic indicators.

6.4.1 Use of Germplasm for Crop Improvement

Different reference sets (core, mini-Core) have been employed over the years for evaluating the allelic diversity in order to trace out variation patterns. Those accessions posing accurate agronomic features particularly how they respond to environmental stresses are incorporated in the core (10% of the complete assortment) and mini-core (40% of the core assortment or −1% of the complete assortment). Millet species analyzed on the basis of morphological and assessment parameters have been found to fit in core and mini-core assemblages. As compared to other control cultivars, both finger and foxtail millet while analyzed on the basis of core assortment of germplasm accessions generate more grain yield. The approximation of finger millet and foxtail millet core assortments ensued in the assortment of germplasm accessions that ripe earlier, yield extra grain or fodder than control cultivars, besides varying in panicle form, size, and seed color, as well as a few drought and salinity-resistant accessions. Different parameters like seed protein and mineral content have also been exploited (ICRISAT 2009).

6.4.2 Enhancement of Drought Tolerance

The discovery of substantial QTL genes to amplify crop harvest and yield index in PRLT under terminal drought strain was made possible by ICRISAT's pearl millet study (Yadav et al. 2002). Under varying moisture stress conditions, it has been observed that the QTL derived topcross hybrids out-yielded the field-based topcross hybrids by a moderate but significant margin. Though, this benefit of low stress came at the expense of a little outcome in the non-stressed setting for the same hybrids. The crosses bloomed in advance, had less successful basal tillers, produced less biomass, and had a higher harvest index. Under post-flowering drought stress circumstances, hybrids upraised on quantitative trait loci introgression lines and generated 21% more grain, with no influence on grain yield in non-stressed conditions, according to field testing in carefully managed field environments. Due to the high panicle harvest index, many of these introgression lines had an expressively favorable combining capacity for grain yield under terminal stress. As a result, these marker-based breeding products are more imperative than any parental line for water resistance and guaranteed moisture environments.

6.4.3 Markers with Agronomic Traits, Abiotic and Biotic Stress Confrontation, and Product Quality

The heritable characters in pearl millet, finger millet, foxtail millet, and tef are ample enough to recognize traits that are beneficial. QTL based agronomic physiognomies have been studied comprehensively in pearl millet, together with resilience to environmental stresses (Jones et al. 2002; Morgan et al. 1998; Hash and Witcombe 2001; Breese et al. 2002; Gulia et al. 2007; Sharma et al. 2007; Yadav et al. 2002).

Kholova et al. (2010) employed the QTL and near isogenic lines introgression comprising a terminal drought-tolerance QTL and evaluate the impact of drought resistant and responsive inbred lines to see if the management of water forfeiture in non-limiting circumstances is intricated in pearl millet drought stress. When the tolerant genotypes were exposed to a water shortage, transpiration started to slow down at a subordinate portion of evaporable soil water, whereas the transpiration degree (Tr) in test crosses of the resistant parental genotypes was subordinate than in those of the responsive parental genotypes under well-watered conditions. Furthermore, Kholova et al. (2010) found that ABA levels in drought-tolerant testcross genotypes, like QTL-NILs, were higher in under adequate water supply than in responsive genotypes. It is important to emphasize at vapor pressure deficit levels, ABA levels did not increase rapidly in relation to water pressure in some of the test crosses, and sufficiently watered Tr was subordinate in resistant genotypes compared to responsive genotypes. These findings upkeep the hypothesis that drought-tolerant pearl millet has water-saving (avoidance) mechanisms (low Tr even at low VPD), which could be related to leaf ABA or sensitive predictor to higher VPD, which limits Tr further. Constitutive variants that did not result in modifications in transpiration efficiency (greater leaf ABA levels lessen Tr) can enhance absolute water conservation. This has the potential to become critical for grain filling in situations where total water availability is reduced, and it deserves to be considered in the development of pearl millet genotypes that are highly resistant to drought stress when developed on soils with a pH of 6.5 or higher.

6.4.4 Molecular Breeding Tools Enhanced by Genomic Sequences

Most genomic and molecular field of investigation has put its focus on cereal crops to meet the demands of world food production while undermining the capabilities and prominence of millet genomic research. In recent decades the genomic research in millets has been intensified, with the goal of assessing their potential for long-term agri-business infacet of climate alteration and global warming. The comparison of different fields of genomics has a remarkable potential in various scientific fields for speeding up the foundation of genomic contrivances in crops using techniques of molecular propagation with other crops. The research advancements have shown that QTL based methods can resolve crucial characters for genetic enhancement of crops specifically by comparative genomics method). Genome conservation is not circumscribed to chromosomes; it can also be comprehended in orthologous regions of different grass family members at the DNA sequence level.

6.5 Applications

Seeds of finger millet have high calcium content in contrast to other cereals and millets (376–515 mg per 100 g) (Barbeau and Hilu 1993). CaM genes obtained from other crops along with different millet species have been recently cloned to establish their structural resemblance and possible role in calcium signaling (Nath et al. 2010; Giussani et al. 2001; Goff et al. 2002; Schnable et al. 2009). Identification of cereal species of different grasses adjoining to the target millet for relative genomic research will be aided by determining the phylogenetic lineage between millets and other cereals. Because of the availability of foxtail millet genome sequences, genome mapping of millet species in conjunction with different breeding utilities would be extremely useful. Analogously, obtaining genome sequences for corn and sorghum will help with sugarcane and Job's tears genome analysis. Comparative genomics is the study of how genes are related to one another and in millets it helped to define the best strategies and tools for isolating genes of agronomic importance from huge and complicated cereal genomes by using accessible grass sequences. Genomic analyses of different crops (sugarcane, barley, wheat, etc.) have been successfully established using such methods (Kilian et al. 1997; Roberts et al. 1999; Asnaghi et al. 2000; Druka et al. 2000). A novel category of DNA marker, single-strand conformational polymorphism (SSCP)-SNP, has been designed in *Pennisetum glaucum* using annotated rice genomic sequences to predict intron–exon borders in millet ESTs and then designing primers to amplify regions across the introns (Bertin et al. 2005). For the genome saturation, genetic assisted maps in conjunction with markers can be utilized in both foxtail and tef millet (Yu et al. 2006; Dida et al. 2008; Arya et al. 2009). Since these crops descended from single *Poaceae* progenitor about 70 million years before, their genomes have a lot in common with foxtail millet and its wild relative green foxtail (*S. viridis*) regarded as model crops (Brutnell et al. 2010). Pearl millet grain possesses many outstanding properties particularly the ethanol production in the USA among millet species. The grains can produce 0.43 L kg^{-1} of ethanol, which is comparable to *Hordeum vulgare* and *Avena sativa* grains but lower than *Zea mays, Oryza sativa, Sorghum bicolor*, and *Triticum aestivum* grains (0.52–0.57 L kg^{-1}). Similarly to maize and sorghum grains, pearl millets have starch-based fermentation efficiency. As a result, in areas which are unfavorable for growth of maize and sorghum like excessive hotness and dryness, pearl millet may be used as a fuel source. Augmented genetic synteny in grass genomes should contribute gene function findings in bioenergy prototypical crops (Carpita and McCann 2008), posing viable genomic characters and have previously had their genomes sequenced (Schnable et al. 2009). The convenience of NGS alongside with genome characterization tools can be used to explore the source of biomass development at molecular level (Varshney et al. 2009), as well as analyzing changes in cell wall employing brown midrib variants that amend cell wall conformation, especially lignin subunit alignment. NGS may also be utilized to discover major-effect genes and QTLs accompanying to food quality and reliability. Microsatellites (Pandey et al. 2013; Zhang et al. 2014; Kumari et al. 2013), intron-length polymorphisms (Muthamilarasan et al. 2014), miRNA-based markers (Yadav et al. 2014), and transposable elements-based markers (Yadav et al. 2014) have all aided from the foxtail millet genome sequence (Yadav et al. 2015). These genomic methods can be used to map QTLs and categorize statistically important phenotype–genotype associations using GWAS and historic linkage disequilibrium (Varshney et al. 2014). Genes of interest can be utilized by greater-resolution genetic plotting, which incorporates sequencing-facilitated mapping with bulked segregant analysis. Major genes intricated in blast resistance (Pii) and QTLs involved in rice seedling vigor have been recognized using this tactic. Following the discovery of the genes, they will be introduced in anticipated food grains by expending genomic markers to monitor progenies regarding the advantageous allele combination. NGS strategies in conjunction with genotyping arrays can efficaciously be utilized for integrated aimed loci into exclusive varieties (Gupta et al. 2010). Transgene-based methods may also be used to do this (Wakasa et al. 2006; Frizzi et al. 2008). The development of elite crop varieties with greater nutrient content will come to a close with this approach.

6.6 Conclusion and Future Research Needs

Genetic and genomic contrivances of millet can be utilized to assess the arrangement and genetic assortment of some millet; however, in the meantime analogous studies in other species are required. Although various agronomic traits related with QTLs have been detected in millet crops; however, fonio and Job's tears are yet to be ascertained in terms of genomic tools (markers and genetic maps) as these two are understudied millets. In the future, for maximizing total biomass and grain yield foxtail millet has a lot of potential by using heterosis. Elite genetic accessions in superior breeds with opposing phenotypes should be resequenced. The variation in the sequence may be responsible for phenotypic variation mostly among hybrids. Furthermore, DNA sequence data open new avenues and make it easier to find genes with various agronomically beneficial traits for crop enhancement. The production of millets has shown a declining trend in locally adapted millet growing areas with respect to other staple crops. Despite the reason a millet grain is suitable for health and has proved to be useful against various health aliments. Reduced consumption has resulted from lower demand, which could be linked to a shift in lifestyle as a result of overall economic growth. Many anthropogenic and few of natural factors play critical role in meager production of millet species. In order to raise the status of this crop one should take possible efforts to erase its underutilized status possibly to aware people about its health benefits, more research is needed specifically for understudied millets and value addition through development of new products that will be of daily use along with government support to include millet in public distribution in accordance with that people will realize its value. A collaborative effort is needed involving certain organization to frame efficient strategy.

References

Alaunyte I, Stojceska V, Derbyshire E, Plunkett A, Ainsworth P (2010) Iron-rich Teff-grain bread: an opportunity to improve individual's iron status. Proc Nutr Soc 69:E105

Aliscioni S, Giussani LM, Zuloaga FO, Kellogg EA (2003) A molecular phylogeny of Panicum (Poaceae: Paniceae): tests of monophyly and phylogenetic placement within the Panicoideae. Am J Bot 90(5):796–821

Arora RK (1977) 'Job's tears' (Coix lacryma-jobi), a minor food and fodder crop of North-Eastern India. Econ Bot 31:358–366

Arya LM, Verma VK, Gupta, Karihaloo JL (2009) Development of EST-SSRs in finger millet and their transferability to pearl millet (*Pennisetum glaucum*). J Plant Biochemist Biotechnol 18:97–100

Asnaghi C, Paulet F, Kaye C, Grivet L, Deu M, Glaszmann JC, D'Hont A (2000) Application of synteny across Poaceae to determine the map location of a sugarcane rust resistance gene. Theor Appl Genet 101:962–969

Barbeau WE, Hilu KW (1993) Protein, calcium, iron and amino acid content of selected wild and domesticated cultivars of finger millet. Plant Foods Hum Nutr 43:97–104

Basavaraj G, Rao PP, Bhagavatula S, Ahmed W (2010) Availability and utilization of pearl millet in India. SAT eJournal 8:1–6

Bennett MD, Bhandol P, Leitch IJ (2000) Nuclear DNA amounts in angiosperms and their modern uses 807 new estimates. Ann Bot (Lond) 86:859–909

Bennetzen JL, Schmutz J, Wang H, Percifield R, Hawkins J, Pontaroli AC, Jenkins J (2012) Reference genome sequence of the model plant Setaria. Nat Biotechnol 30(6):555–561

Bergamini N, Padulosi S, Ravi SB, Yenagi N (2013) Minor millets in India: a neglected crop goes mainstream. In: Fanzo J, Hunter D, Borelli T, Matei F (eds) Diversifying food and diets: using agricultural biodiversity to improve nutrition and health, pp 313–325

Bertin I, Zhu JH, Gale MD (2005) SSCP-SNP in pearl millet-a new marker system for comparative genetics. Theor Appl Genet 110:1467–1472

Breese WA, Hash CT, Devos KM, Howarth CJ (2002) Pearl millet genomics—an overview with respect to breeding for resistance to downy mildew. In: Leslie JF (ed) Sorghum and millets pathology. Iowa State Press, Ames, pp 243–246

Brunken J, De Wet JMJ, Harlan JR (1977) The morphology and domestication of pearl millet. Econ Bot:163–174

Brutnell TP, Wang L, Swartwood K, Goldschmidt A, Jackson D, Zhu XG, Kellogg E, Eck JV (2010) Setaria viridis: a model for C4 photosynthesis. Plant Cell 22(8):2537–2544

Cannarozzi G, Plaza-Wüthrich S, Esfeld K, Larti S, Wilson YS, Girma D, Lyons E (2014) Genome and transcriptome sequencing identifies breeding targets in the orphan crop tef (*Eragrostis tef*). BMC Genomics 15(1):1–21

Carpita NC, McCann MC (2008) Maize and sorghum: genetic resources for bioenergy grasses. Cell 13:415–420

Clayton WD, Renvoze SA (1986) Genera graminum, grasses of the world. Kew Bull Add Ser XIII:389

Colosi JC, Schaal BA (1997) Wild proso millet (*Panicum miliaceum*) is genetically variable and distinct from crop varieties of proso millet. Weed Sci 45:509–518

Dansi A, Adoukonou-Sagbadja H, Vodouhe R (2010) Diversity, conservation and related wild species of Fonio millet (Digitaria spp.) in the northwest of Benin. Genet Resour Crop Evol 57 (6):827–839

de Lumen BO, Thompson S, Odegard WJ (1993) Sulfur amino acid-rich proteins in acha (Digitaria exilis), a promising underutilized African cereal. J Agric Food Chem 41(7):1045–1047

Devos KM, Pittaway TS, Reynolds A, Gale MD (2000) Comparative mapping reveals a complex relationship between the pearl millet genome and those of foxtail millet and rice. Theor Appl Genet 100(2):190–198

Dida MM, Wanyera N, Dunn MLH, Bennetzen JL, Devos KM (2008) Population structure and diversity in finger millet (*Eleusine coracana*) germplasm. Trap Plant Biol 1(2):131–141

Doust AN, Devos KM, Gadberry MD, Gale MD, Kellogg EA (2004) Genetic control of branching in foxtail millet. Proc Natl Acad Sci U S A 101(24):9045–9050

Druka A, Kudrna D, Han F, Kilian A, Steffenson B, Frisch D, Tomkins J, Wing R, Kleinhofs A (2000) Physical mapping of the barley stem rust resistance gene rpg4. Mol Genet Genomics 264:283–290

Dwivedi S, Upadhyaya H, Subudhi P, Gehring C, Bajic V, Ortiz R (2010) Enhancing abiotic stress tolerance in cereals through breeding and transgenic interventions. Plant Breed Rev 33:31–114

Fernandez DR, Vanderjagt DJ, Millson M, Huang Y-S, Chuang L-T, Pastuszyn A, Glew RH (2003) Fatty acid, amino acid and trace mineral composition of Eleusine coracana (Pwana) seeds from northern Nigeria. Plant Foods Hum Nutr 58(3):1–10

Frizzi A, Huang S, Gilbertson LA, Armstrong TA, Luethy MH, Malvar TM (2008) Modifying lysine biosynthesis and catabolism in corn with a single bifunctional expression/silencing transgene cassette. Plant Biotechnol J 6(1):13–21

Gale MD, Devos KM (1998) Plant comparative genetics after 10 years. Science 282:656–659

Giussani LM, Cota-Sanchez JH, Zuloaga FO, Kellogg EA (2001) A molecular phylogeny of the grass subfamily Panicoideae (Poaceae) shows multiple origins of Cs photosynthesis. Am J Bot 88:1993–2012

Goff SA, Ricke D, Lan TH, Presting G, Wang R, Dunn M, Glazebrook J, Sessions A (2002) A draft sequence of the rice genome (*Oryza sativa L. ssp. japonica*). Science 296(5565):92–100

Goron TL, Raizada MN (2015) Genetic diversity and genomic resources available for the small millet crops to accelerate a new green revolution. Front Plant Sci 6:157

Gowda CLL, Rai KN, Reddy BVS, Saxena KB (2006) Hybrid parents research at ICRISAT. International Crops Research Institute for the Semi-Arid Tropics

Gulia SK, Hash CT, Thakur RP, Breese WA, Sangwan RS (2007) Mapping new QTLs for downy mildew (*Sclerospora graminicola* (Sacc.) J. Schroet. J resistance in pearl millet (*Pennisetum glaucum* (L.) R. Br.). pp 373–386

Gupta PK, Langridge P, Mir RR (2010) Marker-assisted wheat breeding: present status and future possibilities. Mol Breed 26(2):145–161

Gupta S, Kumari K, Muthamilarasan M, Parida SK, Prasad M (2014) Population structure and association mapping of yield contributing agronomic traits in foxtail millet. Plant Cell Rep 33 (6):881–893

Harlan JR (1975) Our vanishing genetic resources. Science 188(4188):618–621

Hash CT, Witcombe JR (2001) Pearl millet molecular marker research. Int Sorghum Millets Newsl 42:8–15

Ibrahima OF, Chibani SA, Oran M, Boussaid Y, Karamanos, Raies A (2005) Allozyme variation among some pearl millet (*Pennisetum glaucum L.*) cultivars collected from Tunisia and West Africa. Genet Resour Crop Evol 52:1087–1097

ICRISAT (2009) ICRISAT archival report (2008) MTP project 5: producing more and better food at lower cost of staple cereal and legume hybrids in the Asian SAT (sorghum, pearl millet, and pigeon pea) through genetic improvement. pp 158–159

Ingram AL, Doyle JJ (2003) The origin and evolution of Eragrostis tef (Poaceae) and related polyploids: evidence from nuclear waxy and plastid rps16. Am J Bot 90(1):116–122

Jia G, Huang X, Zhi H, Zhao Y, Zhao Q, Li W, Zhu C (2013) A haplotype map of genomic variations and genome-wide association studies of agronomic traits in foxtail millet (*Setaria italica*). Nat Genet 45(8):957–961

Joint FAO/WHO Expert Committee on Food Additives (1972) Evaluation of mercury, lead, cadmium and the food additives amaranth, diethyl pyro carbonate, and octyl gallate (No. 51). World Health Organization, Geneva

Jones ES, Breese WA, Liu CJ, Singh SD, Shaw DS, Witcombe JR (2002) Mapping quantitative trait loci for resistance to downy mildew in pearl millet: field and glasshouse detect the same QTL. Crop Sci 42:1316–1323

Kellogg EA (2003) What happens to genes in duplicated genomes. Proc Natl Acad Sci U S A 100:4369–4371

Khairwal IS, Rai KN, Yadav OP, Bhatnagar SK (2004) Pearl millet cultivars. All India coordinated pearl millet improvement project, Mandor, Jodhpur, India. Indian Council of Agricultural Research, p 22

Kholova J, Hash CT, Kumar PL, Yadav RS, Koaova M, Vadez V (2010) Terminal drought-tolerant pearl millet (*Pennisetum glaucum* (L.) R. Br.) have high leaf ABA and limit transpiration at high vapour pressure deficit. J Exp Bot 61:1431–1440

Kilian A, Chen J, Han F, Steffenson B, Kleinhofs A (1997) Toward map-based cloning of the barley stem rust resistance gene Rp gland rpg4 using rice as a inter genomic cloning vehicle. Plant Mol Biol 35:187–195

Kumar A, Gaur VS, Goel A, Gupta AK (2015) De novo assembly and characterization of developing spikes transcriptome of finger millet (*Eleusine coracana*): a minor crop having nutraceutical properties. Plant Mol Biol Report 33(4):905–922

Kumari PL, Sumathi S (2002) Effect of consumption of finger millet on hyperglycemia in non-insulin dependent diabetes mellitus (NIDDM) subjects. Plant Foods Hum Nutr 57 (3–4):205–213

Kumari K, Muthamilarasan M, Misra G, Gupta S, Subramanian A, Parida SK, Prasad M (2013) Development of eSSR-markers in Setaria italica and their applicability in studying genetic diversity, cross-transferability and comparative mapping in millet and non-millet species. PLoS One 8(6):e67742

Lata C, Gupta S, Prasad M (2013) Foxtail millet: a model crop for genetic and genomic studies in bioenergy grasses. Crit Rev Biotechnol 33(3):328–343

Li HW, Li CH, Pao WK (1945) Cytological and genetic studies of interspecific cross of the cultivated foxtail millet, *Setaria italica* P. Beauv., and the green foxtail millet, *S. viridis L.* J Am Soc Agron 37:32–54

Malconber ST, Preston JC, Reinheimer R, Kossuth J, Kellogg EA (2006) Developmental gene evolution and the origin of grass inflorescence diversity. Adv Bot Res 44:425–480

Malleshi NG, Klopfenstein CF (1998) Nutrient composition, amino acid and vitamin contents of malted sorghum, pearl millet, finger millet and their rootlets. Int J Food Sci Nutr 49(6):415–422

Mengesha MH, Pickett RC, Davis RL (1965) Genetic variability and interrelationship of characters in Teff, Eragrostis tef (Zucc.) Trotter 1. Crop Sci 5(2):155–157

Mishra AK, Muthamilarasan M, Khan Y, Parida SK, Prasad M (2014) Genome-wide investigation and expression analyses of WD40 protein family in the model plant foxtail millet (*Setaria italica L.*). PLoS One 9(1):e86852

Miura R, Terauchi R (2005) Genetic control of weediness traits and the maintenance of sympatric crop–weed polymorphism in pearl millet (*Pennisetum glaucum*). Mol Ecol 14(4):1251–1261

Morgan RN, Wilson JP, Hanna WW, Ozias-Akins P (1998) Molecular markers for rust and pyricularia leaf spot disease resistance in pearl millet. Theor Appl Genet 96:413–420

Muthamilarasan M, Prasad M (2015) Advances in Setaria genomics for genetic improvement of cereals and bioenergy grasses. Theor Appl Genet 128(1):1–14

Muthamilarasan M, Suresh V, Pandey B, Kumari G, Parida K, Prasad M (2014) Development of 5123 intron-length polymorphic markers for large-scale genotyping applications in foxtail millet. DNA Res 21(1):41–52

Nath M, Goel A, Taj G, Kumar A (2010) Molecular cloning and comparative in silico analysis of calmodulin genes from cereals and millets for understanding the mechanism of differential calcium accumulation. J Proteomics Bioinform 3:294–301

National Research Council (NRC) (1996) National science education standards. National Academy of Sciences

Pandey G, Misra G, Kumari K, Gupta S, Parida SK, Chattopadhyay D, Prasad M (2013) Genome-wide development and use of microsatellite markers for large-scale genotyping applications in foxtail millet [*Setariaitalica* (*L.*)]. DNA Res 20(2):197–207

Phillips S (1995) Flora of Ethiopia and Eritrea. Poaceae (Gramineae), vol 7. National Herbarium, Addis Ababa Univ., Addis Ababa, Ethiopia, and Department of Systematic Botany, Uppsala Univ., Uppsala

Poncet V, Martel E, Allouis S, Devos K, Lamy F, Sarr A, Robert T (2002) Comparative analysis of QTLs affecting domestication traits between two domesticated× wild pearl millet (*Pennisetum glaucum L.*, Poaceae) crosses. Theor Appl Genet 104(6–7):965–975

Porteres R (1976) African cereals: eleusine, fonio, black fonio, teff, Brachiaria, Paspalum, Pennisetum and African rice. In: Harlan J, de Wet JMJ, Stemler A (eds) Origins of African plant domestication. Mouton Publ, The Hague, pp 409–452

Puranik S, Sahu PP, Mandal SN, Parida SK, Prasad M (2013) Comprehensive genome-wide survey, genomic constitution and expression profiling of the NAC transcription factor family in foxtail millet (*Setariaitalica L.*). PLoS One 8(5):e64594

Roberts MA, Reader SM, Dalgliesh C, Miller TE, Foote TN, Fish LJ, Snape JW, Moore G (1999) Induction and characterization of Phi wheat mutants. Genetics 153:1909–1918

Saha D, Gowda MC, Arya L, Verma M, Bansal KC (2016) Genetic and genomic resources of small millets. Crit Rev Plant Sci 35(1):56–79

Schnable PS, Ware D, Fulton RS, Stein JC, Wei F, Pasternak S, Minx P (2009) The B73 maize genome: complexity, diversity, and dynamics. Science 326(5956):1112–1115

Schontz D, Rether B (1998) Genetic variability in foxtail millet, *Setariaitalica (L.)* P. Beauv.-RFLP using a heterologous rDNA probe. Plant Breed 117(3):231–234

Sehgal DT, Napolean L, Skot V, Vadez CT, Yaday RS (2009) Conducting association mapping analysis of global pearl millet germplasm collection to validate candidate genes associated with a major drought tolerance QTL. In: Plant and animal genome 17 conference, San Diego, 10–14 Jan

Sharma YK, Yadav SK, Khairwal IS (2007) Evaluation of pearl millet germplasm lines against downy mildew incited by *Sclerospora graminicola in* western Rajasthan. J SAT Agric 3:2

Shimanuki S, Nagasawa T, Nishizawa N (2006) Plasma HDL subfraction levels increase in rats fed proso-millet protein concentrate. Med Sci Monit 12:BR221–BR226

Till-Bottraud I, Reboud X, Brabant P, Lefranc M, Rherissi B, Vedel F, Darmency H (1992) Outcrossing and hybridization in wild and cultivated foxtail millets: consequences for the release of transgenic crops. Theor Appl Genet 83(8):940–946

Tobias CM, Sarath G, Twigg P, Lindquist E, Nicholas J, McCann C, Carpita N, Lazo G (2008) Comparative genomics in switchgrass using 61, 585 high-quality expressed sequence tags. Plant Genome 1:111–124

Varshney RK, Nayak SN, May GD, Jackson SA (2009) Next generation sequencing technologies and their implications for crop genetics and breeding. Trends Biotechnol 27:522–530

Varshney RK, Terauchi R, McCouch SR (2014) Harvesting the promising fruits of genomics: applying genome sequencing technologies to crop breeding. PLoS Biol 12(6):e1001883

Wakasa K, Hasegawa H, Nemoto H, Matsuda F, Miyazawa H, Tozawa Y, Miyagawa H (2006) High-level tryptophan accumulation in seeds of transgenic rice and its limited effects on agronomic traits and seed metabolite profile. J Exp Bot 57(12):3069–3078

Yadav RS, Hash CT, Bidinger FR, Cavan GP, Howarth CJ (2002) Quantitative trait loci associated with traits determining grain and stover yield in pearl millet under terminal drought-stress conditions. Theor Appl Genet 104(1):67–83

Yadav CB, Muthamilarasan M, Pandey G, Khan Y, Prasad M (2014) Development of novel microRNA-based genetic markers in foxtail millet for genotyping applications in related grass species. Mol Breed 34(4):2219–2224

Yadav CB, Bonthala VS, Muthamilarasan M, Pandey G, Khan Y, Prasad M (2015) Genome-wide development of transposable elements-based markers in foxtail millet and construction of an integrated database. DNA Res 22(1):79–90

Yu JY, Kantety RV, Graznak E, Benscher D, Tefera H, Sorrells ME (2006) A genetic linkage map for tef [Eragrostis tef (Zucc.) Trotter]. Theor Appl Genet 113:1093–1102

Zhang C, Zhang H, Li J (2007) Advances of millet research on nutrition and application. J Chin Cereals Oils Assoc 22:51–55

Zhang L, Liu R, Niu W (2014) Phytochemical and antiproliferative activity of proso millet. PLoS One 9(8):e104058

Genomic Approaches for the Improvement and Conservation of Buckwheat

7

Shazia Mukhtar, Zaffar Bashir, Rakeeb Ahmad Mir, and Sajad Majeed Zargar

Abstract

Buckwheat (*Fagopyrum* sp.) is a vital pseudocereal annual crop grown widely in temperate regions and can strongly acclimatize to adverse conditions. The crop has gained immense importance in recent times because of its higher medicinal and nutraceutical values, as it is rich in vitamins, fats, proteins, minerals, polyphenols, and bioflavonoids such as rutin and quercetin. However, the crop is suffering from several barriers responsible for its underutilization which includes, low yield, self-incompatibility, increased seed shattering, low seed set, lodging, frost susceptibility, etc. To overcome such barriers, biotechnological tools such as molecular markers, genomics, plant tissue culture, and mutagenesis approaches have played crucial role for genetic improvement of buckwheat trait descriptors. Till date, only few of them have been implemented in this crop which are discussed in this chapter.

S. Mukhtar (✉) · Z. Bashir
Proteomics Laboratory, Division of Plant Biotechnology, Sher-e-Kashmir University of Agricultural Sciences and Technology of Kashmir (SKUAST-K), Kashmir, Jammu and Kashmir, India

R. A. Mir
Department of Biotechnology, School of Biosciences & Biotechnology, Baba Ghulam Shah Badshah University, Rajouri, Jammu and Kashmir, India

S. M. Zargar
Proteomics Laboratory, Division of Plant Biotechnology, Sher-e-Kashmir University of Agricultural Sciences and Technology of Kashmir, Srinagar, Jammu and Kashmir, India
e-mail: smzargar@skuastkashmir.ac.in

S. M. Zargar et al. (eds.), *Neglected and Underutilized Crops - Towards Nutritional Security and Sustainability*, https://doi.org/10.1007/978-981-16-3876-3_7

Keywords

Buckwheat · Pseudocereal · Nutraceutical · Bioflavonoids · Polyphenols · Rutin and Quercetin

7.1 Introduction

Buckwheat, an underutilized crop belongs to the genus *Fagopyrum* and family *Polygonaceae* (Hou et al. 2016). It is considered as a pseudocereal crop, because of its phylogenetic divergence from major cereal crops like maize, wheat, and rice (Zhang et al. 2017). The crop originated first in China and was later on introduced in Europe and America in the fifteenth and seventeenth centuries, respectively (Ohnishi and Matsuoka 1996; Guo et al. 2007). At present, it is cultivated mainly in the USA, Asia, Canada., Russia, East Europe, and other cold arid regions across the world (Koyama et al. 2013; Yasui et al. 2016). Buckwheat is cultivated in wide variety of soil types and is well suited to shady, moist, and cool environment. About 27 species of buckwheat have been reported till date of which *Fagopyrum esculentum* (common buckwheat), *Fagopyrum tartaricum* (tartary buckwheat), and *Fagopyrum dibotrys* (perennial buckwheat) are three important species of buckwheat (Tang et al. 2016; Jing et al. 2016). *Fagopyrum esculentum* (common buckwheat) is an annual Asian herb possessing triangular edible seeds and group of white and pink flowers, while *Fagopyrum tartaricum* (tartary buckwheat) is an annual erect herb possessing seeds of smaller size. Besides, *Fagopyrum dibotrys*, also known as *Fagopyrum cymosum* is an erect perennial herb with edible leaves and seeds (Jing et al. 2016). Among the above three mentioned species of buckwheat, *Fagopyrum esculentum* and *Fagopyrum tartaricum* are the two major species in the buckwheat gene pool cultivated in various parts of the world (Krkoskova and Mrazova 2005). These two crop plants are having diverse properties as *Fagopyrum esculentum* is characterized by sweet taste, easy dehulling of seed coat, and larger seed size, while *Fagopyrum tartaricum* is bitter in taste and seed size is small with tight seed coat (Jiang et al. 2007; Kim et al. 2008). Despite the above limitations, *Fagopyrum tartaricum* has been preferred by researchers in the recent years as it is reported to possess higher phenolic content in comparison to *Fagopyrum esculentum* (Sun et al. 2008; Guo et al. 2010; Zheng et al. 2012). In addition to the above mentioned major species of buckwheat, various other species were reported which includes *Fagopyrum crispatifolium, Fagopyrum pugense, Fagopyrum wenchuanense, Fagopyrum qiangcai, Fagopyrum hailuogouense,* and *Fagopyrum luojishanense* (Liu et al. 2008; Tang et al. 2010; Shao et al. 2011; Zhou et al. 2015; Hou et al. 2015).

Buckwheat holds remarkable potential owing to its short life cycle, its ability to be cultivated at higher altitudes, and presence of higher protein content in its grains. The flour of buckwheat is free of gluten and is utilized as an alternative to wheat flour for the people suffering from gluten intolerance (Comino et al. 2013). Buckwheat seeds are consumed as potential functional food material because of its higher nutraceutical and medicinal value (Kim et al. 2008; Inglett et al. 2011). It is rich in starch, dietary fiber, essential vitamins, biological proteins, minerals, antioxidants, flavonoids, exhibits prebiotic activities, and contains essential well balanced amino acids (Krkoskova and Mrazova 2005; Alamprese et al. 2007; Gimenz-Bastida and Zielinski 2015). Besides, several health benefits are offered by buckwheat species: *Fagopyrum esculentum, Fagopyrum tartaricum, and Fagopyrum dibotrys. Fagopyrum esculentum* seeds have the ability to remove stagnation of food, revitalize spleen, possess anti-hemorrhagic, anti-inflammatory activities, therapeutics for reducing the fever in Korean folk medicine and detoxification purposes (Kim et al. 2003). Conversely, *Fagopyrum tartaricum* is known for its richness in a bioflavonoid called as rutin (quercetin 3-*O*-β-rutinoside) which provides the species anti-carcinogenic, antioxidant, anti-inflammatory, and antihypertensive properties (Fabjan et al. 2003; Hou et al. 2016). Moreover, its roots are utilized to treat various incurable and chronic diseases like rheumatic disorders, treat general debility, relieve pain, drain dampness, and rejuvenate spleen. Additionally, it is utilized for hypolipidemic and hypoglycemic purposes (Guo et al. 2007; Yao et al. 2008). Besides, *Fagopyrum dibotrys* also imparts certain health benefits as its rhizome is utilized for detoxification purposes, to eliminate pus and expel blood stasis, treats cancer, lung diseases, dysmenorrheal, snakebite, rheumatism, trauma, and inflammation (Chan 2003).

Despite the above mentioned advantages of buckwheat, it is suffering from various constraints. One of the main constraints in the cultivation of buckwheat involves low grain yield and low seed set (Woo et al. 2010). Other limitations responsible for underutilization and lower yield of buckwheat include seed shattering during harvesting period, lodging, flower abortion, and abiotic stress conditions such as frost, temperature, drought, photoperiod, and water scarcity (Farooq et al. 2016). To overcome such limitations, various crop improvement programs were applied for utilization of buckwheat. Improving seed yield is one of the main objectives in buckwheat crop improvement programs. In addition, various other crop breeding programs were initiated to address other problems of buckwheat which include increased seed shattering resistance, increased seed size, early maturity, determinate flowering, zero allergenic proteins, increased groat percentage, increased flour shelf life, lodging resistance, and easier dehulling ability (Dar et al. 2018). Achieving these objectives through breeding techniques is facing certain problems of which one of the peculiar problems in buckwheat is self-incompatibility (Chrungoo et al. 2012; Farooq et al. 2016). Other factors restricting effective breeding involve sterility, shattering, low seed set, and apomixes which are determined by genotype and environmental ($G \times E$) interactions. This poses serious problem in effective breeding and makes selection of genotypes tedious (Chrungoo et al. 2012).

In order to advance the utilization of buckwheat, integrated approach involving conventional and nonconventional biotechnological and genomic approaches are required which includes: marker assisted selection of genotypes exhibiting quality traits, improvement of suitable trait descriptors and mutagenic approach which aims at exclusion of dominant immunological allergenic protein in whole buckwheat gene pool and in its wild species (Dar et al. 2018). Over 2500 accessions of buckwheat belonging to different have been identified from diverse regions of the world which provide large wealth of genes for a variety of quality traits (International Plant Genetic Resources Institute and the Consultative Group on International Agriculture).

7.2 Biotechnological Interventions for Combating Underutilization of Buckwheat

Various efforts have been taken to overcome the breeding barriers in buckwheat till date. Although conventional breeding was employed to hybridize *Fagopyrum esculentum* with *Fagopyrum tataricum*, but due to the prevailing problem of self-incompatibility in buckwheat, these hybridization approaches did not prove fruitful (Adachi et al. 1989; Samimy 1991; Asaduzzaman et al. 2009). Further, interspecific hybridization was carried between the diploid genus *Fagopyrum esculentum* and tetraploid genus *Fagopyrum cymosum* to improve the buckwheat crop (Ujihara et al. 1990; Suvorova et al. 1994; Hirose et al. 1995; Woo et al. 1999; Asaduzzaman et al. 2009). The success of interspecific hybridization was limited by the resultant hybrids being sterile. However, one of the major success in buckwheat crop improvement was production of self-pollinated buckwheat through hybridization of *Fagopyrum esculentum* with *Fagopyrum homotropicum* (Campbell 1997). Moreover, biotechnological techniques have large scope in the crop for combating its underutilization. One of the most important biotechnological tools involves the use of molecular markers which have played pivotal role in understanding diversity of buckwheat crop. In this context, earlier Javornik and Kump (1993) employed RAPD (random amplified polymorphic DNA) markers to analyze diversity in populations of *Fagopyrum esculentum* and *Fagopyrum tartaricum* and in limited number of individuals belonging to diverse cultivars. The results revealed higher polymorphism within populations and cultivars. Thus, RAPD markers have proved to be an efficient technique for analyzing genetic diversity in buckwheat. Other nonconventional methods which were implemented for the genetic improvement of buckwheat include embryo rescue, plant tissue and cell culture, protoplast fusion, development of haploids, and utilization of variation in plastid DNA like chloroplast DNA (Fig. 7.1).

Embryo rescue has proved to be an indispensable technique for accelerating production of interspecific hybrids between buckwheat species which otherwise was impossible through normal hybridization technique. Successful attempts were made to produce fertile interspecific hybrids between the diploid genus *Fagopyrum esculentum* and tetraploid genus *Fagopyrum cymosum* (Ujihara et al. 1990; Hirose et al. 1993; Suvorova et al. 1994; Woo et al. 1995; Wang and Campbell 1998). It has been revealed that *Fagopyrum esculentum* is more related with *Fagopyrum cymosum* rather than with *Fagopyrum tartaricum* (Hirose et al. 1995). Interspecific hybrids between *Fagopyrum cymosum* and *Fagopyrum tataricum* were successfully produced by performing embryo rescue technique (Woo et al. 1999). Moreover, in

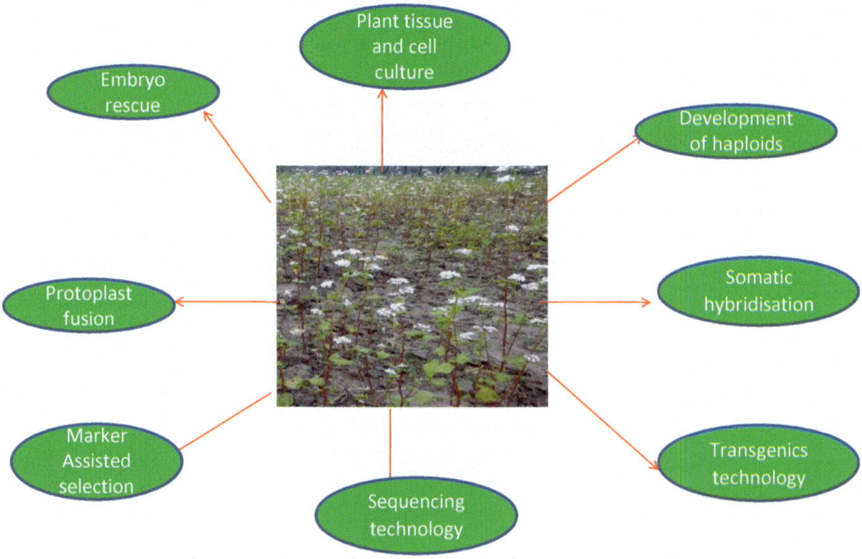

Fig. 7.1 Biotechnological tools and techniques employed for improvement of buckwheat

another attempt, *Fagopyrum esculentum* and *Fagopyrum tartaricum* were hybridized and the resulting ovules were grown on media to overcome hybrid sterility (Woo et al. 2010).

Another pivotal tool for improving genetics of buckwheat crop involves plant tissue and cell technique. The hypocotyl segments of *Fagopyrum esculentum* and *Fagopyrum homotropicum* were identified to have regenerative capabilities through organogenesis and somatic embryogenesis techniques (Woo et al. 2012). Conversely, protoplasts were identified to regenerate callus. Further, fusion techniques were harnessed in buckwheat to combat the breeding barriers by elaborating the simple procedures. In an experiment, mesophyll protoplasts of *Fagopyrum esculentum* were fused with the hypocotyl protoplasts of *Fagopyrum tartaricum* (carrier) by polyethylene glycol (PEG) fusion (Lachmann 1991). By utilizing the manifestation of *Fagopyrum tartaricum* calli and intolerance of mesophyll protoplasts, an easy pre-selection method of two steps was established and the resultant hybrid was confirmed by RFLP marker analysis (Samimy et al. 1996). In addition, development of haploids is gaining more attention currently in buckwheat crop improvement programs. Anthers of three polishbred common buckwheat cultivars were developed in vitro, where the most prominent class of regenerates was tetraploids. No haploids were identified among the regenerates obtained from diploid cultivars, likewise the tetraploid cultivars did not yield any diploid regenerate (Berbec and Doroszewska 1999). Based on breeding behavior and morphological characterization of *Fagopyrum* species, it was identified that *Fagopyrum tartaricum* and *Fagopyrum cymosum* are not related to each other. However, based on the analysis of chloroplast DNA, it was identified that *Fagopyrum tartaricum* and *Fagopyrum cymosum* are more related to each other in contrast to *Fagopyrum esculentum* (Kishima et al. 1995). Conversely, *Fagopyrum tartaricum* and *Fagopyrum cymosum* were identified to differ in their origin in the Tibet-Himalayan belt through the analysis of three chloroplast DNA (Yamane et al. 2003). Moreover, by making analysis of cpDNA, nucleotide sequence variation and allozyme of Adh gene, it has been revealed that interspecific genetic variation is more in *Fagopyrum cymosum* and low in *Fagopyrum tartaricum* which provides *Fagopyrum cymosum* as a valuable genetic resource for cultivated *Fagopyrum tartaricum* (Dar et al. 2018).

Apart from above mentioned approaches, other approaches which have emerged recently and have solved various agro-economical problems include transgenics approaches and genome editing tools. These approaches have proved to be fruitful as only some specific genes are transferred to develop particular genotypes in comparison to the traditional breeding approach where two complete parental genomes are mixed and subsequently undesirable genes are removed by back crossing method.

7.3 Molecular Marker System and Its Application in Buckwheat

The use of molecular markers has been widely adapted in crops at rapid pace for crop improvement programs. Various molecular marker systems such as RAPD, AFLP, SSRs, ISSRs, ESTs, and array based markers have been developed for understanding the genetics and to address various problems of buckwheat such as self-incompatibility and its allergenic nature (Yasui et al. 2004; Konishi and Ohnishi 2006; Hara et al. 2011; Yabe et al. 2014; Heffler et al. 2014). Identification and elimination of allergenic proteins in buckwheat is another area which requires attention of researchers. Generation of nucleotide information for the development of hypo-allergenic buckwheat has been considered most effective in this regard (Nair and Adachi 1999). Thus, eliminating genes responsible for allergens and self-incompatibility and improving nutritional quality of the grain are the important objectives in buckwheat breeding programs. These molecular markers have also been applied for understanding genetic diversity and genotyping of buckwheat. With the help of RAPD (random amplified polymorphic DNA) marker technique, *Fagopyrum tartaricum* was identified to be closely related to its wild ancestor *Fagopyrum tartaricum* ssp. *potaninii Batalin* by analyzing 28 diverse accessions belonging to two subspecies and 14 species of *Fagopyrum* (Sharma and Jana 2002). Phylogenetic relationships were studied among 42 accessions of *Fagopyrum tartaricum* from 21 populations through AFLP markers which revealed the species originated most likely in northwestern Yunnan in China or in eastern Tibet (Tsuji and Ohnishi 2001). About 195 tartary buckwheat accessions were also studied for rutin content variation by utilizing AFLP (amplified fragment length polymorphism) fingerprinting technique which exhibited grouping of germplasm into two diverse clusters, one cluster having genotypes with lower rutin content, while other cluster was having genotypes with higher rutin content (Gupta et al. 2012). In another study, ISSR (inter-simple sequence repeats) technique was also applied to 15 germplasms of *Fagopyrum tartaricum* belonging to Western Himalayas to analyze genetic diversity. It was identified that genetic diversity of *Fagopyrum tartaricum* and altitude are significantly correlated with each other (Kishore et al. 2013). Simple sequence repeat (SSR) markers are extensively used for map construction and high-throughput genotyping in crops because of their random distribution, stable co-dominance, greater abundance within the genome, higher reproducibility, and high polymorphism information content (PIC) (Zietkiewicz et al. 1994; Iwata et al. 2005). However, SSR markers have limited application in deciphering genetic diversity of buckwheat due to lack of genome sequence data, high cost of its development, difficulty in amplifying specific loci, increased genetic diversity between EST marker systems and buckwheat cultivars (Woo et al. 2010; Dar et al. 2018). Despite the above limitations, SSR markers were also developed for common buckwheat (Iwata et al. 2001). By utilizing SSR markers, significant genetic diversity was observed among Japanese common buckwheat cultivars (Iwata et al. 2005). Further, array based markers were also developed to construct novel genome map of buckwheat which has enough markers to cover the whole genome. However, to interpret the fluorescence signals of the arrays, a specialized instrument is required which is a major limitation of this marker system (Yabe et al. 2014).

One of the most important applications of molecular marker system is marker assisted selection and creation of genetic maps. Marker assisted selection (MAS) emphasizes on selection of plants possessing genomic fragments involved in expression of desired traits through the application of molecular marker technology. With the availability and development of various molecular markers and genetic maps in crop plants, marker assisted selection (MAS) has become successful for the traits conferred both by major genes and by quantitative trait loci (QTLs). Various prospective advantages are being offered in breeding programs by utilizing molecular markers linked to genes of interest in breeding programs, thus changing the scenario from phenotype to genotype based selection (Dar et al. 2018). Creation of genetic maps has proved to be an important tool for plant breeders and for genetic researchers as it aids in determination of chromosomal locations of both complex and simple traits (Paterson et al. 1991; Khan 2015). With the help of MAS technique, various efforts have been taken for construction of genetic maps in buckwheat. Earlier, RAPD marker system was employed to develop a linkage map of buckwheat for self-incompatibility trait and the map produced was comparatively dense with a 0.6 cM distance between markers. Mapping of self-incompatibility gene was an initial step for better comprehension of sexual reproduction and these markers are beneficial for introgression of self-compatibility in buckwheat (Aii et al. 1998). Further, three RAPD fragments tightly linked to self-compatibility gene (Sh gene) were cloned and sequenced in buckwheat to develop SCAR markers. For PCR amplification of SCAR markers, 21-mer oligonucleotide primers were designed by utilizing the sequence information obtained from each end of the three linked RAPD markers. In *Fagopyrum homotropicum,* a single fragment was amplified by two of these SCAR primers, SCB14 1250 and SCP8 1000, and the same fragment was absent in common buckwheat, *Fagopyrum esculentum*. Another SCAR marker, SCQ7 800, showed diverse PCR fragments between *Fagopyrum homotropicum* and *Fagopyrum esculentum*. This co-dominant marker proved advantageous for distinguishing both types of homozygous and heterozygous forms (Aii et al. 1998). Among the PCR based markers, various other powerful molecular markers were used for construction of linkage maps, which includes AFLP and SSR markers. In an investigation, AFLP (amplified fragment length polymorphism) and bulked segregate analysis were employed to develop molecular markers associated with self-compatibility (*Sh*) gene. AFLP marker provides many genomic bands for buckwheat. About 62 bands were detected which were exclusive for homomorphic flowered plants. Thus, AFLP is useful for marker assisted selection for detailed analysis of plant genomes as well as introgression of self-compatibility into common buckwheat (Aii et al. 1998). More recently, SSR markers were also utilized for construction of genetic maps in buckwheat in both common wheat and tartary buckwheat. The first genetic linkage map in buckwheat was constructed in wild species of *Fagopyrum tartaricum* by using the SSR marker technique. Development of this genetic map helped in elucidation of the MAS, QTL mapping, evolution, genomics, map-based cloning which provided basis for molecular breeding and location of important genes in buckwheat (Xiaolei et al. 2013). For the analysis of genetic diversity of *Fagopyrum tataricum*, a genomic survey was taken by using simple sequence repeat (SSR) markers where about 21.9 GB raw sequence reads were produced which were assembled into 348.34 Mb genome sequences including 204,340 contigs (Hou et al. 2016). Further, analysis of SSR marker was conducted on 64 accessions of *Fagopyrum tartaricum* collected from diverse regions of the world. The clustering exhibited two distinguished sub-groups of *Fagopyrum tataricum*. Among the two groups, one group belonged to Bhutan, Nepal, and Yunnan-Guizhou Plateau regions, while other group belonged to Hunan, Loess Plateau regions, and Hubei of China, and the USA. This undoubtedly revealed significant genetic diversity present among the accessions of *Fagopyrum tartaricum* and this diversity can be advantageous for improving Tartary buckwheat breeding programs by utilizing resources of its germplasm (Hou et al. 2016). Dense linkage map has also been successfully constructed in common buckwheat (*Fagopyrum esculentum)*. The map comprised of 8884 markers and 751 loci. The average distance between adjacent loci was 2.13 cM. This linkage map was identified to be very beneficial for analyzing the genome of other populations of buckwheat populations (Yabe et al. 2014). Although the linkage map of buckwheat was previously available, but it had inadequate applicability due to its great genetic variation (Konishi and Ohnishi 2006; Pan and Chen 2010; Yasui et al. 2004).

7.4 Sequencing Status of Buckwheat

With the advent of next-generation sequencing (NGS) technology, genome sequences of both model and non-model crop species have been elucidated. The technique has proved to be robust for database creation revealing gene information and comparative analysis of both model and non-model crops, including buckwheat (Yasui et al. 2016). These efforts have led to the foundation of sequencing in crop plants including buckwheat to understand the molecular mechanism regulating agronomically important traits. Thus in buckwheat, a reference genome had served a high value database for backing molecular breeding, molecular genetics, comparative genomics, and advancing eudicot phylogenetic reconstructions. Further, whole genome sequencing data has revealed that tartary buckwheat is the most suitable species among all related species of buckwheat owing to its low heterozygosity and small size (Zhang et al. 2017). Based on NGS technology, a high-throughput genotyping method known as GBS (genotyping by sequencing) technology was developed in recent past (Elshire et al. 2011). GBS is a rapid and robust approach for reduced-representation sequencing of multiplexed samples that combines genome-wide molecular marker discovery and genotyping for any species, including non-model organisms (Andrews et al. 2016). The GBS system has been harnessed to sequence buckwheat genome and a database for buckwheat was developed with the name Buckwheat Genome Database (BGDB; http://buckwheat.kazusa.or.jp). This database will be helpful for rapid facilitation of homologous sequences previously deposited in other plant databases. For example, in an investigation, short reads produced from NGS technology were utilized to generate a draft assembly of the buckwheat genome which subsequently led to the construction of Buckwheat Genome Database. Post assembling of short reads, about 387,594 scaffolds were determined as the draft genome sequence (FES_r1.0). The N50 of the scaffolds was 25,109 bp and the total length of draft genome sequence (FES_r1.0) was 1,177,687,305 bp. About 286,768 coding sequences (CDSs; FES_r1.0_cds) including those associated with transposable elements were revealed through gene prediction analysis. The N50 of cds was 1101 bp and total length of FES_r1.0_cds was 212,917,911 bp. Of all these coding sequences, the function of about 35,816 CDSs was annotated by BLAST analysis except for transposable elements. The database proved useful for identification of new candidate genes in buckwheat such as genes encoding granule-bound starch synthases (GBSSs), 2S albumin-type allergens, and genes regulating heteromorphic self-incompatibility and flavonoid biosynthesis (Yasui et al. 2016). In another investigation, an important neutraceutical gene known as PAL (phenylalanine ammonia lyase) gene was sequence characterized from three species of buckwheat (*Fagopyrum. tataricum, Fagopyrum esculentum,* and *Fagopyrum dibotrys*), for comprehension of its genomics. The gene plays a crucial role in the biosynthesis of bioactive compounds, such as rutin and quercetin, which are neutraceutically important found in buckwheat. The results revealed presence of three SNPs (single nucleotide polymorphisms) and four indels at inter- and intraspecific levels. The frequency of SNPs was less in case of *Fagopyrum tartaricum* in comparison to other two species (*Fagopyrum esculentum* and *Fagopyrum dibotrys*). The SNPs identified in *Fagopyrum tartaricum* did not result in any change of amino acid, while the SNPs identified in other two species exhibited both conserved and non-conserved variations. The information obtained through the sequence characterization of PAL gene will be helpful for understanding its evolutionary importance and can be efficiently harnessed for the genetic improvement of buckwheat in relevance to its medicinal value (Thiyagarajan et al. 2016). In tartary buckwheat (*Fagopyrum tataricum*), a high quality reference genome sequence of 489.3 Mb was unveiled by combining the technologies of whole genome shotgun sequencing of both Illumina short reads and single-molecule real-time (SMRT) long reads, Hi-C sequencing, BioNano genome maps and sequence tags of a large DNA insert fosmid library. Based on expression evidence, about 33, 366 protein-coding genes of high confidence were annotated. Further, comparative analysis of buckwheat and sugar beet genome revealed independent whole genome duplication in the lineage of buckwheat, which would have occurred after the divergence of buckwheat from its common ancestor. The reference genome of tartary buckwheat (*Fagopyrum tartaricum*) facilitated identification of various novel genes involved in aluminum stress resistance, cold stress, drought responses, regulation and biosynthesis of rutin. The higher tolerance of tartary buckwheat (*Fagopyrum tartaricum*) genome to abiotic stress is due to widening of multigene families involved in gene regulation, membrane transport, and signal transduction. The accessibility of these genomic resources will assist in genetic improvement of tartary buckwheat and in discovery of nutritionally and agronomically important genes (Zhang et al. 2017). All the genes of buckwheat sequenced and identified are summarized in Table 7.1.

Apart from the sequencing of nuclear genome of buckwheat, its chloroplast genome was also sequenced to understand its evolutionary trend and mutational patterns. To explore the phylogenetic relationships of buckwheat, variable regions were identified in sufficient number. Further, two complete chloroplast genomes of buckwheat were sequenced which includes *Fagopyrum dibotry*s and *Fagopyrum luojishanense.* The chloroplast genome size of *Fagopyrum* d*ibotrys* and *Fagopyrum luojishanense* was 159,320 bp and 159,265 bp, respectively. Also 48 SSRs were identified in *Fagopyrum dibotrys,* while 61 SSRs were identified in *Fagopyrum Luojishanense.* Sequencing analysis revealed close association of *Fagopyrum dibotrys* with *Fagopyrum tataricum.* The genetic information is highly valuable for identification, molecular breeding, phylogenetic studies, and taxonomy of *fagopyrum* species (Wang et al. 2017).

Table 7.1 Candidate genes identified and sequenced in buckwheat (*Fagopyrum* sp.)

Buckwheat species	Genes identified and sequenced	References
Fagopyrum esculentum	Genes encoding GBSSs, 2S albumin-type allergens, and genes regulating heteromorphic self-incompatibility and flavonoid biosynthesis	Yasui et al. (2016)
Fagopyrum esculentum, Fagopyrum tataricum, and *Fagopyrum dibotrys*	PAL gene	Thiyagarajan et al. (2016)
Fagopyrum tataricum	Genes involved in regulation and biosynthesis of rutin, abiotic stress resistance such as aluminum stress resistance, cold stress, and drought responses	Zhang et al. (2017)

GBSSs granule-bound starch synthases, *PAL* phenylalanine ammonia lyase

7.5 Future Prospectus

Various efforts have been made in buckwheat crop improvement programs, where most of the stress has been laid on development of high yielding varieties, increased rutin concentration, determinate growth habits, resistance in seed shattering, lodging, early maturity, and removal of allergenic proteins. Apart from all the mentioned traits, crop improvement in buckwheat should also focus on other agronomic traits such as frost resistance, insect resistance, improving nutritional value by lowering the concentration of anti-nutritional factors, homostyly, and self-compatibility. The need of the hour is to enhance the research perspective in this crop in order to identify the potential varieties of buckwheat gene pool, so that the valuable genes could be identified to address the key issues related with this crop. Further, both conventional and nonconventional approaches should be employed to improve the genetic potential of buckwheat. Thus, in order to make the buck improvement a great success, research efforts are required through interdisciplinary collaborations especially between the agriculturists, local growers, and research institutes.

References

Adachi T, Yamaguchi A, Miike Y, Hoffman F (1989) Plant regeneration from protoplasts of common buckwheat (*Fagopyrum esculentum*). Plant Cell Rep 8:247–250

Aii J, Nagano M, Penner GA, Campbell CG, Adachi T (1998) Identification of RAPD markers linked to the *homostylar* (*Ho*) gene in buckwheat. Breed Sci 48:59–62

Alamprese C, Casiraghi E, Pagani MA (2007) Development of gluten-free fresh egg pasta analogues containing buck wheat. Eur Food Res Technol 225:205–213

Andrews KR, Good JM, Miller MR, Luikart G, Hohenlohe PA (2016) Harnessing the power of RADseq for ecological and evolutionary genomics. Nat Rev Genet 17:81–92

Asaduzzaman M, Minami M, Matsushima K, Nemoto K (2009) An in-vitro ovule culture technique for producing interspecific hybrid between tartary buckwheat and common buckwheat. J Biol Sci 9(1):1–11

Berbec A, Doroszewska T (1999) Investigations on androgenetic induction in three varieties buckwheat species: phylogenetic relationships, origin of the reproductive systems and buckwheat. Fagopyrum 13:35–39

Campbell GC (1997) Buckwheat Fagopyrum esculentum Moench. Promoting the conservation and use of underutilized and neglected crops 19. Institute of Plant Genetics and Crop Plant Research, Gatersleben, Germany and the IPGRI, Rome, Italy

Chan PK (2003) Inhibition of tumor growth in vitro by the extract of *Fagopyrum cymosum* (fago-c). Life Sci 72:1851–1858

Chrungoo NK, Kreft I, Sangma SC, Devadasan N, Dohtdong L, Chetri U (2012) Genetic diversity in Himalayan buckwheats: a perspective for use in crop improvement programmes. In: Proceedings of the 12th International Symposium on Buckwheat, Laško, Aug. 21–25. Pernica: *Fagopyrum*, pp 198–211

Comino I, de Lourdes Moreno M, Real A, Rodríguez-Herrera A, Barro F, Sousa C (2013) The gluten-free diet: testing alternative cereals tolerated by celiac patients. Nutrients 5:4250–4268

Dar FA, Pirzadah TB, Malik B, Tahir I, Rehman RU (2018) Molecular genetics of buckwheat and its role in crop improvement. In: Zhou M, Kreft I, Suvorova G, Tang Y, Woo SH (eds) Buckwheat germplasm in the world. Academic Press, Chennai, pp 271–286

Elshire RJ, Glaubitz JC, Sun Q (2011) A robust, simple genotyping-by-sequencing (GBS) approach for high diversity species. PLoS One 6:e19379

Fabjan N, Rode J, Kosir IJ, Wang Z, Zhang Z, Kreft I (2003) Tartary buckwheat (*Fagopyrum tartaricum* Gaertn.) as a source of dietary rutin and quercitrin. J Agric Food Chem 51:6452–6455

Farooq S, Rehman RU, Pirzadah TB, Malik B, Dar FA, Tahir I (2016) Cultivation, agronomic practices, and growth performance of buckwheat. Academic Press, Oxford, pp 299–320

Gimenz-Bastida JA, Zielinski H (2015) Buckwheat as a functional food and its effects on health. J Agric Food Chem 63:7896–7913. https://doi.org/10.1021/acs.jafc.5b02498

Guo YZ, Chen QF, Yang LY, Huang YH (2007) Analyses of the seed protein contents on the cultivated and wild buckwheat *Fagopyrum esculentum* resources. Genet Resour Crop Evol 54:1465–1472

Guo X, Zhu K, Zhang H, Yao H (2010) Anti-tumor activity of a novel protein obtained from tartary buckwheat. Int J Mol Sci 11:5201–5211

Gupta N, Sharma SK, Rana JC, Chauhan RS (2012) AFLP fingerprinting of tatary buckwheat accessions (Fagopyrum tataricum) displaying rutin content variation. Fitoterapia 83:1131–1137

Hara T, Iwata H, Okuno K, Matsui K, Ohsawa R (2011) QTL analysis of photoperiod sensitivity in common buckwheat by using markers for expressed sequence tags and photoperiod-sensitivity candidate genes. Breed Sci 61:394–404

Heffler E, Pizzimenti S, Badiu I, Guida G, Rolla G (2014) Buckwheat allergy: an emerging clinical problem in Europe. J Aller Ther 5:168

Hirose T, Ujihara A, Kitabayashi H, Minami M (1993) Morphology and identification by isozyme analysis of interspecific hybrids in buckwheats. Fagopyrum 13:25–30

Hirose T, Ujihara H, Kitabayashi, Minami M (1995) Pollen tube behavior related to self-incompatibility interspecific crosses of *Fagopyrum*. Breed Sci 45:65–70

Hou LL, Zhou ML, Zhang Q, Qi LP, Yang XB, Tang Y (2015) *Fagopyrum luojishanense*, a new species of polygonaceae from Sichuan, China. Novon 24(1):22–26

Hou S, Sun Z, Linghu B, Xu D, Wu B, Zhang B (2016) Genetic diversity of buckwheat cultivars (*Fagopyrum tartaricum* Gaertn.) assessed with SSR markers developed from genome survey sequences. Plant Mol Biol Reprod 34:233–241

Inglett GE, Chen D, Berhow M, Lee S (2011) Antioxidant activity of commercial buckwheat flours and their free and bound phenolic compositions. Food Chem 125:923–929

Iwata H, Imon K, Yoshimura K, Tsumura Y, Oshawa R (2001) Genetic diversity of common buckwheat varieties in Japan based on microsatellite markers. In: Proceedings of the 8th international symposium on buckwheat, Chuncheon, Korea, pp 240–247

Iwata H, Imon K, Tsumura Y, Ohsawa R (2005) Genetic diversity among Japanese indigenous common buckwheat (*Fagopyrum esculentum*) cultivars as determined from amplified fragment length polymorphism and simple sequence repeat markers and quantitative agronomic traits. Genome 48:367–377

Javornik B, Kump B (1993) Random amplified polymorphic DNA (RAPD) markers in buckwheat. Fagopyrum 13:35–39

Jiang P, Burczynski F, Campbell C, Pierce G, Austria JA, Briggs CJ (2007) Rutin and flavonoid contents in three buckwheat species *Fagopyrum esculentum*, *F. Tataricum*, and *F. Homotropicum* and their protective effects against lipid peroxidation. Int J Food Res 40:356–364

Jing R, Li HQ, Hu CL, Jiang YP, Qin LP, Zheng CJ (2016) Phytochemical and pharmacological profiles of three *fagopyrum* buckwheats. Int J Mol Sci 17:589. https://doi.org/10.3390/ijms17040589

Khan S (2015) QTL mapping: a tool for improvement in crop plants. Res J Recent Sci 4:7–12

Kim CD, Lee WK, No KO, Park SK, Lee MH, Lim SR, Roh SS (2003) Anti-allergic action of buckwheat (Fagopyrum esculentum moench) grain extract. Int Immunopharmacol 3:129–136

Kim SJ, Zaidul ISM, Suzuki T, Mukasa Y, Hashimoto N, Takigawa S, Takahiro N, Chie ME, Hiroaki Y (2008) Comparison of phenolic compositions between common and tartary buckwheat (*Fagopyrum*) sprouts. Food Chem 110:814–820

Kishima YK, Ogura KM, Mikami T, Adachi T (1995) Chloroplast DNA analysis in buckwheat species: phylogenetic relationships, origin of the reproductive systems extended inverted repeats. Plant Sci 108:173–179

Kishore G, Pandey A, Dobhal R, Gupta S (2013) Population genetic study of *Fagopyrum tartaricum* from Western Himalaya using ISSR markers. Biochem Genet 51:750–765

Konishi T, Ohnishi O (2006) A linkage map for common buckwheat based on microsatellite and AFLP markers. Fagopyrum 23:1–6

Koyama M, Nakamura C, Nakamura K (2013) Changes in phenols contents from buckwheat during growth stage. J Food Sci Technol 50:86–93

Krkoskova B, Mrazova Z (2005) Prophylactic components of buckwheat. Food Res Int. 38:561–568

Lachmann S (1991) Plant cell and tissue culture in buckwheat: an approach towards genetic improvements by means of unconventional breeding techniques. In: Adachi T (ed) Proceedings of an international colloquium on overcoming breeding barriers by means of plant biotechnology, Miyazaki, Japan, pp 145–154

Liu JL, Tang Y, Xia ZM, Shao JR, Cai GZ, Luo Q (2008) *Fagopyrum crispatifolium* Liu, J L., a new species of Polygonaceae from Sichuan, China. J Syst Evol Res 46:929–932

Nair A, Adachi T (1999) Immunoblotting and characterization of allergenic proteins in common buckwheat (*Fagopyrum esculentum*). Plant Biotechnol 16:219–224

Ohnishi O, Matsuoka Y (1996) Search for the wild ancestor of buckwheat II. Taxonomy of *Fagopyrum* (Polygonaceae) species based on morphology, isozymes and cpDNA variability. Genes Genet Syst 72:383–390

Pan SJ, Chen QF (2010) Genetic mapping of common buckwheat using DNA, protein and morphological markers. Hereditas 147:27–33

Paterson AH, Tanksely SD, Sorrells ME (1991) DNA markers in plant improvement. Adv Agron 46:39–90

Samimy C (1991) Barrier to interspecific crossing of Fagopyrum esculentum with *Fagopyrum tataricum*: I. Site of pollen tube arrest: II. Organogenesis from immature embryos of *F. tataricum*. Euphytica 54:215–219

Samimy C, Bjorkman T, Siritunga D, Blanchard L (1996) Overcoming the barrier to interspecific hybridization of *Fagopyrum esculentum* with *Fagopyrum tataricum*. Euphytica 91:323–330

Shao JR, Zhou ML, Zhu XM, Wang DZ, Bai DQ (2011) *Fagopyrum wenchuanense* and *Fagopyrum qiangcai*, two new species of Polygonaceae from Sichuan, China. Novon 21:256–261

Sharma R, Jana S (2002) Species relationships in Fagopyrum revealed by PCR-based DNA fingerprinting. Theor App Gene 105:306–312

Sun BH, Wu YQ, Gao HY, Huang J, Wu LJ (2008) Chemical constituents of *Fagopyrum tartaricum* (L.) gaertn. J Shenyang Pharm Univ 25:541–543

Suvorova GN, Fesenko NN, Kosturbin MM (1994) Obtaining of interspecific buckwheat hybrid (Fagopyrum esculentum Moench × Fagopyrum cymosum Meissn.). Fagopyrum 14:13–16

Tang Y, Zhou ML, Bai DQ, Shao JR, Zhu XM, Wang DZ et al (2010) Fagopyrum pugense (Polygonaceae), a new species from Sichuan, China. Novon 20:239–242

Tang Y, Ding M-Q, Tang Y-X, Wu Y-M, Shao J-R, Zhou ML (2016) Germplasm resources of buckwheat in China. In: Zhou M, Kreft I, Woo SH, Chrungoo NK, Wieslander G (eds) Molecular breeding and nutritional aspects of buckwheat. Academic Press, The Netherlands, pp 13–20

Thiyagarajan K, Vitali F, Tolaini V, Galeffi P, Cantale C, Vikram P (2016) Genomic characterization of phenylalanine ammonia lyase gene in buckwheat. PLoS One 11(3):e0151187. https://doi.org/10.1371/journal.pone.0151187

Tsuji K, Ohnishi O (2001) Phylogenetic relationships among wild and cultivated Tartary buckwheat (Fagopyrum tartaricum Gaertn.) populations revealed by AFLP analyses. Genes Genet Syst 76:47–52

Ujihara A, Nakamura Y, Minami M (1990) Interspecific hybridization in genus *Fagopyrum*-properties of hybrids (F. esculentum Moench 3 F. cymosum Meissner) through ovule culture. Gamma Field Radiation Breeding, NIAR, MAFF, Japan, pp 45–51

Wang YJ, Campbell CG (1998) Interspecific hybridization in buckwheat among *Fagopyrum esculentum, F. homotropicum* and *F. tataricum*. Adv Buckwheat Res 7:11–12

Wang CL, Ding MQ, Zou CY, Zhu XM, Tang Y, Zhou ML, Shao JR (2017) Comparative analysis of four buckwheat species based on morphology and complete chloroplast genome sequences. Sci Rep 7:6514

Woo SH, Tsai QS, Adachi T (1995) Possibility of interspecific hybridization by embryo rescue in the genus *Fagopyrum*. Curr Adv Buckwheat Res 6:225–237

Woo SH, Adachi T, Jong SK, Campbell CG (1999) Inheritance of self-compatibility and flower morphology in interspecific buckwheat hybrids. Can J Plant Sci 79:483–490

Woo SH, Kamal AHM, Tatsuro S, Campbell CG, Adachi T, Yun YH, Chung KY, Choi JS (2010) Eur J Plant Sci Biotechnol 4:1–16

Woo SH, Suzuki T, Mukasa Y, Morishita T, Yun HY, Park HC (2012) Present status, future breeding strategy and prospects for buckwheat. In: Proceedings of the 12th international symposium on buckwheat, Laško, Aug. 21–25. Pernica: Fagopyrum. pp 25–26

Xiaolei D, Zongwen Z, Bin W, Yanqin L, Anhu W (2013) Construction and analysis of genetic linkage map in Tartary buckwheat (*Fagopyrum tataricum*) using SSR. Chin Agric Sci Bull 29 (21):61–65

Yabe S, Hara T, Ueno M, Enoki H, Kimura T, Nishimura S (2014) Rapid genotyping with DNA micro-arrays for high-density linkage mapping and QTL mapping in common buckwheat (*Fagopyrum esculentum*). Breed Sci 64:291–299

Yamane K, Yasui Y, Ohnishi O (2003) Interspecific cpDNA variations of diploid and tetraploid perennial buckwheat, *Fagopyrum cymosum* (Polygonaceae). Am J Bot 90:339–346

Yao Y, Shan F, Bian J, Chen F, Wang M, Ren GD (2008) Chiro-inositol-enriched Tartary buckwheat bran extract lowers the blood glucose level in KK-Ay mice. J Agric Food Chem 56:10027–10031

Yasui Y, Wang Y, Ohnishi O, Campbell CG (2004) Amplified fragment length polymorphism linkage analysis of common buckwheat (*Fagopyrum esculentum*) and its wild self-pollinated relative *Fagopyrum homotropicum*. Genome 47:345–351

Yasui Y, Hirakawa H, Ueno M, Matsui K, Katsube-Tanaka T, Yang SJ (2016) Assembly of the draft genome of buckwheat and its applications in identifying agronomically useful genes. DNA Res 23(3):215–224. https://doi.org/10.1093/dnares/dsw012

Zhang L, Li X, Ma B, Gao Q, Du H, Han Y, Li Y, Cao Y, Qi M, Zhu Y, Lu H, Ma M, Liu L, Zhou J, Nan C, Qin Y, Wang J, Cui L, Liu H, Liang C, Qiao Z (2017) The Tartary buckwheat genome provides insights into rutin biosynthesis and abiotic stress tolerance. Mol Plant 10:1224–1237

Zheng CJ, Hu CL, Ma XQ, Peng C, Zhang H, Qin LP (2012) Cytotoxic phenylpropanoid glycosides from *Fagopyrum tartaricum*(L.) gaertn. Food Chem 132:433–438

Zhou ML, Zhang Q, Zheng YD, Tang Y, Li FL, Zhu XM (2015) *Fagopyrum hailuogouense* (Polygonaceae), one new species from Sichuan, China. Novon 24(2):222–224

Zietkiewicz E, Rafalski A, Labuda DF (1994) Genome fingerprinting by simple sequence repeat (SSR)-anchored polymerase chain reaction amplification. Genomics 20:176–183

Genetic and Genomic Resources of Chestnut

8

Shahid Qayoom Dar, Sumaya Mumtaz, Shabnam Ahad,
Abdul Waheed Wani, and Tawseef Rehman Baba

Abstract

Chestnuts (*Castanea* sp.) belong to the family *Fagaceae*, comprising more than 900 species of trees and shrubs. The family *Fagaceae* is dominated by temperate forest trees like oaks (*Quercus* spp.) and beeches (*Fagus* sp.), which are found mainly in the Northern Hemisphere countries of the world (Manos et al. 2008). From time to time coevolution among different *Castanea* sp. and biodiversity found among these species have ensued the extent of these opulent and diverse *Castanea* sp. to spread all over the world, particularly in high altitude areas. The elasticity and adaptability of various chestnut species to different agro-climates and the extensive genetic base of these species determined the range of various ecotypes in the wild. During the course of many decades, human population has selected, maintained, and utilized these diverse genotypes for many uses, such as garden-fresh consumption, production of flour, timber production, and physical nourishment, thus promisingly making contribution towards the preservation of natural biodiversity of these species and giving an outstanding model for horticultural crop conservation. However, presently the genomic diversity of the species is in danger of extinction and huge number of cultivars are liable of being vanished as a result of various phytosanitary issues like *Pseudomonas syringae* causing canker blight, *Phytophthora palmivora* causing ink disease, gall wasp, *Cryphonectria parasitica*, *Phytophthora* spp., and *Dryocosmus kuriphilus*. These issues have resulted in the decline and desertion of chestnut cultivation and loss of male chestnut as well. In recent times, numerous research programs have strived to improve policies for the preservation of *Castanea* sp. biodiversity. The main goal of this chapter is to present the research studies

S. Q. Dar (✉) · S. Mumtaz · S. Ahad · A. W. Wani · T. R. Baba
Division of Fruit Science, Sher-e-Kashmir University of Agricultural Sciences and Technology, Shalimar, Srinagar, Jammu and Kashmir, India

S. M. Zargar et al. (eds.), *Neglected and Underutilized Crops - Towards Nutritional Security and Sustainability*, https://doi.org/10.1007/978-981-16-3876-3_8

on genomic variability and preservation of *Castanea* sp. germplasm to pave way for conservation and development of resilient chestnut varieties.

Keywords

Biodiversity · Castanea · Chestnut · Genetics · Variability · Conservation

8.1 Introduction

Chestnuts (*Castanea* species) belong to family *Fagaceae*, sub-family *Quercoideae*, and order *Fagales*. The members of this family represent various essential forest tree species worldwide. *Castanea* comprises seven deciduous tree and shrub species, grouped into three divisions, Eucastanon, Balanocastanon, and Hypocastanon (Lang et al. 2007). Five species of chestnut, viz. *Castanea dentata* (Marsh.) Borkh., *C. sativa* Mill., *C. screnata* (Sieb and Zucc), *C. mollissima* (Blume), and *C. seguinii* (Dode) are best researched particularly in relation to resistance for various diseases and pests and other biotechnological aspects. In Asian and European countries the chestnut shows significant importance for nut production. In North American countries they serve as versatile forest trees, supplying a varied choice of harvests for native populations in Appalachian mountain region. In addition, *Castanea sativa* Mill. found in Mediterranean region is primarily used for household products like honey, timber, nuts, and tannin. Besides, *Vitis* spp. and *Olea* spp. tracing the natural diversity of the most ancient cultured species of sweet chestnut are impractical, as being greatly affected by human activities to a large extent (Conedera et al. 2004). The continuous challenges faced by forest trees due to industrialization, large scale deforestation, devastating pests/pathogens, and a threat of increased global warming have made their existence difficult. Managing such constraints dynamically with approaches based on scientific knowledge to endure and advance the forest tree vigor mostly depends on having extremely categorized varied germplasm constituents, genomic possessions, and art of breeding techniques for apprehending and transferring significant characters (e.g. disease resistance) into enhanced plant material for replant and rejuvenation exertions.

8.2 Genetics of Chestnut

The extreme genetic variability at neutral loci, tiny population assembly, genomic variation among populations, population hereditary characters shared by several other long-lived outcrossing forest trees are the main characteristics of *Castanea* species (Petit and Hampe 2006). Various morphological, physiological, and phenological traits are used for characterizing the genetic diversity of chestnut. However, variation in physical characteristics could be used to differentiate amid generations or ecotypes of a species, their association to environmental changes is not clear, thus focus is given on phenological traits like bud burst, bud set, and time of anthesis that

Table 8.1 Origin and distribution of *Castanea* spp.

Species	Origin	Common name	Orchards	Predominant Use
Europe	*C. sativa* Mill.	European or sweet Chestnut	Europe, Asia Minor, North Africa	Nut, timber
Asia	*C. crenata* Sieb. et Zucc.	Japanese chestnut	Japan, Korea	Nut
	C. mollissima Blume	Chinese chestnut	China	Nut
	C. seguinii Dode	Willow leaf or pearl Chestnut	China	Firewood
	C. davidii Dode		China	Firewood
	C. henryi (Skan) Rehd. et E. H. Wils		China	Timber
America	*C. dentata* (Marsh.) Borkh.	American chestnut	North America	Timber
	C. pumila Mill. var. pumila	Allegheny chinkapin	Southeast USA	Nut
	C. pumila Mill. var. ozarkensis (Ashe) Tucker	Ozark chinkapin	USA (AR, MI, OK)	Timber
	C. floridana (Sarg.) Ashe	Florida chinkapin	Southeast USA	Ornamental
	C. ashei (Sudw.) Ashe	Ashe chinkapin	Southeast USA	Ornamental
	C. alnifolia Nutt	Creeping chinkapin	Southeast USA (AL–FL)	–
	C. paucispina Ashe		Southeast USA (TX–LA)	–

have close relation with adaptation. In American and Chinese chestnuts, 26 morphometric variables were calculated on the basis of their leaf, stipule, twig, and bud variations, and four hybrid chestnut were characterized by means of an index of species identity (ISI) (Diskin et al. 2006). The origin, distribution, and predominant use of different *Castanea* spp. are shown in Table 8.1.

American chestnuts (*Castanea dentata*) characterized by upright growth form, small and sweet nuts, but extremely prone to chestnut blight.

European chestnuts (*Castanea sativa*) have upright tree form, nuts good from some cultivars; however, nuts from seedlings are bitter in taste and difficult to peel, prone to blight, thus cultivated in areas free from blight diseases.

Chinese chestnuts (*Castanea mollissima*) have good resistance to blight and yield high quality, trees show upright form of growth, but few have multiple branches near the ground and are spreading type.

Japanese chestnuts (*Castanea crenata*) are dwarf, tolerant to blight, spreading type but yield nuts of poor quality and are typically employed for hybridization.

Chinkapins (several species) yield tasty nuts of small size (one per burr), have shrubby form, and are not commercially used for nuts.

Chestnut species being diploid (2*n*¼2 × ¼24) can be hybridized freely with its correspondents; however, the resulting progenies are less vigorous and show male sterility (Jaynes 1975). Chestnut flowers mature in a manner defined by Stout (1928) as duodichogamy, wherein male flowers emerged on long, slender catkins become receptive first, subsequently by female blossoms, and lastly male flowers on bisexual catkins. Chestnut is monoecious, predominantly wind pollinated, and largely carrying three massive seeds in each capsule. A genetically controlled mechanism preventing the self-fertilization is called as self-incompatibility. Since chestnut carries binucleate pollen grains, thus showing gametophytic type of self-incompatibility system and several pollen tubes enter to the base of the style preceding fertilization (Burnham et al. 1986a, b). About 12–18 years are required by chestnuts to reach reproductive maturity under favorable environments (Weeks et al. 2005).

8.3 Genomic Approach to Disease Resistance in American Chestnut

Castanea species preferably define the nature of coevolution and biology of host/pathogen genome, as Chinese chestnut (*C. mollissima*) coexists and developed antagonism against duo offensive infectious microorganisms, *Cryphonectria parasitica* and *Phytophthora cinnamomi*, which are accountable for the widespread loss of most vulnerable American chestnut (*C. dentata*). By hybridizing the Chinese and American chestnut the disease resistance and other important traits of interest can be segregated in the resulting interspecies hybrid families. Moreover, it is promising to induce profuse flowering and reduce juvenility, thereby extensively decreasing breeding phase for fundamental genetics as well as genetic engineering procedures for restoration programs of forest trees. Increased attention towards developing resistance against Phytophthora root rot has urged researchers for interspecific crossing along with assortment for tolerance in synthetic immunized experiments (Jeffers et al. 2009). These efforts are generally based on the occurrence of inherent opposition in the mutually developed susceptible types and of them soundly increased interspecific hybridization with the extremely prone irrelevant kinds (Graves 1926). North America is found to be rich in germplasm possessions of *C. dentata* (Kubisiak and Roberds 2006) due to the chestnuts' capability to put forth new growth after dieback (Mattoon 1909; Graves 1926; Paillet 2002) and inability of *P. Cinnamomi* to spread in cooler situations (Balci et al. 2007; Griffin et al. 2009).

Wide range of marker systems have been used for analyzing genetic diversity, structure, and relation of phylogenes within and among *Castanea* species (Kubisiak and Roberds 2006; Tanaka et al. 2005; Lang et al. 2006; Han et al. 2007). Comparative genomics enables us to understand the mechanism of gene networks responsible

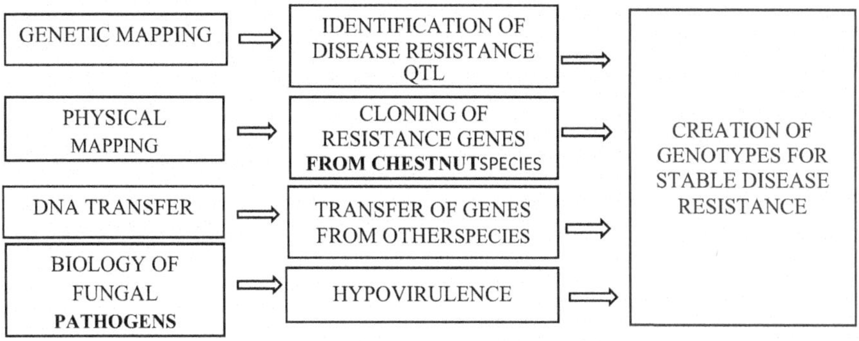

Fig. 8.1 Genomic approach to disease resistance in American Chestnut (Lecape et al. 2007)

for resistance systems in forest trees. However, in perennial trees there is a limited awareness about the level of maintenance of the physical association of the genotypes between close or distantly connected types. Improvement in the gene pool of several forest flora has increased the magnitude of genome conservation among these species. The genetically anchored EST sequence markers that are physically mapped from Chinese chestnut chromosomal diagram and the Chinese chestnut BAC end sequences (BES) offer perfect base for genetic assessments with extremely considered DNA sequence, like peach (*Prunus persica*) and poplar (*Populus trichocarpa*). Recognized preserved masses in these genetic pool may possibly help in locating the gene of interest together by chestnut characteristic plotting exertions. Moreover, such nucleotide sequence evaluations offer a former fine resolution preview of the degree of chromosomal alteration, which has been persistent as long as splitting of the families. Thus to exploit relative methodologies with complete organized and collected tree orientation genes as a means to recognize probable gene of interest in distinct intervals of less better sequence described species, it is necessary to evaluate the degree of conservation among tree DNA sequence at peak resolution, viz. in 1 centimorgan (cM) windows. Besides this, to develop disease resistance (Barakat et al. 2012) and biocontrol (Dawe and Nuss 2013), modern biotechnological approaches are employed. Later approach includes chromosomal plotting, sequencing, asexual multiplication, micropropagation, and genomic variation of chestnut trees (Zhebentyayeva et al. 2014) along with virulence diminishing viruses and bacteria (Dawe and Nuss 2013). Genomic approach to disease resistance in American Chestnut is shown diagrammatically in Fig. 8.1.

8.4 Conservation of Chestnut Germplasm

In 1904, the first infestation of Chestnut blight (*Cryphonectria parasitica*) was witnessed in New York State. The quick dissemination of this disease shrunk the whole genus to an insignificant underwood bush within 50 years (Anagnostakis 1987). Inexhaustible stub growing has permitted *C. dentata* to continue above maximum of its inherent variety since 50 years. Large number of genetic resources still survives, but experiences constant loss, as left behind root structures fail to grow (Paillet 1988). Due to the continuous increasing stress of natural environment demolition the genotypes may be endangered through elimination. A management plan is important for *C. dentata* to survive (Irwin 2003). In this regard, American Chestnut Foundation (TACF) employed the breeding approaches to transmit the genetic material against blight resistance from the Asian species into *C. dentata* though preserving the inherent genomic variability (Burnham et al. 1986a, b; Hebard 2004).

8.4.1 In Situ Conservation

In situ conservation is conservation of genetic resources within their natural habitat. It exists as a highly preferable form of conservation due to ease in implication of biological operations and protocols (McIlwrick et al. 2000). If fruitful, this method safeguards not merely the germplasm, but the species, their environment, and the biological processes that endure them without growth or flowering (Paillet 1988). Sometimes, canopy interferences, viz. uprooted plants by fast moving winds or road-breaks let buds to develop as well as flower within a few years beforehand vanished by blight (Griffin 2000). The inaccessible chestnut locations could be achieved in order to shrink antagonism via bringing out of neighboring undergrowth. Persisting *C. dentata* is usually hedged to guard from animal consumption and physical damage. Furthermore, in situ conservation approach includes natural restraint of chestnut to blight disease. Hypovirulence exhibits its effective biotic management in Europe on *C. sativa.*, in Michigan on *C. dentata,* and on grafted plants of long living American chestnut in Lesesne State Forest, However, this method is not that significant on uncultivated (vulnerable) American chestnut as that for grafted cultivated chestnut clones (Griffin 2000). Within the native range of *Castanea dentata* in North America, hypovirulence has usually not stood operative as a biocontrol (Griffin 2000; MacDonald and Fulbright 1991).

8.4.2 Ex Situ Conservation

Ex situ conservation denotes the methods employed to protect genomic possessions exterior to their area of natural habitat in a controlled manner. It involves procedures like organ and tissue banks, pedigree plantations, and nurseries. Furthermore, ex situ conservation approaches aim to maintain genetic variation to a lesser extent. Seeds of *C. dentata* are extremely perishable due to excess water loss under average moisture, thereby decreasing viability. While dealing with storage of pollens and seeds, due to some unreliability's, conservation the gene constitution of indigenous *C. dentate* community in progeny plantations is preferred. Locally adapted back-cross hybrids are important for preserving adaptive gene complexes that are crucial for the attainment of success in any breeding program (Bounous et al. 1993). Such existing repositories can further be utilized in breeding programs for producing blight impervious nuts. For carrying out successful multiplication programs favorable growing conditions like management of unwanted plants, disease and insect control, defense from animal consumption, natural control through hypovirulence (Alexander 2003), irrigation, fertilization, and other soil amendment practices are sustained.

8.4.3 In Vitro Conservation

Tissue culture is type of ex situ conservation method employed for somatic and zygotic embryos, haploid megagametophytic tissue, meristem tissues, and cell suspensions stored for a long time; however, their restoration is challenging (Bajaj 1985). This technique shows promising results in germplasm conservation especially of threatened plant species (Fay 1992; Amo-Marco and Lledo 1996; Benson 1999; Martin and Pradeep 2003). In vitro multiplication of full grown *C. dentata* plants was attained by using buds of softwood sprouts enforced from suckers gathered in winter season, scultured on woody plant medium (WPM) with 0.5 mg/l benzyl adenine to enhance new growth. Axillary shoot production can be achieved by embryo or nodal explant culturing of *C. dentate* on nutrient media augmented with benzyl adenine (Keys and Cech 1979; Serres et al. 1990). Xing et al. (1997) carried out the culturing of shoots from *C. dentata,* which was then subjected to elongation of shoots supplemented with cytokinin rich medium, followed by its excision and then rooting in auxin augmented medium along with activated charcoal and lastly for hardening to decrease shoot tip necrosis. More recently, Oakes et al. (2013) reported that the rooting and survival percentage can be boosted by controlling the light exposure, active carbon, and duration on root proliferation media prior shifting to other medium.

8.5 Conclusion

Castanea species are facing a severe threat of germplasm extinction due to the decline of cultivated lands, the extent of a restricted number of particular genotypes, and the amassed destruction caused by biotic agents regardless of the present exertions to conserve them, their genetic diversity is crucially threatened; moreover, large number of genotypes and selections are at the verge of getting vanished due to a number of plant diseases like canker blight, ink disease, *Phytophthora* spp., gall wasp (*Dryocosmus kuriphilus*). *P. cinnamomi* is one of the most damaging pathogens related with the decline of this fruit species. Various initiatives like creation of the arboretum, breeding, integrated plant conservation, and molecular approaches have been taken by few international organizations to signify the initial phase in preventing the decline of biodiversity; nevertheless, extensive research exertions are still required to fully apply the diverse plans in the recalcitrant upkeep of *Castanea* species and development of resilient chestnut plants.

References

Alexander SH (2003) Host pathogen interactions in a segregating population of BC2F2 hybrid Castanea dentata and other chestnut hybrids after exposure to hypovirus containing and hypovirus-free strains of *Cryphonectria parasitica*. Thesis Presented for the Master of Science Degree at The University of Tennessee at Chattanooga

Amo-Marco JB, Lledo MD (1996) In vitro propagation of Salix tarraconensis Pau ex Font Quer, an endemic and threatened plant. In Vitro Cell Dev Biol Plant 32:42–46

Anagnostakis SL (1987) Chestnut blight—the classical problem of an introduced pathogen. Mycologia 79:23–37

Bajaj YPS (1985) Cryopreservation of embryos. In: Katha KK (ed) Cryopreservation of plant cell and organs. CRC Press, Inc, Boca Raton, pp 227–242

Balci Y, Balci S, Eggers J, MacDonald WL, Juzwik J, Long RP, Gottschalk (2007) *Phytophthora spp.* associated with forest soils in eastern and north-central U.S. oak ecosystems. Plant Dis 91:705–710

Barakat A, Staton M, Cheng CH, Park J, Yassin NBM, Ficklin S, Yeh CC, Hebard F, Baier K, Powell W, Schuster SC, Wheeler N, Abbott A, Carlson JE, Sederoff R (2012) Chestnut resistance to the blight disease: insights from transcriptome analysis. BMC Plant Biol 12:38

Benson EE (1999) An introduction to plant conservation biotechnology. In: Benson EE (ed) Plant conservation biotechnology. Taylor and Francis, London, pp 3–10

Bounous G, Paglietta R, Craddock JH, Bellini E (1993) An overview of chestnut breeding. In: Proceedings of international congress on chestnut, Spoleto, Italy, 20–23 October. pp 251–269

Burnham CR, Rutter PA, French DW (1986a) Breeding blight-resistant chestnuts. Plant Breed Rev 4:347–397

Burnham CR, Rutter PA, French DW (1986b) Breeding blight-resistant chestnuts. Plant Breed chestnut trees. Phytopathology 6:615–621

Conedera M, Krebs P, Tinner W, Prandella M, Torriani D (2004) The cultivation of *Castanea sativa* (Mill.) in Europe, from its origin to its diffusion on a continental scale. Veg Hist Archaeobot 13:161–179. https://doi.org/10.1007/s00334-004-0038-7

Dawe AL, Nuss DL (2013) Hypovirus molecular biology: from Koch's postulates to host self-recognition genes that restrict virus transmission. Adv Virus Res 86:109–147

Diskin M, Steiner K, Hebard FV (2006) Recovery of American chestnut characteristics following hybridization and backcross breeding to restore blight ravaged *Castanea dentata*. For Ecol Manag 223:439–447

Fay MF (1992) Conservation of rare and endangered plants using in vitro methods. In Vitro Cell Dev Biol Plant 28:1–4

Graves AH (1926) The cause of the persistent development of basal shoots from blighted chestnut trees. Phytopathology 16:615–621

Griffin GJ (2000) Blight control and restoration of the American chestnut. J For 98:22–27

Griffin GJ, Reaver D, Osborne CK, Yancey MM (2009) Root recovery rates for *Phytophthora cinnamomi* and rate of symptom development from root rot on Abies fraseri trees over 7 years. For Pathol 39:15–27

Han JC, Wang GP, Kong DJ, Liu QX, Zhang XY (2007) Genetic diversity of Chinese chestnut (*Castanea mollissima*) in Hebei. Acta Hortic 760:573–577

Hebard FV (2004) Notes from Meadowview research farms 2003-2004. J Am Chestnut Found 18:38–45

Irwin H (2003) The road to American chestnut restoration. J Am Chestnut Found 16:6–13

Jaynes RA (1975) Chestnut. In: Janick J, Moore J (eds) Advances in fruit breeding. Purdue University Press, West Lafayette, pp 590–503

Jeffers SN, James JJ, Sisco PH (2009) Screening for resistance to *Phytophthora cinnamomi* in hybrid seedlings of American chestnut. In: Goheen EM, Frankel SJ (eds) Phytophthoras in forests and natural ecosystems. Proceedings of the fourth meeting of the International Union of Forest Research Organizations (IUFRO) working party S07.02.09. Monterey, CA, 334 p

Keys RN, Cech FN, (1979) Growth and differentiation of American chestnut callus cultures. In. Proc. 26th north-eastern forest tree improvement, Penn. State University. pp 13–24

Kubisiak T, Roberds JH (2006) Genetic structure of American chestnut populations based on neutral DNA markers : in restoration of American chestnut to forest lands. National Park Service, Washington, pp 109–122

Lang P, Dane F, Kubisiak TL (2006) Phylogeny of *Castanea* (Fagaceae) based on chloroplast trnT–L–F sequence data. Tree Gen Genom 2:132–139

Lang P, Dane F, Kubisiak TL, Huang H (2007) Molecular evidence for an Asian origin and a unique westward migration of species in the genus Castanea *via* Europe and North America. Mol Phylogenet Evol 43:49–59. https://doi.org/10.1016/j.ympev.2006.07.022

Lecape J-M, Nguyen T-B, Hau B, Giband M (2007) Targeted introgression of cotton fibre quality quantitative trait loci using molecular markers. In: Guimaraes EP, Ruane J, Schert BD, Sonnino A, Dargie JD (eds) Marker-assisted selection: current status and future perspectives in crops, livestock, forestry and fish. FAO, Rome

MacDonald WL, Fulbright DW (1991) Biological control of chestnut blight: use and limitations of transmissible hypovirulence. Plant Dis 75:656–661

Manos PS, Cannon CH, Oh S-H (2008) Phylogenetic relationships and taxonomic status of the paleoendemic Fagaceae of western North America: recognition of a new genus Notholithocarpus. Madrono 55:181–190

Martin KP, Pradeep AK (2003) Simple strategy for the in vitro conservation of *Ipsea malabarica* an endemic and endangered orchid of the Western Ghats of Kerala, India. Plant Cell Tissue Organ Cult 74:197–200

Mattoon WR (1909) The origin and early development of chestnut sprouts. Forest Q 7:34–37

McIlwrick K, Wetzel S, Beardmore T, Forbes K (2000) Ex situ conservation of American chestnut (*Castanea dentata* (Marsh.) Borkh.) and butternut (*Juglans cinera* L.), a review. For Chron 76:765–774

Oakes AD, Powell WA, Maynard CA (2013) Doubling acclimatization survival of micropropagated American chestnuts with darkness and shortened rooting induction time. J Environ Hortic 31 (2):77–83

Paillet FL (1988) Character and distribution of American chestnut sprouts in southern New England woodlands. Bull Torrey Bot Club 115:32–44

Paillet FL (2002) Chestnut: history and ecology of a transformed species. J Biogeogr 29:1517–1530

Petit RJ, Hampe A (2006) Some evolutionary consequences of being a tree. Annu Rev Ecol Evol Syst 37:187–214

Serres R, Read P, Hackett W, Nissen P (1990) Rooting of American chestnut microcuttings. J Environ Hort 8:86–88

Stout AB (1928) Dichogamy in flowering plants. Bull Torrey Bot Club 55:141–153

Tanaka T, Yamamoto T, Suzuki M (2005) Genetic diversity of Castanea crenata in northern Japan assessed by SSR markers. Breed Sci 55:271–277

Weeks S, Weeks H, Parker GR (2005) Native trees of the Midwest: identification, wildlife values, and landscape use. Purdue Univ. Press, West Lafayette

Xing Z, Satchwell M, Powell WA, Maynard CA (1997) Micro propagation of American chestnut: increasing rooting rate and preventing shoot-tip necrosis. In Vitro Cell Dev Biol Plants 33:43–48

Zhebentyayeva T, Staton ME, Olukolu B, Chandra A, Jeffers SN, James J, Sisco PH, Hebard FV, Georgi L, Nelson CD, Abbott AG (2014) Genetic and genomic resources for mapping resistance to root rot disease in chestnut. Acta Hortic 1019:263–270

Genetic and Genomic Resources in Rice Bean (*Vigna umbellata* Thunb.): Availability, Advancements, and Applications

9

Showkat A. Zargar, Humara Fayaz, Aijaz A. Wani, M. I. S. Saggoo, Rakeeb A. Mir, and Reyazul Rouf Mir

Abstract

Rice bean (*Vigna umbellata* Thunb.) is an important pulse crop grown in Asia mainly for dried grain or fodder or as green manure and plays an important role in human and animal nutrition. The protein present in rice bean is rich in lysine, which makes it as an excellent addition to the cereal-based diet. The seeds are also rich in minerals and vitamins including thiamine, riboflavin, niacin, and ascorbic acid. The rice bean landraces constitute a rich source of biodiversity. The conservation needs these genetic resources to be unraveled judiciously. Expediting the genetic improvement of rice bean, genomic resources, and tools for molecular breeding is important for the crop improvement programs and the

S. A. Zargar (✉)
Cytogenetics and Reproductive Biology Laboratory, Department of Botany, University of Kashmir, Srinagar, Jammu and Kashmir, India

Department of Botany, Punjabi University, Patiala, India

H. Fayaz · A. A. Wani
Cytogenetics and Reproductive Biology Laboratory, Department of Botany, University of Kashmir, Srinagar, Jammu and Kashmir, India
e-mail: aawani@uok.edu.in

M. I. S. Saggoo
Department of Botany, Punjabi University, Patiala, India

R. A. Mir
Department of Biotechnology, School of Bio-sciences & Biotechnology, Baba Ghulam Shah Badshah University, Rajouri, Jammu and Kashmir, India

R. R. Mir
Division of Genetics and Plant Breeding, Sher-e-Kashmir University of Agricultural Sciences and Technology of Kashmir, Wadura, Sopore, Jammu and Kashmir, India
e-mail: rrmir@skuastkashmir.ac.in

S. M. Zargar et al. (eds.), *Neglected and Underutilized Crops - Towards Nutritional Security and Sustainability*, https://doi.org/10.1007/978-981-16-3876-3_9

breeding activities. The global food security is constrained by several factors including the heavy reliance on very few key staple crops. The need of the hour is to reduce the heavy reliance on the major staple crops. With this regard it becomes imperative to unravel the genetic and genomic resource availability in rice bean which would serve as a major backbone to the underprivileged part of the global population. The genetic diversity studies have been carried out using number of molecular marker systems including, RAPD, AFLP, RFLP, SSR.

Keywords

Rice bean · Genetic resources · Genomic resources · Underutilized crop

9.1 Introduction

Rice bean (*Vigna umbellata* Thunb.) is a warm seasonal annual vine legume, also referred as red bean, oriental bean, and mountain climbing bean, is widely cultivated in South and Southeast Asia (Khadka and Acharya 2009; Pattanayak et al. 2019). It belongs to the family Fabaceae, tribe *Phaseoleae,* genus *Vigna,* and species *umbellate.* It is diploid with chromosome number of $2n = 2x = 22$ (Asha et al. 2015). The subgenus *Ceratotropis* contains six cultivated species, among which four species are economically important in Asian countries, namely rice bean (*V. umbellata*), mungbean (*Vigna radiata*), azuki bean (*V. angularis*), and black gram (*V. mungo*) (Isemura et al. 2010). Rice bean is typically an annual and short-lived legume. It shows different growth habits as it is erect, semi-erect or twined. It normally measures between 30 and 100 cm in height, but can grow up to 200 cm (Ecoport 2014). Its taproot system can go 100–150 cm deep and the stem is branched; leaves trifoliate, flowers bright yellow, up to 5–10 cm long are papillonaceous, a characteristic feature of family fabaceae. Fruits are usually cylindrical, 7–12.5 cm long pods with 6–8 mm seeds having concave hilum. About 16 different grade colors/market classes mainly black, red, cream, violet, purple, maroon, brown, chocolate, or mottled grains with greenish, brownish, or ash gray background, are available in rice bean grains. Typically, rice bean is ready to be harvested in 4–5 months after sowing. It can be harvested when about 75% of the pods turn brown. Seed shattering results in considerable decrease of crop yield (Khadka and Acharya 2009).

The cultivation of rice bean is primarily confined to the hills of tribal regions in Northeastern, the western and the Eastern Ghats of India (Arora et al. 1980). The crop is also cultivated in other parts of India including Punjab, Himachal Pradesh, central Indian plains, hills of Chhattisgarh, southern hills of Odisha and Kerala. The center of domestication for rice bean is presumed to be Indo-China belt (Tomooka et al. 1991). The wild species *V. umbellata* var. *gracilis* of this crop is found in the Himalayas and Central China to Malaysia (Seehalak et al. 2006). Rice beans are now commonly cultivated in tropical Asia and to a smaller degree in Fiji, USA, Australia, South-Western Asia, tropical Africa and America. From an area of 1804 ha, the rice bean world production is approximately 1407 metric tons with an approximate productivity of 2.25q/ha. As far as marketing in the international trade is concerned meager information is available. Japan is the major importer of rice bean, whereas the major exporters include Thailand, Myanmar, China, and Madagascar (Wadikar and Bepary 2020).

Rice bean is a non-conventional underutilized crop which has a potential to generate livelihood for the underprivileged sections of the population particularly in south and South East Asia. Rice bean is nutritionally rich and genetically diverse agricultural crop, having potential to withstand extreme environmental conditions. It shows resistance to various diverse biotic and abiotic stress factors. Every part of rice bean plant is edible and used for cooking. Rice bean seeds are consumed by boiling and eating dry seeds with rice or substituting the rice in stew or soup. In Madagascar, the seeds are grounded to produce nutrient flour as food for children. Moreover, boiled leaves, young pods, and sprouted seeds are eaten as vegetables, while as young pods can be eaten raw (Rajerison 2006). Contrary to other pulses, rice beans are not easily transformed into dhal, because of their fibrous mucilage, which prevents hulling and cotyledon separation (Rajerison 2006; Ecoport 2014; Van Oers 1989). For livestock feeding, rice bean is very useful, as the vegetative parts of the plant can be fed freshly as well as made into hay, while as the stems, empty pods, leafy portions, and some seeds are used as rice bean straw (Göhl 1982). In addition, plants of rice bean are grown as a green manure, cover crop as well as a living fence or biological barrier (Ecoport 2014). Therefore, rice bean can be considered both as a grain and a fodder legume.

Rice bean helps to improve the nitrogen status and fertility of soil, due its capability to efficiently fix atmospheric nitrogen. Apart from its normal physiological functions, the remaining taproots in the soil act as good source to enhance the soil fertility, reduction of soil erosion and provide strong soil cover (Dahipahle et al. 2017). Maize is frequently cultivated as a strategical intercrop in rice bean fields. It was found that the crop yield of maize was enhanced by rice bean intercropping (Buergelt et al. 2009). The critical limitations of rice bean include higher levels of labor and low crop yield as compared to other closely associated crops like green grams (*V. radiata*) and heck (*V. mungo*) (Buergelt et al. 2009). In addition, the rice bean is also categorized as underutilized crop due it is very low utilization as vegetable crop. Later limitation may be due to failure to initiate agricultural practices for development of higher yielding varieties by modern tools. Other reason for its low consumption includes its tough seed coat, which remains intact even after boiling. A very faint scientific knowledge of this crop is available, even though it is believed that this crop has potential to alleviate economy and helps in reducing nutritional deficiency (Buergelt et al. 2009).

Therefore, precise cataloging and frequent dissemination of available rice bean germplasm is crucial. The development of rice bean breeding varieties with different desirable characteristics is necessary for sustainable agricultural development and global food security. In addition, crop may be possibly cultivated in less fertile soil in arid regions of developing countries due its resilient features (Dhillon and Tanwar 2018). Current chapter aims to highlight the genomics and genetic resources of rice bean to pave way for initiation of conservation programs, production of improved varieties and its utilization at large scale.

9.2 Genomic Resources: Advancement in Marker Technology

In the last few decades, a trend in the population growth as well as in the pattern of consumption in developing countries indicates a continuing rise of global demand for food. In addition to improve quality and food availability, it is important to use newer technologies to enhance nutritional values of rice bean (Godfray et al. 2010). The molecular breeding approaches combined with phenotyping and genotyping methods can possibly offer new perspectives to address challenges associated with this crop (Varshney and Dubey 2009). Genomic resources, such as molecular markers, genetic maps, and data pertaining to sequence information play important role to augment the molecular breeding programs (Varshney 2010). Current data reveals that very less molecular breeding studies of minor crops like rice bean is carried globally, even though these crops play important role in local food security. The utilization and conservation of rice bean crop landraces are of dire importance, as they are regarded as rich source of biodiversity (Muthusamy et al. 2008). The rice bean is primarily characterized based on morphology, agronomics, and biochemical characteristics. Even though these traditional methods of characterization are merely used for closely related landraces due to low polymorphism (Lumpkin and McClary 1994), modern genomic studies have led to study landraces very deeply, helping to uncover the hidden distinctions based on DNA sequences. The understanding of genetic characterization and variability among species has been successfully employed as basis for conservation of the landrace germplasm (Paterson et al. 1991). Molecular markers have been mostly employed for assessment of genetic variation among different genotypes of landraces. Later studies can be utilized to construct the linkage maps, in turn to screen the desired varieties of crop plants (Paterson et al. 1991).

Several PCR based markers have been employed to characterize the crop plants. Among these molecular markers like simple sequence repeats (SSRs), inter-simple sequence repeats (ISSRs), single nucleotide polymorphism (SNPs), and diversity array technologies (DArTs) are frequently employed to have insights in genomic variation of crop plants (Jones et al. 2009). The SSR makers have been frequently used for breeding of crop plants (Gupta and Varshney 2000). In comparison to SSRs, ISSRs are frequently used for assessing genetic variation, diversity analysis, and molecular breeding of crop plants due to its higher level of polymorphism (Reddy et al. 2002). The SNPs are leading markers employed in studying the genetic variation of humans and other animals, due to the fact that they are readily available for high performance genotyping and cost efficiency (Varshney 2010). Later features make SNP potentially a dominating marker to characterize the crop plants.

The DArT marker system has been shown to be very prominent in revealing genetic variation in germplasm collections and is well suitable for molecular breeding studies (Kilian et al. 2003). These resources will accelerate the discovery of genes and unravel a clear path towards the molecular breeding of this crop, thereby enhancing crop productivity in developing countries in order to guarantee food safety (Varshney et al. 2010). It has been stated that genetic variation can be resolved more specifically in accordance with the number of polymorphisms identified by the marker system (Sivaprakash et al. 2004). Genomic studies of rice bean, however, have declined owing to the absence of polymorphic DNA markers as compared to other legume plants like soybean, mung bean, azuki bean, and common bean. There are very few studies on the genetic diversity of rice bean using AFLP, RAPD, SSR, and ISSR markers (Bajracharya et al. 2008; Muthusamy et al. 2008; Thakur et al. 2017; Tian et al. 2013; Iangrai et al. 2017). Genetic variation among 10 landraces of rice bean using random RAPD and ISSR markers were reported by Muthusamy et al. (2008). From among these markers, 987 amplification products were generated by RAPD with 719 polymorphic bands while as ISSR markers generated 479 bands with 296 polymorphic bands. They also reported that RAPD markers detected more polymorphism (70.30%) than ISSR markers (61.79%) with average PIC (polymorphism information content) for each (0.243 for RAPD and 0.203 for ISSR) marker which indicated that both marker systems were equally successful for evaluating polymorphism. Genetic diversity in 13 rice bean varieties and their 11 narrow leaf crosses was studied using RAPD markers (Shafiqul et al. 2017).

In another study, 35 polymorphic markers selected from azuki bean were used to study genetic diversity in a set of 112 rice bean germplasm collected from a wide geographical area of Nepal and India (Bajracharya et al. 2008). Furthermore, the characterization of genotypes was done for diversity indices, and it was proposed that azuki bean SSR markers can be utilized for evaluating genetic diversity in rice bean. Tian et al. (2013) gave the first comprehensive SSR analysis on a total of 472 diverse rice bean accessions (388 cultivated and 84 wild) from 16 Asian countries. This extensive study carried out by Tian et al. (2013) also utilized the 13 azuki bean SSR markers retrieved from the available literature of Wang et al. (2004). In cultivated populations, gene diversity was approximately 83% of the wild populations. The most genetically diverse cultivars were from Vietnam, Myanmar, Nepal, and India. Sampled from Thailand, North and South Korea, Japan, and China and the East Asian accessions formed a separate cluster. The most similar genetic structure to wild accessions was in Indonesian-cultivated rice bean accessions, showing strong genetic divergence from other cultivated accessions. Most of the accessions from Western regions were different and created a particular group in Nepalese-cultivated accessions (Bajracharya et al. 2008). Chen et al. (2016) selected 300 SSR loci randomly for validation using PCR amplification. Out of these loci, they observed only 23 primer pairs polymorphic for 32 rice bean accessions. A set of microsatellite markers from a SSR-enriched genomic library of rice bean were also characterized and developed by Wang et al. (2004). Genetic diversity of rice beans from northeast India was also analyzed by 28 SSR marker developed by Iangrai et al. (2017). They found high PIC value for SSR markers that detected a high genetic diversity among 65 accessions. Apart from the genetic diversity related studies, there are many reports that are carried out on rice bean, which have focused on its high level of resistance to the major storage pest bruchid beetles (*Callosobruchus* spp.) (Tomooka et al. 2000; Kashiwaba et al. 2003; Somta et al. 2006).

QTL analysis carried out in an interspecific mapping population derived from rice bean and *V. nakashimae* led to the mapping of bruchid resistance (Somta et al. 2006). Four putative bruchid-resistant QTLs with 18–60% effects on phenotype were identified. One putative QTL for resistance to each of these bruchid species was located on different LGs. Some other putative QTLs linked to both *C. chinensis* and *C. maculates* resistance were found on the linkage group 1. In addition, using intra-specific mapping population QTLs were identified for different domestication related traits in rice bean (Isemura et al. 2010). A draft genome sequencing study by Kaul et al. (2019) revealed that rice beans have relatively rich biochemical pathways, enzyme distribution and reaction types, active plant compound-specific gene hubs, genomic structures, protein-coding genes, and the composition of relatively non-coding RNA. Rice beans, which focus on their palatability and flowering potential, gene response to stress and genetic disease resistance, have had irrelevant effects on their genetic diversity and their subsequent breeding. The genetic enhancement of rice bean can potentially contribute to the sustainable production of food, nutrition; can provide favorable farmer-accessible flowering conditions, and consumer-accessible palatable crop. According to the genetic information drawn from the draft of the *Vigna umbellate,* it is indicated that this species is suitable for domestication and cultivation in large areas and potentially presents high value commercial crop because of its close relation to mung bean, azuki bean, and cowpea species, a high-density genetic map in rice bean, could help to trace important economic traits and meet the long-standing challenge of its cultivation. This bids an opportunity to scientists for improving the traits via CRISPR/Cas genome editing tool. Moreover, this led to the cultivation of legume crop varieties enduring huge nutritional value for wide consumption, to meet up with the century old challenge of its consumption and cultivation.

The huge diversity maintained around its small geographical range and the fact that there are few marketing outlets means the genetic advancement of rice beans has great potential. There are many characteristics of rice beans which need to be focused by scientists before they can be extensively used for regular consumption. These include extreme photoperiod sensitivity, delayed flowering, high vegetative growth issues, twining habit, hard seeds, and susceptibility to shatter (Smartt 2012). Rice beans are vulnerable to pests, leading to severe harvest losses and reduced crop quality. Restricted number of genomic tools has prevented the increase in rice bean yield. Domestication and genetic improvement of rice beans is urgently required to produce high yielding varieties with good organoleptic properties and lower rates of anti-nutrients. Although, the majority of the literature available shows the rice bean possess resistance to insect and aluminum toxicity (toxicity tolerance), not much concern is shown to understand its mechanism of flowering, palatability, and disease resistance profiles which are important for the full potential of the crop (Kaul et al. 2019).

9.3 Genetic Resources in Rice Bean

Systematic plant exploration and collection in India initiated in the early 1940s with the creation of a plant introduction unit in the Botany Division, Indian Agriculture Research Institute (IARI) New Delhi for the management of genetic resources, which finally became the National Bureau of Plant Genetic Resources (NBPGR). In the late 1960s and early 1970s under the IARI PL-480 project, intensive efforts were made to gather all the available landraces of *Vigna* species from the different states of India. Approximately 2600 germplasm accessions were collected from diverse agro-ecological areas. The explored areas include the Rajasthan and Gujarat; parts of Madhya Pradesh, Bihar, Uttar Pradesh, Punjab, Himachal Pradesh, Jammu and Kashmir, Andhra Pradesh, Orissa, and Haryana except the central zone; the northeastern, western, and coastal regions of Maharashtra; southern districts of Karnataka and eastern Kerala; and major areas of Tamil Nadu (Malik et al. 2001). Majority of the explored accessions showed determinate growth habit. As far as seed size is concerned wide variation was observed. The bold seeded accessions were collected from Maharashtra and Madhya Pradesh, whereas the accessions having more number of seeds per pod were collected from Rajasthan, Gujarat, Haryana, Bihar, and Jammu and Kashmir.

In the last three decades, NBPGR introduced significant number accessions of pulse crops including trial material from more than 50 countries (Singh et al. 2001). In addition, successful introduction of rice bean has been made from the USA (Gautam et al. 2000). Some rice bean accessions from Taiwan were high in yield, wide adaptability, and resistant to drought (Bisht and Singh 2013). Various national and international gene banks maintaining wide diversity of *Vigna* species include International Center for Agriculture Research in the Dry Areas (ICARDA), Aleppo, Syria; Asian Vegetable Research and Development Centre (AVRDC), Shanhua, Taiwan; International Institute for Tropical Agriculture (IITA), Ibadan, Nigeria; Bogor Research Institute for Food Crops (BORIF), Bogor, Indonesia; Common wealth Scientific and Industrial Organization (CSIRO), Canberra, Australia; Malang Research Institute for Food Crops (MARIF), Malang, Indonesia; National Plant Genetic Research Institute (NPGRI), University of Philippines, Los Banos, Philippines, and US Department of Agriculture (USDA), Southern Regional Plant Introduction Station, Georgia, and NBPGR (Bisht and Singh 2013). *Vigna* species have orthodox seeds, which can be dried and stored with a minimum loss of viability over a long time period. The long-term repository at the NBPGR in New Delhi contained a total of 10,551 accessions of different *Vigna* species, such as urd bean (3131), azuki bean (185), mung bean (3704), rice bean (2045), and moth bean (1486) at -18 °C. In various institutes, worldwide germplasm collections including moth beans, mung bean, rice beans, urd bean, and suki beans have been maintained.

The Taiwan-based World Vegetable Center has 351 rice bean accessions from 24 different countries of the world with Nepal, India, Brazil, Philippines, and Congo as major contributors (Table 9.1) (Pattanayak et al. 2019); (http://seed.world veg. org/search/passport).

Table 9.1 Rice bean germplasm held at some key centers

S. No.	Center	Country	No. of accessions present
1.	ICAR—National Bureau of plant genetic resources	India	2071
2.	Institute of Crop Genetic Resources	China	1363
3.	National Agriculture and Food Research Organization	Japan	399
4.	World Vegetable Centre	Taiwan	351
5.	Nepal Agricultural Research Council	Nepal	300
6.	National Plant Genetic Resource Laboratory, UPLB, Laguna	Philippines	161
7.	Centre of Biology, Indonesian Institute of Science Research and Development, Bogor	Indonesia	100
8.	ARS–GRIN	USA	147

Adopted from Pattanayak et al. (2019)

Among the Indian accessions, IC 7506, EC 18771, IC 7588, and IC 8229 are stated to be least to photoperiods (Kim et al. 2015). In India, there are 300 accessions maintained at the Indian Institute for Pulses Research and the National Bureau of Plant Genetic Resources stations at Uttar Pradesh and Bhowali. Besides, there are a total of 2071 rice bean accessions in the National Gene bank maintained by NBPGR of which 96 are exotic collections and 1993 are indigenous collections (http://www.nbpgr.ernet.in:8080/PGRPortal). Among these 2071 accessions, 656 accessions were collected from Northeast India, 240 accessions from Northwest, 187 accessions from Eastern parts, 40 accessions from Southern, and 7 accessions from Central parts. Most of the ethnic collections were taken from NE India (656 accessions) Moreover, a collection of some 300 accessions from different parts of the country in Nepal are maintained at the Plant Genetic Resources Unit of the NARC (Pattanayak et al. 2019). For commercial cultivation of rice bean in India, a large number of rice bean varieties are being released from time to time on the basis of some selected salient features highlighting the zones suitable for their cultivation (Table 9.2).

9.4 Conclusion and Future Perspective

Rice bean is a multi-purpose legume, having great potential to provide nutrition at a cheaper cost. Rice bean landrace genome harbors desirable traits coding for pest resistance, tolerance to a variety of abiotic stress factors, easy to cultivate, and adaptation to different cropping systems and climatic conditions. Several factors are responsible for its underutilization include over-reliance on major cultivated crops; abrupt climatic change and land degradation, in turn hampers sustainable agricultural productivity. The challenge posed to several crop landraces are undoubtedly unnoticed, leading to increase in rural poverty, malnutrition, and global food insecurity. In order to meet global food needs, urgent emphasis should be initiated to

Table 9.2 Varieties of rice bean released in India for fodder and grain

S. No.	Variety	Host institute of development	Year of release	Salient features	Yield (q/ha)	Recommendation Zone
Grain type						
1.	RBL-1	PAU, Ludhiana	1987	Normal duration[a] with high yield	15.0	Punjab and northern plains
2.	PRR-1	GBPUAT, Ranichauri	1995	Normal duration with high yield	16.0	Uttarakhand Hills
3.	PRR-2	GBPUAT, Ranichauri	1997	Normal duration with high yield	15.0	Northwest Hills
4.	RBL-6	PAU, Ludhiana	2000	Early type	15.0	Northern plains
5.	RBL-50	PAU, Ludhiana	2003	Normal duration with high yield	15.5	Northern plains
6.	RBL-35	PAU, Ludhiana	2003	Normal duration with green color grain	15.0	Northern plains
7.	BRS-1	ICAR-NBPGR, Bhowali	2003	Normal duration, black seed	17.0	Northwest Hills
8.	Him Shakti (VRB-3)	ICAR-VPKAS, Almora and ICAR-NBPGR, Shimla	2013	Normal duration, light green seed	17.1	Northeast and Northwest Hills
Fodder type						
9.	Konkan rice bean 1	BSKKV, Dapoli	1997	High fodder yield	220.0	Konkan region
10.	Bidhan rice bean 1	BCKV, West Bengal	2000	High fodder yield, insect resistant	350.0	East and Northeast zone
11.	Bidhan rice bean 2	BCKV, West Bengal	2005	High fodder yield, insect resistant	300.0	East and Northeast zone
12.	Suravi	KAU, Thrissur	2013	High fodder yield	220.0	Kerala

(continued)

Table 9.2 (continued)

S. No.	Variety	Host institute of development	Year of release	Salient features	Yield (q/ha)	Recommendation Zone
13.	Bidhan rice bean 3	BCKV, West Bengal	2015	High fodder yield, insect resistant	350.0	East and Northeast zone

Adopted from Pattanayak et al. (2019)
[a]Normal duration is 135–145 days

promote the cultivation and use of other crops, which are neglected and underutilized. Due to its higher genetic diversity, rice beans are highly suitable for crop breeding. However, the precise identification of germplasm resources, the characterization of trait donors, and the development of mapping populations are the first priorities for rice bean advancement. The use of comparative genomics for determining genetics of crop resistance and chief agronomic traits can pave way for the identification of genes/QTLs. The correct use of genetic diversity in the collection of germplasm resources requires a good understanding of its characteristics. Additionally, characterizing regional diversity of crops is necessary to understand the direction of farmer's selection and location-specific need for improvement.

References

Arora RK, Chandel KPS, Joshi BS, Pant KC (1980) Rice bean: tribal pulse of eastern India. Econ Bot 34:260–263

Asha RK, Koundinya AVV, Das A, Chattopadhyay SB (2015) A review on an underutilised multipurpose legume: rice bean. In: III International symposium on underutilized plant species 1241, pp 57–64

Bajracharya J, Singh S, Dangol B, Hollington PA, Witcombe JR (2008) Food security through rice bean research in India and Nepal (FOSRIN) report 2. Identification of polymorphic markers in ricebean (*Vigna umbellata*)

Bisht IS, Singh M (2013) Asian *Vigna*. In: Genetic and genomic resources of grain legume improvement. Elsevier, pp 237–267

Buergelt D, von Oppen M, Yadavendra JP (2009) Quality parameters in relation to consumer's preferences in rice bean. In: Presentation at the international conference on grain legumes: quality improvement, value addition and trade

Chen H, Chen X, Tian J, Yang Y, Liu Z, Hao X, Wang L, Wang S, Liang J, Zhang L, Yin F (2016) Development of gene-based SSR markers in rice bean (*Vigna umbellata* L.) based on transcriptome data. PLoS One 11(3):e0151040

Dahipahle AV, Kumar S, Sharma N, Singh H, Kashyap S, Meena H (2017) Rice bean—a multipurpose, underutilized, potential nutritive fodder legume—a review. J Pure Appl Microbiol 11:433–439

Dhillon PK, Tanwar B (2018) Rice bean: a healthy and cost-effective alternative for crop and food diversity. Food Secur 10(3):525–535

Ecoport (2014) Ecoport database. http://www.ecoport.org/

Gautam PL, Sharma GD, Srivastava U, Kumar A, Saxena RK, Srinivasan K (eds) (2000) 20 Glorious years of NBPGR, 1976. National Bureau of Plant Genetic Resources, New Delhi

Godfray HCJ, Beddington JR, Crute IR, Haddad L, Lawrence D, Muir JF, Pretty J, Robinson S, Thomas SM, Toulmin C (2010) Food security: the challenge of feeding 9 billion people. Science 327(5967):812–818

Göhl B (1982) Les aliments du bétail sous les tropiques: données sommaires et valeurs nutritives. Organisation des Nations Unies pour l'alimentation et l'agriculture

Gupta PK, Varshney RK (2000) The development and use of microsatellite markers for genetic analysis and plant breeding with emphasis on bread wheat. Euphytica 113(3):163–185

Iangrai B, Pattanayak A, Khongwir DEA, Pale G, Gatphoh EM, Das A, Chrungoo NK (2017) Development and characterization of a new set of genomic microsatellite markers in rice bean (*Vigna umbellata* (Thunb.) Ohwi and Ohashi) and their utilization in genetic diversity analysis of collections from North East India. PLoS One 12(7):e0179801

Isemura T, Kaga A, Tomooka N, Shimizu T, Vaughan DA (2010) The genetics of domestication of rice bean, *Vigna umbellata*. Ann Bot 106(6):927–944

Jones N, Ougham H, Thomas H, Pašakinskienė I (2009) Markers and mapping revisited: finding your gene. New Phytol 183(4):935–966

Kashiwaba K, Tomooka N, Kaga A, Han OK, Vaughan DA (2003) Characterization of resistance to three bruchid species (*Callosobruchus* spp., Coleoptera, Bruchidae) in cultivated rice bean (*Vigna umbellata*). J Econ Entomol 96(1):207–213

Kaul T, Eswaran M, Thangaraj A, Meyyazhagan A, Nehra M, Raman NM, Bharti J, Gayacharan DR, Badapanda C, Balamurali B (2019) Rice bean (*Vigna umbellata*) draft genome sequence: unravelling the late flowering and unpalatability related genomic resources for efficient domestication of this underutilized crop. bioRxiv 2019:816595

Khadka, K., & Acharya, B. D. (2009). Cultivation practices of ricebean. Research and Development (LI-BIRD), Local Initiatives for Biodiversity, Pokhara

Kilian A, Huttner E, Wenzl P, Jaccoud D, Carling J, Caig V, Evers M, Heller-Uszynska K, Cayla C, Patarapuwadol S, Xia L (2003) The fast and the cheap: SNP and DArT-based whole genome profiling for crop improvement. In: Proceedings of the international congress in the wake of the double helix: from the green revolution to the gene revolution, pp 443–461

Kim SK, Nair RM, Lee J, Lee SH (2015) Genomic resources in mungbean for future breeding programs. Front Plant Sci 6:626

Lumpkin TA, McClary DC (1994) Azuki bean: botany, production and uses. CAB International, Wallingford

Malik SS, Srivastava U, Tomar JB, Bhandari DC, Pandey A, Hore DK, Dikshit N (2001) Plant exploration and germplasm collection. In: National Bureau of Plant Genetic Resources: a compendium of achievements. National Bureau of Plant Genetic Resources, New Delhi, pp 31–68

Muthusamy S, Kanagarajan S, Ponnusamy S (2008) Efficiency of RAPD and ISSR markers system in accessing genetic variation of rice bean (*Vigna umbellata*) landraces. Electron J Biotechnol 11 (3):32–41

Paterson AH, Tanksley SD, Sorrells ME (1991) DNA markers in plant improvement. In: Advances in agronomy, vol 46. Academic Press, pp 39–90

Pattanayak A, Roy S, Sood S, Iangrai B, Banerjee A, Gupta S, Joshi DC (2019) Rice bean: a lesser known pulse with well-recognized potential. Planta 2019:1–18

Rajerison R (2006) *Vigna umbellata* (Thunb.) Ohwi & H. Ohashi. Cereals and pulses. PROTA1, Wageningen, Pays Bas, Netherlands

Reddy MP, Sarla N, Siddiq EA (2002) Inter simple sequence repeat (ISSR) polymorphism and its application in plant breeding. Euphytica 128(1):9–17

Seehalak W, Tomooka N, Waranyuwat A, Thipyapong P, Laosuwan P, Kaga A, Vaughan DA (2006) Genetic diversity of the *Vigna* germplasm from Thailand and neighboring regions revealed by AFLP analysis. Genet Resour Crop Evol 53(5):1043–1059

Shafiqul I, Dutta M, Shachi S, Kumar P, Bhatt KV (2017) Assessment of genetic diversity of ricebean [(*Vigna umbellate*) (Thunb.) Ohwi & Ohashi] varieties and their narrow leaf cross derivatives using RAPD markers. Int J Agric Environ Biotechnol 10(4):415–421

Singh RV, Chand D, Brahmi P, Verma N, Tyagi V, Singh SP (2001) Germplasm exchange. In: Dhillon BS, Varaprasad KS, Singh M, Archak S, Srivastava U, Sharma GD (eds) National Bureau of Plant Genetic Resources: a compendium of achievements. NBPGR, New Delhi, pp 69–89

Sivaprakash KR, Prashanth SR, Mohanty BP, Parida A (2004) Genetic diversity of black gram (*Vigna mungo*) landraces as evaluated by amplified fragment length polymorphism markers. Curr Sci 86(10):1411–1416

Smartt J (2012) Morphological, physiological and biochemical changes in Phaseolus beans. In: Gepts P (ed) Genetic resources of Phaseolus beans: their maintenance, domestication, evolution and utilization, vol 6. Springer, Dordrecht, p 143

Somta P, Kaga A, Tomooka N, Kashiwaba K, Isemura T, Chaitieng B, Srinives P, Vaughan DA (2006) Development of an interspecific *Vigna* linkage map between *Vigna umbellata* (Thunb.) Ohwi & Ohashi and V. nakashimae (Ohwi) Ohwi & Ohashi and its use in analysis of bruchid resistance and comparative genomics. Plant Breed 125(1):77–84

Thakur S, Bhardwaj N, Chahota RK (2017) Evaluation of genetic diversity in ricebean [*Vigna umbellata* (Thunb.) Ohwi and Ohashi] germplasm using SSR markers. Electron J Plant Breed 8 (2):674–679

Tian J, Isemura T, Kaga A, Vaughan DA, Tomooka N (2013) Genetic diversity of the rice bean (*Vigna umbellata*) genepool as assessed by SSR markers. Genome 56(12):717–727

Tomooka N, Lairungreang C, Nakeeraks P, Egawa Y, Thavarasook C (1991) Mungbean and the genetic resources, the subgenus Ceratotropis. Tropical Agriculture Research Center, Tsukuba

Tomooka N, Kashiwaba K, Vaughan DA, Ishimoto M, Egawa Y (2000) The effectiveness of evaluating wild species: searching for sources of resistance to bruchid beetles in the genus *Vigna* subgenus Ceratotropis. Euphytica 115(1):27–41

Van Oers CCCM (1989) *Vigna aconitifolia*. In: Plant resources of South-East Asia: I. Pulses. Pudoc, Wageningen, pp 66–67

Varshney RK (2010) Gene-based marker systems in plants: high throughput approaches for marker discovery and genotyping. In: Molecular techniques in crop improvement. Springer, Dordrecht, pp 119–142

Varshney RK, Dubey A (2009) Novel genomic tools and modern genetic and breeding approaches for crop improvement. J Plant Biochem Biotechnol 18(2):127–138

Varshney RK, Glaszmann JC, Leung H, Ribaut JM (2010) More genomic resources for less-studied crops. Trends Biotechnol 28(9):452–460

Wadikar DD, Bepary RH (2020) Rice bean. In: Pulses. Springer, Cham, pp 297–331

Wang XW, Kaga A, Tomooka N, Vaughan DA (2004) The development of SSR markers by a new method in plants and their application to gene flow studies in azuki bean [*Vigna angularis* (Willd.) Ohwi & Ohashi]. Theor Appl Genet 109(2):352–360

10 Advancements in High-Throughput Omics-Technologies for Understanding the Biology of Neglected and Underutilized Crops

Ashish Kumar Choudhary, Riyazuddin Riyazuddin, Arun Kumar Maurya, and Ravi Gupta

Abstract

The world's population depends heavily only on a limited number of crops for its basic nutrition, viz., carbohydrates, proteins, and fats. As of today, three unique and universal (almost pan-cultural) crops including wheat, rice, and maize are the main sources of carbohydrates for human dietary consumption. Besides these three staple crops, several neglected and underutilized crops (NUCs) are documented as rich sources of essential micro and macronutrients, however, these have neither been widely consumed nor exploited at a large scale for human consumption. One of the major bottlenecks in the large-scale utilization of these crops is the lack of scientific literature that limits our understanding of how these NUCs behave during various biotic and abiotic stressors. Development of various omics-based platforms including transcriptomics, proteomics, and metabolomics have facilitated the understanding of various model plants, such as Arabidopsis and rice, and the same can also be used in the future to decipher

A. K. Choudhary
Department of Botany, University of Delhi, Delhi, India

R. Riyazuddin
Department of Plant Biology, Faculty of Science and Informatics, University of Szeged, Szeged, Hungary

Doctoral School in Biology, Faculty of Science and Informatics, University of Szeged, Szeged, Hungary

A. K. Maurya
Department of Botany, Multanimal Modi College, Modinagar, Uttar Pradesh, India

R. Gupta (✉)
Department of Forestry, Environment, and Systems, College of Science and Technology, Kookmin University, Seoul, South Korea
e-mail: dr.ravigupta@jamiahamdard.ac.in

S. M. Zargar et al. (eds.), *Neglected and Underutilized Crops - Towards Nutritional Security and Sustainability*, https://doi.org/10.1007/978-981-16-3876-3_10

the NUCs responses to environmental fluctuations. In this chapter, we are presenting the technological advancements in the omics tools and downstream data analysis to understand the biology of these NUCs in greater detail. The information derived from these studies could be used in the future to exploit NUCs for their large-scale cultivation to address the issue of food security in the near future.

Keywords

Transcriptomics · Metabolomics · Proteomics · Data integration · Neglected and underutilized crops · Next-generation sequencing · Bioinformatics tools

10.1 Introduction

Owing to the increasing global population, hunger, starvation, and malnutrition are the major concerns nowadays. As per the State of Food Security and Nutrition, over 820 million people are suffering from hunger globally (FAO et al. 2019). Similarly, micronutrient deficiencies have been observed in two billion people due to insufficient intake of vital minerals and vitamins (Bailey et al. 2015; World Health Organization 2018). It has been estimated that more than 1600 million people suffer from iron deficiency, approximately 500 million people from protein deficiency, and more than 200 million people from vitamin-A deficiency (World Health Organization 2009). Furthermore, the world's 17.3% population suffers from zinc deficiency resulting in the deaths of over 400,000 children (of $\leq$5 years of age) annually (Hefferon 2019). The Sustainable Development Goal using alternative food sources to completely overcome the hunger and malnutrition problem by 2030 have been declared by the world food program (Kilonzi et al. 2017; Davari et al. 2018; FAO et al. 2019). At present, approximately 80,000 plants are exploited for food industries, fiber, and medicines (Kermali et al. 1997), of which, over 12,000 crop species are recognized for human consumption (Adhikari et al. 2017). Among these, around 150 plant species are being cultivated for food purposes, and of these only 30 crops are highly exploited to meet the global food demands (Arora 2014; Shelef et al. 2017). Of these 30 species of crop plants, only 20 species currently fulfill 90% of the global food requirements (Collins and Hawtin 1999). Wheat, rice, and maize are known as "staple crops," because these comprise 60% of the human diet (Collins and Hawtin 1999; Chivenge et al. 2015). There are many reasons for global food scarcity and the major problem is the limited number of food crops being utilized to meet the global food requirements. In addition, these food crops fail to produce higher yields under various stress conditions, such as drought, salinity, excess moisture, and diverse biotic stresses. Because of this scarcity, emphasis on neglected and underutilized crops (NUCs) can help to fix global food problems, due to their resilient feature to withstand various stress conditions.

"Neglected crops" are grown by local communities/traditional farmers at their centers of origin (Mayes et al. 2012), however, "Underutilized crops" are those crops which once grew more intensively, but are now falling into disuse due to various economic, genetic, and cultural reasons (IPGRI 2002). These underutilized crop species are also known as "minor," "promising," "orphan," or "little-used" crops (Thakur 2014). Several unexplored/less explored plant groups come under the neglected and underutilized crops, which include cereals, legumes, oilseeds, vegetables, roots and tubers, fruits and nuts, and aromatic and medicinal plants (Barman et al. 2017; Chandra et al. 2020). Therefore, NUCs have huge potential to become nourishing food sources and at the same time provide relief to the pressing food scarcity problems in low-income groups and developing countries. NUCs are contributing to our dietary health, because these crops provide important macronutrients (carbohydrates, proteins, and fats), micronutrients (vitamins and minerals), and bioactive non-nutrient (Toledo and Burlingame 2006; Blasbalg et al. 2011; Fanzo et al. 2013; Dulloo et al. 2014; Romanelli et al. 2015; Bailey 2016). The NUCs are wild crop relatives and are an important resource of germplasm, as they contain useful traits for stress tolerance (abiotic and biotic) and higher nutritional values (Castañeda-Álvarez et al. 2016).

Several NUCs are abiotic (drought, heat, and frost), biotic (pests and diseases) stress-tolerant, and also adapted to semi-arid and arid environmental conditions (Padulosi et al. 2011; Mayes et al. 2012; Chivenge et al. 2015; Chimonyo et al. 2016; Mabhaudhi et al. 2017; Hadebe et al. 2017; Chibarabada et al. 2017). Since the NUCs are nutritionally rich food sources, utilization of these crops can potentially resolve the global food scarcity and poverty problems. However, these NUCs have not been exploited at the large scale, because of the limited information available on these crops. In this chapter, we are providing various omics-based high-throughput tools that can be used for the in-depth understanding of NUCs, which in turn, will facilitate the large-scale utilization of these crops for fulfilling the dietary and nutritional requirements of humans and livestock in the future.

10.2 Technical Advancement in Transcriptomics to Study NUCs

Advancements in the omics-based technologies have led to the identification of several thousands of transcripts, proteins, and metabolites, facilitating our understanding of how an organism responds to various internal and external signals. Since transcription is the first and key regulatory step for gene expression in the cell (Crick 1970), malfunctioning at the level of transcription has a disastrous effect on the life of an organism (Churko et al. 2013). Therefore, researchers have always been interested to study and analyze the global changes in the gene expression by identification of corresponding RNAs. All the RNA molecules expressed from the genetic material of an organism are collectively constitute the transcriptome. Transcriptomics has worked as a tool to unravel the information and functionality of the genome as gene expression dynamics, spatio-temporal pattern, and knowing the functions of cell types. The transcriptome has a high degree of complexity and encompasses multiple types of coding and non-coding RNA molecules (Mattick and Makunin 2006; Wang et al. 2009).

Due to emergence of low-cost sequencing technologies, the field of transcriptomics has revolutionized and diversified with time, and a great rise in the application of transcriptomics became available after the completion of genome sequencing projects of various model and non-model plants (Table 10.1) (Ekblom and Galindo 2011). These projects have generated a huge amount of sequenced data and information about the genome of these organisms. Transcriptomic analysis has also supported the downstream research in post-transcriptional and proteomic research (Manzoni et al. 2018). Parallel to these applications, the transcriptomics has also exposed the structural and functional studies on non-coding RNAs playing a crucial role in gene regulation. The hybridization-based microarray technique offers relatively low cost and high-throughput output and became instrumental for the first transcriptomic study (Schena et al. 1995). Transcriptomic studies can be carried out by various techniques, such as microarray, EST, SAGE, CAGE, RNA-Seq/NGS, and *de novo* assembly. Apart from well-established and sequenced model plant species like *Arabidopsis* (2000), Rice (2002), Papaya (2008), and Maize (2009); these techniques have played a vital role in the studies on NUCs, such as *Petunia axillaris*, *P. inflata*, *Olea europaea*, *Camellia sinensis*, *Vitis vinifera*, *Solanum lycopersicoides*, *Prunus triloba*, *Beta vulgaris*, *Lens esculenta*, *Saccharum officinarum*, *Cannabis sativa*, and *Medicago truncatula* (Zenoni et al. 2011; Alagna et al. 2009; Li et al. 2017a, b; Chen et al. 2015; Liu et al. 2017; Moliterni et al. 2015; Singh et al. 2017; Vital et al. 2017; Van Bakel et al. 2011; Trindade et al. 2010).

10.2.1 Hybridization-Based Technology

10.2.1.1 DNA Microarray/DNA Chips

A microarray or DNA chip is a molecular technique that allows automated side-by-side analysis of multiple DNA sequences. It comprises of hundreds or thousands of DNA fragments (called probes/reporter) arrayed on small glass slides. Affymetrix was the first company in California, USA in the early 1990s and is based on the hybridization principle. The microarray glass slide contains thousands of labeled spots with single-strand DNA/RNA molecules labeled with a fluorescent dye. When another solution containing ssDNA or RNA fragment (called target) are poured on the chip, hybridization between probe and target fragments occurs that gives florescence. This fluorescence is scanned and analyzed to generate the data for further uses. At present, two main types of microarray chips are available, short oligonucleotide and cDNA based microarray. DNA chips may be pre-fabricated or synthesized directly onto the slide surface by modification through the phosphoramidite method. The reverse approach has also been used to develop a DNA microarray, where microarray targets instead of probes, arrayed on a glass slide and uses a labeled specific gene as a probe. The technique has been named as a metagenome microarray and is widely used in the rapid screening of metagenomic libraries (Roh et al. 2010; Missoum 2018; Miller and Tang 2009). Wide applications of DNA microarray include studying gene expression profiling, comparative genomic hybridization, SNP detection, and alternative splicing detection (Govindarajan et al. 2012) and besides, it is widely employed for genomic studies of model organisms, such as *Arabidopsis thaliana*, Rice (*Oryza sativa*), *Brassica napus* (Arabidopsis Genome

Table 10.1 List of NUCs sequenced using NGS technology

S. No	Plant common name (botanical name)	Reference
1	Commerson's nightshade (*Solanum commersonii*)	Aversano et al. (2015)
2	Eggplant (*Solanum melongena*)	Hirakawa et al. (2014)
3	Hops (*Humulus lupulus*)	Natsume et al. (2014)
4	False Flax (*Camelina sativa*)	Kagale et al. (2014)
5	Cultivated cotton (*Gossypium hirsutum* TM-1)	Li et al. (2014)
6	Horseweed (*Conyza canadensis*)	Peng et al. (2014)
7	American Cranberry (*Vaccinium macrocarpon*)	Polashock et al. (2014)
8	*Jujube* (*Ziziphus jujuba*)	Liu et al. (2014)
9	African Rice (*Oryza glaberrima*)	Wang et al. (2014b)
10	Greater duckweed (*Spirodela polyrhiza*)	Wang et al. (2014c)
11	Amborella (*Amborella trichopoda*)	Albert et al. (2013)
12	Lupin (*Lupinus angustifolius*)	Yang et al. (2013)
13	Wild rice (*Oryza brachyantha*)	Chen et al. (2013)
14	Pear (*Pyrus bretschneideri*)	Wu et al. (2013)
15	Sacred lotus (*Nelumbo nucifera*)	Ming et al. (2013)
16	Corkscrew plant (*Genlisea aurea*)	Leushkin et al. (2013)
17	Kiwifruit (*Actinidia chinensis*)	Huang et al. (2013)
18	Dwarf birch (*Betula nana*)	Wang et al. (2013)
19	Water melon (*Citrullus lanatus*)	Guo et al. (2013)
20	Einkorn wheat (*Triticum urartu*)	Ling et al. (2013a, b)
21	Sugar beet (*Beta vulgaris*)	Dohm et al. (2014)
22	Cotton (*Gossypium raimondii*)	Wang et al. (2012a)
23	Flax (*Linum usitatissimum*)	Wang et al. (2012b)
24	Chinese Plum/Mei (*Prunus mume*)	Zhang et al. (2012)
25	Glaucophyte (*Cyanophora paradoxa*)	Price et al. (2012)
26	Cannabis (*Cannabis sativa*)	Van Bakel et al. (2011)
27	Woodland strawberry (*Fragaria vesca*)	Shulaev et al. (2011)
28	Date Palm (*Phoenix dactylifera*)	Al-Dous et al. (2011)
29	Prince-of-wales feather (*Amaranthus hypochondriacus*)	Lightfoot et al. (2017)
30	Para rubber tree (*Hevea brasiliensis*)	Pootakham et al. (2017)
31	Barley (*Hordeum vulgare* L.)	Mascher et al. (2017)
32	Lettuce (*Lactuca sativa*)	Reyes-Chin-Wo et al. (2017)
33	Sugarcane (*Saccharum spontaneum* L.)	Zhang et al. (2018)
34	Cassava (*Manihot esculenta*)	Bredeson et al. (2016)
35	Rose (*Rosa chinensis*)	Raymond et al. (2018)
35	Loblolly pine (*Pinus taeda*)	Wegrzyn et al. (2014)
37	Cavendish Banana (*Musa acuminate*)	Martin et al. (2016)
38	Japanese climbing fern (*Lygodium japonicum*)	Aya et al. (2015)
39	Bracken fern (*Pteridium aquilinum*)	Der et al. (2011)
40	*Ceratopteris* spp.	Bushart et al. (2013)
41	Physcometrella moss (*Physcomitrella patens*)	Rensing et al. (2008)
42	Dumortier's liverwort (*Dumortiera hirsute*)	Singh et al. (2015)
43	Green algae (*Bathycoccus prasinos* BBAN7)	Moreau et al. (2012)
44	Red Algae (*Chondrus crispus*)	Collén et al. (2013)

Initiative 2000; Van Zhong and Burns 2003; Kreps et al. 2002; Li et al. 2005) as well as various NUCs. The use of cross-oligonucleotide microarrays to profile the response of the banana (*Musa* spp.) leaf transcriptome to drought stress using a genomic DNA (gDNA)-based probe-selection strategy to improve the efficiency of detection of differentially expressed *Musa* transcripts showed promising results (Davey et al. 2009). Although microarray-based RNA-profiling is the best tool to study complex and diverse responses in plant systems during biotic and abiotic stress, the limited availability of commercial microarrays for few plant species with great input cost makes it less applicative in plant research. Microarray analysis based studies have been carried out to identify the differentially expressed genes engaged in fruit development between table and wine grape (Wang et al. 2014a), *Prunus mume* and *Prunus armeniaca* (Li et al. 2012), using a high-density oligomicroarray under drought stress in various genotypes of Cassava (Utsumi et al. 2012), Sunflower (*Helianthus annuus* L.) (Fernandez et al. 2012) and the roles played by abscisic acid and auxins in the development and ripening of strawberry fruits (Medina-Puche et al. 2016).

Moreover, the technique output is not easily accessible and disruptive as it requires the physical disruption of a cell to obtain genetic material for study. The quality and quantity of data obtained from organisms pose a challenge to determine correct results, as false microarray data can be generated from degraded mRNAs (Russo et al. 2003), sensitive to cross-hybridization of probes to non-target sequences due to the similarity of sequences possess high background noise making a great challenge to determine correct results. It has been tried to overcome by the use of multiple probes for each target. Probes specificity in DNA microarray is affected by its lengths and found that as length increases specificity decreases (Wren 2002; Loy and Bodrossy 2006; Jayaraman et al. 2006). Apart from these limitations, microarray techniques show a limited dynamic range of detection owing to both background and saturation signals and comparing expression levels across different experiments and require normalization methods and a priori knowledge of the sequences being interrogated (Casneuf et al. 2007; Shendure and Ji 2008).

10.2.2 Sequencing-Based Technologies

10.2.2.1 Expressed Sequence Tags

Following microarray, sequence-based technologies came into existence for the better elaboration of transcriptome by directly determining the transcript sequence. For example, the generation of expressed sequence tags (EST) libraries by Sanger sequencing of complementary DNA (cDNA) for gene expression studies. ESTs are the fragments of mRNA sequences derived through single sequencing reactions performed on randomly selected clones from cDNA libraries and also uses plasmid vectors. The cDNA libraries are constructed from mRNAs isolated from specific tissues in such a way that the 5′ and 3′ ends of the message can be identified by sequencing one or both vector-insert joins of the plasmid. In contrast to full-length cDNA sequences, ESTs typically give only partial coverage of the coding sequence and 3′ untranslated regions of the genes. EST was pioneered by Craig Venter and his co-workers (Adams et al. 1991). EST has been used extensively for the comparative transcriptomic analysis of NUCs including comparative transcriptomic analysis of male and female flowers of monoecious *Quercus suber* (Rocheta et al. 2014), prediction and analysis of GH14 family β-amylases in oat (*Avena sativa* L.) seedling extract (Ben Halima 2018), characterization of a transcriptome from *Cladonia rangiferina*, the grey reindeer lichen, using high-throughput next-generation sequencing and EST sequence data (Junttila and Rudd 2012), generation, functional annotation and comparative analysis of black spruce (*Picea mariana*) ESTs (Mann et al. 2013).

EST generation provided a quick and easy platform to discover the new gene, which is important for functional genomics and to help understand the molecular mechanism for sustainable agriculture production (Afzal et al. 2020), however, detection of only more abundant transcripts, relatively low throughput and unable to quantify transcripts are some of the drawbacks of this technique (Adams et al. 1991). dbEST of NCBI is the specialized division of GenBank, containing all publically available EST collections from different organisms (Boguski et al. 1993).

10.2.3 Serial Analysis of Gene Expression (SAGE) and Cap Analysis Gene Expression (CAGE)

To overcome issues posed by EST, tag-based methods, such as serial analysis of gene expression (SAGE) and cap analysis of gene expression (CAGE) were developed. SAGE allows quantitative estimation of mRNA expression levels by measuring short (10–14 mer) sequences of transcribed messages, and using these to infer the identity of specific transcripts. It is used to detect and quantify the gene activities in the desired cells and provides clues to discovering novel genes. CAGE, on the other hand, collects 21 bp tags from the 5′ end of cDNAs to determine the 5′ end of transcripts and the location of the promoter (Kim and Wang 2009). These two techniques give better results, but still not free from challenges, such as splice isoforms levels showed assay insensitivity, unable to discover new gene, a large amount of RNA required for the experiment, laborious and high cost of sequencing. Furthermore, gene predictions and use of EST evidence led to the enhancement of knowledge of transcriptomes, but these were partial and biased. These limitations were overcome by NGS/RNA-Seq technology that revealed a larger and complex landscape of transcriptomes with greater sensitivity and accuracy (Martin and Wang 2011). The Array Express Archive and the Gene Expression Omnibus (GEO) are two main public collections of functional genomics data currently available and helpful in transcriptomic studies (Ambrosino et al. 2020). SAGE has been successfully applied to multiple NUCs including *Musa acuminata* (Coemans et al. 2005) and lentil (*Lens culinaris* Medik.) (Barrios et al. 2017). Similarly, CAGE has also been applied for the genome wide identification of transcription start site in maize.

10.2.4 Next-Generation Sequencing (NGS)/RNA Sequencing (RNA-Seq)

High-throughput next-generation sequencing (NGS), also known as RNA sequencing (RNA-Seq), is an advanced technology used to sequence the RNA of a biological sample to determine the relative abundance of each transcript with great sequencing depth (100–1000 reads per base pair of a transcript) (Wang et al. 2009). RNA-Seq has contributed in overcoming the problems faced by previous technologies to study transcriptome and revealed many molecular functions at the single-cell level, the identification of novel cell types, genes and the study of global patterns of stochastic gene expression (Stegle et al. 2015), presented a high-resolution view of the global transcriptional landscape, alternative splicing, and allele-specific expression. NGS also contributes to scale-up transcriptome sequencing, de novo assembly, and sequence-based marker resources for crop development that is less costly than phenotyping (Unamba et al. 2015). Comparative to microarray technology, the RNA-Seq technique gives good base-pair-level resolution, much higher dynamic range of expression levels, and also capable of de novo annotations (Wang et al. 2009). However, RNA-Seq analysis presents computational challenges not encountered in other sequencing-based analyses and requires specific consideration to the biases inherent in expression data. The key experimental limiting factors in the detection of transcripts in single cells are cDNA synthesis and PCR amplification. A human cell contains <1 pg of mRNA (Kawasaki 2004), whereas most sequencing protocols, such as Illumina's TruSeq RNA-Seq kit recommends 400 ng to 1 μg of input RNA material. Therefore, the efficiency of RNA-to-cDNA conversion is imperfect and estimated to be as low as 5–25% of all transcripts (Islam et al. 2012). In case of model organisms, the microarray is still a good choice, but high-quality annotation and appropriate arrays do not exist especially in NUCs, in these plants sequencing technology offers the best method to understand the transcriptomes (Islam et al. 2011), making possible to catalogue and measure gene expression from a wider range of organisms. NGS-based whole genome sequencing approaches were essential to reduce costs and time (Schwarze et al. 2018). Successful genome sequencing has been accomplished for several NUCs like chickpea (Varshney et al. 2013a, b, c), pigeon pea (Varshney et al. 2012), *Trefoil* spp. (Sato et al. 2008), soybean (Schmutz et al. 2010), *Medicago* (Young et al. 2011), *Streptocarpus rexii* (Chiara et al. 2013), *Hevea brasiliensis* (Xia et al. 2011; Varshney et al. 2009). Multiple transcriptomics analysis of NUCs has given great results than traditional studies, such as annotation, taxonomy, and phylogeny (Pai et al. 2018), *de novo* characterization of the gametophyte transcriptome in bracken fern, *Pteridium aquilinum* (Der et al. 2011), integrated RNA-seq and sRNA-seq analysis identifies chilling and freezing responsive key molecular players and pathways in tea plant (*Camellia sinensis*) (Zheng et al. 2015), RNA sequencing analysis helped in the identification of floral scent biosynthesis-related genes in *Cymbidium goeringii* (Ramya et al. 2019) and metagenomic data reveal diverse fungal and algal communities associated with the lichen symbiosis (Smith et al. 2020).

Mapping RNA-Seq reads to the genome is considerably more challenging than mapping DNA sequencing reads, because many reads map across splice junctions, but its wider uses have led to several splicing-aware mapping tools, such as GSNAP, MapSplice, RUM, STAR, and TopHat) have been developed for mapping transcriptome data. The NCBI Sequence Read Archive (SRA) is the reference resource public database obtains global NGS collections including data from Illumina Genome Analyzer, Applied Biosystems SOLiD System, Roche 454 GS System, and Helicos Heliscope (Ambrosino et al. 2020).

10.2.5 Transcript Assembly

The study of transcriptome requires the transcript assembly of data obtained from NGS. The transcript assembly could be of three types: (a) the genome-guided/ reference-based/ "ab-initio" assembly, (b) de novo methods based on the requirement of a reference genome or not, and (c) combined strategy.

10.2.5.1 Reference-Based Assembly Method

The reference-based assembly method involves three steps. First, RNA-seq reads are aligned to a reference genome using a splice-aware aligner, such as Blat, TopHat, SpliceMap, MapSplice, or GSNAP. Second, overlapping reads from each locus are clustered to build a graph representing all possible isoforms. Third, this step involves traversing the graph to resolve individual isoforms and examples are Cufflinks and Scripture. The pros of this technique include assembling of transcripts of low abundance, transformation a large assembly problem (millions of reads) into many smaller assembly problems, using parallel computing with lesser RAM, and no contamination problem with great sensitivity. The technique is unable to perform in the plants, where reference genome is unavailable and cannot easily assemble *trans*-spliced genes (Martin and Wang 2011).

10.2.5.2 De Novo Method

The "de novo" transcriptome assembly uses the redundancy of short-read sequencing to find overlaps between the reads and assembles them into transcripts and does not use a reference genome. This strategy is followed by The Rnnotator, Multiplek, Trans-ABySS, Trinity, and Oases using a De Bruijn graph-based approach (Martin and Wang 2011; Pevzner et al. 2001). Because it is independent from the reference genome, it is useful in those organisms where a high-quality finished genome is not available. In addition, it does not depend on the correct alignment of reads to known splice sites (Burset et al. 2000) or the prediction of novel splicing sites and trans-spliced transcripts and similar transcripts originating from chromosomal rearrangements can be assembled. The limitations of de novo assembler are high sensitivity to sequencing errors and the presence of chimeric molecules in the data set which usually requires more than 30× coverage for the same task in comparison to reference-based strategy and need much higher sequencing depth for full-length transcript assembly (Martin and Wang 2011).

10.2.5.3 Combined Strategy

The above two strategies can be amalgamated and called a combined strategy that helps to create a more comprehensive transcriptome. By doing this, the advantage of the sensitivity of reference-based assemblers and the ability of de novo assemblers to detect novel and trans-spliced transcripts are used together. The combined assembly strategy can be carried out either by align-then-assemble or by assemble-then-align strategy (Martin and Wang 2011; Haas and Zody 2010). These assemblies are dependent upon algorithms and implementation purposes. PCR amplification, sequencing techniques, assembly algorithm, and genome characters also decide the quality of assembled transcripts (Deng et al. 2015).

Transcriptomics is progressing at a rapid pace and is getting complimented with the advancement in high-performance computing (HPC) thereby greatly reducing the time required to assemble a large transcriptome data set. Since more quality reference genomes will be available with time, it would be feasible to understand the transcriptome of various NUCs in near future. Cloud computing and improved sequencing technology and protocol will give better results and the future of transcriptome assembly will be "no assembly required."

10.3 Progress Made in Proteomics Technologies to Understand the NUCs

The proteins are the functional products of the genomes and are directly involved in the regulation of the metabolic reactions; it is this fact that analysis of proteome has gained considerable interest in recent years (Gupta et al. 2015). The genomes are static and fixed, whereas proteomes of an organism are dynamic and changes from time to time with respect to cells and environmental factors posed to organisms (Meng et al. 2019). Therefore, the analysis of the proteomes is way more complex than the analysis of the genomes or transcriptomes (Min et al. 2019a). Successful proteome analysis of an organism depends on the availability of the genome sequence that complements the protein identification. Since the genome of only a limited number of plants have been sequenced, proteome analysis of other plants, including NUCs, are challenging and all the steps starting from protein extraction to database searching and downstream data processing needs special attention and modifications (Kim et al. 2017).

10.3.1 Advancements in Protein Extraction and Sample Clean-Up

The protein extraction is one of the most crucial steps in the proteome analysis. Various protein extraction methods have been developed and optimized for the isolation of a high amount of proteins from the neglected or underutilized crops. Phenol (Hurkman and Tanaka 1986) and TCA-acetone (Damerval et al. 1986) methods are the most commonly used methods for the precipitation of isolated proteins from the plants, however, the efficacy of these methods varies among different tissues and different plants (Wu et al. 2014). In order to make a universal protein precipitation method from the plant tissues, Wang et al. (2003) combined both TCA-acetone and phenol methods to make a TCA/acetone/phenol method and the efficacy of this method was tested in the olive leaves (Wang et al. 2003), and many other highly recalcitrant plants including cotton (Xie et al. 2009). Moreover, a relatively less-popular chloroform-methanol precipitation method has been found to be more effective for the protein isolation from the plant tissues rich in lipids, such as seeds (Wessel and Flügge 1984).

Plants contain a number of interfering compounds that if not removed, severely affect can interfere with the protein resolution of SDS-PAGE and 2-D gels and protein identification by mass spectrometry (Ghatak et al. 2017). These compounds majorly include secondary metabolites, lipids, polysaccharides, nucleic acid, and organic acids. Medicinal plants, in particular, are highly rich in secondary metabolites including alkaloids and different types of phenolics, such as tannins, lignin, coumarin, and flavonoids, among others, that co-precipitates with the proteins (Kim et al. 2017). Therefore, while isolating the proteins from different plant tissues, some additional components are included in the protein extraction buffer that differentially precipitates the proteins and these interfering compounds. Among different components, polyvinylpolypyrrolidone (PVPP) and its water-soluble form polyvinylpyrrolidone (PVP) are most commonly included in the extraction buffers to remove the phenolic compounds (Gupta and Deswal 2012). In addition, reducing agents, such as DTT or β-ME, sodium ascorbate, and thiourea are used to retain high-quality purified proteins (Wang et al. 2008; Wu et al. 2014).

10.3.2 Proteolytic Digestion and Database Searching for Gel-Based and Gel-Free Proteomics Approaches

The isolated proteins can directly be subjected to the trypsin digestion to proceed for the gel-free proteome analysis or these can be resolved on the high-resolution polyacrylamide gels, most commonly on the two-dimensional gel electrophoresis (2DE) to proceed for the gel-based proteome analysis (Abdallah et al. 2012; Min et al. 2017). At present, three major trypsin digestion based methods commonly employed include in-solution trypsin digestion, in-gel trypsin digestion, and on-filter trypsin digestion or filter-assisted sample preparation (FASP) (Wiśniewski et al. 2009). In the case of in-solution trypsin digestion, isolated proteins are directly subjected to proteolytic digestion and are majorly used for the gel-free proteome analysis, while in-gel digestion is carried out after the resolution of proteins on the polyacrylamide gels and individual spots/bands of interest are excised from the gels and are subjected to the proteolytic digestion after destaining (Gupta et al. 2016). However, several components of polyacrylamide gels, in the case of in-gel digestion and extraction buffer hinders the trypsin digestion and results in only partial digestion of the isolated proteins that affect the quality of the downstream data. To overcome this problem, the FASP method was devised, where the proteolytic digestion is being carried out in the filter units and multiple washing steps of this method results in complete removal of interfering substances from the proteins prior to their proteolytic digestion, thereby increasing the overall efficiency of trypsin digestion (Wiśniewski et al. 2009). However, a comparative study showed that the efficacy of the in-gel trypsin digestion method is higher over in-solution trypsin digestion and FASP method, while digesting the plasma membrane-localized proteins (Choksawangkarn et al. 2012). Although the comparative efficacies of these methods have not been tested yet for the plant samples rich in secondary metabolites, the FASP method can be efficiently used for such samples, due to removal of salts and other interfering substances.

Since the genome sequence for the majority of the neglected crops is not available, identification of proteins after the MS run is a major bottleneck. In such cases, either the genome of related plants can be used for the database searching or the sequences from all the green plants, Viridiplantae, can be used (Somayajula and Desai 2019). However, the confidence for the protein identification in these cases is usually low, because of the protein sequence variations in different plants. Therefore, to increase the authenticity of protein identification in the neglected crops, nowadays databases are derived from EST or RNA-Seq data and used for protein identification. Since these EST and RNA-Seq libraries are made from the same plant, the confidence for the protein identification increases with an improved proteome coverage (Kim et al. 2019).

For proteome analysis of neglected crops, gel-based proteomics is preferred over the gel-free techniques, as it is confirmed that all the peptides are derived from a single protein/spot during MS analysis, thereby increasing the confidence of protein identification in the former case (Abdallah et al. 2012). However, the use of even gel-free proteomics is mushrooming these days for the neglected crops, leading to the identification of thousands of proteins. As an example, a label-free quantitative proteome analysis of *Cenchrus polystachion* (L.) Schult led to the identification of 135407 proteins using the Poaceae database (Somayajula and Desai 2019). Similarly, gel-free proteome analysis has also been applied in various other neglected crops, such as ginseng (Kim et al. 2019), and Jerusalem artichoke tubers (Min et al. 2019b), among others for the identification of differential proteins in response to various abiotic and biotic stress conditions.

10.4 Technical Advancements in Metabolomics to Facilitate the Metabolome Profiling of NUCs

The plant kingdom can be looked as Nature's factories for the production of diverse metabolites having been exploited as flavorings, fragrances, insecticides, pigments, therapeutics, and other fine chemicals for ages (Facchini et al. 2012). Approximately, 200,000–1,000,000 metabolites are known to be synthesized by plants (Dixon and Strack 2003; Afendi et al. 2012), and these numbers represent only a small portion of the global repertoire of plant metabolites (Hartmann 2007), because it is estimated that approximately 15,000 metabolites could be synthesized by a single plant (Fernie 2007). These metabolites possess diverse properties with benefits to humans that are manufactured as nutraceuticals, drugs (traditional and modern drugs), and cosmetics (Shahaf et al. 2016). In metabolomics studies, analytical techniques that provide very high analytical accuracy and high detection sensitivity in the combination of separation technology, namely, gas chromatography (GC) or liquid chromatography (LC), with mass spectrometry (MS) technique. Therefore, gas chromatography-mass spectrometry (GC–MS) and liquid chromatography-mass spectrometry (LC-MS) are widely utilized for both targeted and untargeted metabolite profiling (Lei et al. 2011; Jorge et al. 2016).

Mass spectrometry-gas chromatography (GC-MS) is a highly sensitive technique and the most frequently used analytical method in metabolic profiling. The GC-MS is extensively used for non-targeted metabolic analysis (Dutta et al. 2012). Till now, the GC–MS technique has been used for several metabolomic studies of diverse living organisms including plants, animals, and microorganisms (Fiehn et al. 2000; Roessner et al. 2001; Fernie et al. 2004; Shellie et al. 2005; Ralston-Hooper et al. 2008; Kusano et al. 2011). The GC-MS technique has been widely utilized to generate a wide variety of plant metabolite analytics from the pathways of primary metabolisms of sugars, amino acids, fatty acids, sugar alcohols, polyamines, and organic acids (Schauer and Fernie 2006; Choudhary 2018). Besides the above-mentioned examples, the variations in metabolites in genetically modified plants (Roessner et al. 2001; Jacobs et al. 2007) and changes in metabolites due to biotic (Broz et al. 2010) and abiotic stresses (Roessner et al. 2006; Urano et al. 2009; Gogna et al. 2020) have been also analyzed by GC-MS. Various specific metabolites of NUCs have been investigated using the GC-MS platform, for example, fatty acid profiles including allenic unusual fatty acids (e.g. laballenic and phlomic acid) of several *Leucas* species have been analyzed for pharmaceutical, nutritional, and chemotaxonomic significance (Choudhary et al. 2017). Similarly, wax, fatty acid profiles, and vasicine content have been analyzed in two morpho types of *Justicia adhatoda* using GS-MS analyses (Berry et al. 2020). *Justicia adhatoda* is an important source of vasicine, vasicine is known to have many pharmaceutical properties, namely, respiratory stimulant, bronchodilatory, and uterine stimulant (Gupta et al. 1977) and has also shown antimicrobial activity against gram-negative and gram-positive bacterial strains—*Klebsiella pneumoniae, Staphylococcus epidermidis,* and *Staphylococcus aureus* (Duraipandiyan et al. 2015).

Liquid chromatography—Gas chromatography (LC-MS) also has been widely used for both targeted and non-targeted metabolic analyses in plants (Turner et al. 2016). This technique detects high mass metabolites (<1500 Da) including both primary and secondary metabolites (Turner et al. 2016). Generally, LC-MS analysis is utilized for detection of a wide range of metabolites along with secondary metabolites, such as complex lipids (e.g., sphingolipids and glycerolipids), alkaloids, flavonoids, phenylpropanoids, terpenes, glucosinolates, and saponins (Verdonk et al. 2003; Naoumkina et al. 2007; De Vos et al. 2007; Farag et al. 2008; Bottcher et al. 2008; Matsuda et al. 2009: Okazaki et al. 2011). Moreover, the coupling of Quadrupole Time-of-Flight Mass Spectrometry (Q-TOF-MS) with gas and liquid chromatography instruments can provide comprehensive and rapid detection of metabolites (Allen and McWhinney 2019). Interestingly, Q-TOF instruments can detect masses within 5 ppm that has been used for the verification of the elemental composition (Ferrer and Thurman 2003). In addition, Q-TOF-MS has been used in pharmaceutical and biological research for peptide analysis, metabolites identification, and drug discoveries (Poon and Johnson 2001; Xie et al. 2012; Ranasinghe et al. 2012). Moreover, the ultra-performance liquid chromatography (UPLC) combined with QTOF-MS identified several metabolites rapidly by enhancing their peak resolution and mass accuracy (Chawla and Ranjan 2016).

The Medicinal Plant Metabolomics Resource (http://www.medicinalplantmetabolimics.org/) is one of the earliest and most comprehensive databases, which represent the specialized metabolites of 14 medicinal plants. The NIST, EPA, and NIH mass spectral databases are the largest databases and provide mass spectra that have been used for the identification of both primary and secondary metabolites (Stein 1999). Recently, several plant resource databases have been constructed and made available as a public domain, for example, PlantDB (Exner et al. 2008), PRIMe (Akiyama et al. 2008), Metabolome Express (Carroll et al. 2010), At MetExpress (Matsuda et al. 2010), MassBank (Horai et al. 2010), the KNApSACK database (Afendi et al. 2012), MeRy-B (Deborde and Jacob 2014), the Plant Metabolite Network (Dreher 2014), MeKO (Fukushima et al. 2014), and Medicinal Plant Metabolomics Resources (MPMR) (http://metnetweb.gdcb.iastate.edu/mpmr_public/). Besides, some metabolite databases have also been constructed for plants belonging to specific geographical regions to represent the structural diversity of metabolites, for instance, AfroDB contains information for more than 1000 natural products of African plants (Ntie-Kang et al. 2013). Similarly, the TCM Database@Taiwan is a traditional Chinese medicines database and provides information for more than 20,000 metabolites (Chen 2011). To date, several metabolic networks are constructed for various model plants, such as Arabidopsis, rice, corn, barley, oilseed rape, and *Chlamydomonas* (Seaver et al. 2012; Collakova et al. 2012). The Plant Metabolome Database (PMDB) is one of the largest secondary metabolites databases of plants and is a freely available database (Udayakumar et al. 2012). Some plant-specific databases, namely, Metabolome Tomato Database (MoTo DB) (Grennan 2009) and Plant Metabolomics (PM) database for Arabidopsis (Bais et al. 2012) are also available now. Several non-model plants have been found to produce significant metabolites, for example, paclitaxel has anti-cancer properties (which is extracted from the bark of *Taxus brevifolia*), artemisinin has been used in the treatment of malarial (which is isolated from *Artemisia annua*), and vincristine and vineblastine have been used in the treatment of leukemia and lymphoma (extracted from *Catharanthus roseus*), and morphine has been used in severe pain (isolated from *Papaver somniferum*) (Rai et al. 2017). Therefore, the screening of metabolites and construction of the metabolite library of non-model plants, undoubtedly emerge as an essential pre-requisite for the identification of novel metabolites to be utilized for human welfare.

10.5 Softwares and Other Bioinformatics Platforms to Facilitate the Studies on NUCs

Plant scientists have seen a spectacular boost in the use of high-throughput omics platforms, driven by the latest technological advancements. The growing accessibility of multi-omic platforms provides new tasks to data assessment. There are many tools and techniques to investigate the metabolite and transcriptome profiling data.

10.5.1 Softwares for the Functional Annotation and Integration of Omics Data

10.5.1.1 Pathview Web: User-Friendly Pathway Visualization and Data Integration

In omics, pathway analysis is an extensively studied phenomenon in which data integration and visualization is a crucial part of the assessment. Relatively, a newly developed R package based to address the need for data integration and visualization called Pathview developed by Luo and Brouwer. However, Pathview Maps integrates and renders a broad variety of biological data on relevant pathway graphs. Pathview Web delivers simple communicating access and creates high quality, hyperlinked graphs. Without any special computer skills or resources, a scientist from all over the world can access and get a comprehensive workflow for both regular and integrated pathway analysis of multiple omics data. This Pathview Web is open and can be accessed without any charges on the following links: https://pathview.uncc.edu/ (Luo et al. 2017). Transcriptomic data analysis for *Zea mays* and *A. flavus* was performed using KEGG Pathview package R by Musungu et al. (2020). Similarly, mapping of abundant proteins identified in susceptible cultivar of *Solanum lycopersicum* to TSWV infection (Gupta et al. 2020).

10.5.1.2 PaintOmics 3: A Web Resource for the Pathway Analysis and Visualization of Multi-Omics Data

PaintOmics 3 is a novel approach for data integration and visualization of various kinds of omics studies. Interestingly, on a single click, it gives you a comprehensive assessment of multi-omics data types onto the Kyoto Encyclopedia of Genes and Genomes (KEGG) pathway diagrams. It provides data analysis with the capability of modern web resources for data visualization, offering researchers a robust outline for interactive evaluation of their multi-omics report. Although, PaintOmics 3 offers a powerful exploratory tool of complex data and facilitates a visual assessment of biological activities and provides a helpful method of understanding functional relationships across different molecular layers. In contrast, it covers a comprehensive pathway assessment road map, containing automatic feature name/identifier conversion, multi-layered feature matching, pathway enrichment, network analysis, interactive heatmaps, trend charts, and further. This is also free of cost and available at www.paintomics.org (Hernández-de-Diego et al. 2018). Identification of deregulated metabolic pathway in Arabidopsis was performed using paintomics and then mapping of differentially regulated genes and metabolites onto the KEGG pathway was done by using Paintomics 3 as well (Hernández-de-Diego et al. 2018).

10.5.1.3 KaPPA-View: Integrating Transcript and Metabolite Data

In 2005, Tokimatsu developed a web-based KaPPA-View tool which helps to provide quantitative data for individual transcripts and/or metabolites on plant metabolic pathway maps (Tokimatsu et al. 2005). The KaPPA-View tool is complementary to other metabolic pathway tools and databases offer a direction to retrieve all the information contained on the map that the user is seeing. However, the visualization tool of KaPPA-View is easy to use, spontaneous, and gives a comprehensive outcome of complex results of transcription analyses (Tokimatsu et al. 2005). It was used by Zhao et al. (2010), in Arabidopsis for transcriptomic profiling of the major carbohydrates and all the related pathways in the roots when treated with rhizotoxin ions. It was also used in *I. lacunosa* for mapping metabolic pathway (Maroli et al. 2018).

10.5.1.4 MapMan: A User-Driven Tool to Functionally Annotate the Identified Genes/Proteins in Various Pathways

In 2004, Thimm and his co-worker developed the MapMan tool especially for plants to facilitate the visualization of omics data on plant pathways (Thimm et al. 2004). MAPMAN is a user-driven tool that exhibits huge genomics data sets onto diagrams of metabolic pathways or other biological processes (Thimm et al. 2004; Peltier et al. 2004). With the help of the data visualization and comparison of multiple omics data from plants, the MapMan tool provides help to decide the biologically appropriate data or give a prompt overview of various processes in large datasets. Firstly, it was only developed for *Arabidopsis thaliana,* but later it was extended and applied on several other plant species by similarity transfer and manual curation (Ling et al. 2013a, b). MapMan has been applied to other plants, such as maize (*Zea mays*) (Xing et al. 1996; Peltier et al. 2004), *Brassica napus*, *Glycine max*, *Hordeum vulgare*, *Medicago truncatula*, *Nicotiana tabacum*, *Oryza sativa* spp. japonica, *Populus trichocarpa*, and *Triticum aestivum* (Jung and An 2012). Due to some limitations, MapMan4 was introduced with a revised version of the associated online Mercator annotation tool, which allows a plant genome to be annotated in few minutes (Schwacke et al. 2019). In comparison to the original MapMan, the advanced MapMan4 tool has been upgraded with Mercator4, which contains the updated knowledge across the land plants to provide protein annotations for the entire embryophytes through comparably high quality. MapMan4 was used to investigate the deduced proteomes of two *Cuscuta* species (Schwacke et al. 2019).

10.5.2 Subcellular Localization

10.5.2.1 CELLO: Subcellular Localization Predictive System

CELLO, predicts the protein's location in subcellular compartments, mainly in the cytoplasm, the periplasm, the inner and outer membrane, and the extracellular spaces. It uses a single module (SVM) to predict the subcellular localization based on multiple *n*-peptide composition. The data set used is of Gardy et al. (2003), extracted from SWISS-PROT release 40.29, consisting of 1443 protein sequences. The result accuracy of CELLO is about 89%, with cytoplasmic location prediction accuracy of 91% and that of periplasmic location is 87% (Yu et al. 2004). An updated version of CELLO is also available, CELLO II, which combines the two-level SVM classifier and the homology search method, as a general tool for the sequence annotation of subcellular localization. The cytoplasmic localization prediction accuracy is 94.6% and that of periplasmic location is 92.8% (Yu et al. 2006). It has been applied to tomato (Cao and Li 2015), wheat (Tounsi et al. 2019), *Brassica rapa*, *Glycine max*, *Oryza sativa*, *Zea mays*, and *Gossypium raimondii* (Vatansever et al. 2016).

10.5.2.2 PSORTb (http://www.psort.org)

It is an update of PSORT, which searches a given protein for amino acid composition, similarity to proteins of known localization, presence of a signal peptide, transmembrane α-helices and motifs corresponding to specific localizations. The results are obtained for 5 possible locations of the proteins in gram-negative bacteria, namely cytoplasm, inner and outer membrane, periplasm, and extracellular space. The query sequence is analyzed in six analytical modules and the result is obtained as the cumulative result of each module, by using Bayesian Network. In its multi-component approach, it uses SCI-BLAST for analyzing the homology, a PROSITE motif-based analysis for detection of localization-specific motifs, HMMTOP for detection of transmembrane alpha-helices, a novel outer membrane protein motif analysis, a type II secretion signal peptide predictor, and a variation of SubLoc. The major advantage of this system is the high precision (97%) and the ability to send the result as "unknown" if an assured prediction is not available (Gardy et al. 2003).

10.5.2.3 TMHMM Server

It is a membrane protein topology prediction method created using the Markov model. The major advantage of this method is that it is able to distinguish between non-membrane and membrane proteins based on standards such as the number of predicted transmembrane helices, the expected number of residues in transmembrane helices along with the expected number of transmembrane helices. The specificity and sensitivity of the server to distinguish between a soluble and membrane protein is about 99% but is lowered in the presence of signal peptide (http://www.cbs.dtu.dk/services/TMHMM/) (Krogh et al. 2001). The server has been used for multiple plants including *Brassica rapa*, *Glycine max*, *Oryza sativa*, *Zea mays*, and *Gossypium raimondii* (Vatansever et al. 2016), *Sesamum indicum* (Dehghan Nayeri and Yarizade 2014).

10.5.2.4 SignalP 3.0 (http://www.cbs.dtu.dk/services/SignalP/)

It is a method of predicting classically secreted proteins that work on the predictions based on neural network and hidden Markov model. This update shows improved accuracy (6–17%) of the cleavage site predictions from gram-negative to gram-positive bacteria to eukaryotes in comparison to the previous version. The dataset contains 1192, 334, and 153 sequences for eukaryotes, Gram-negative and Gram-positive bacteria, respectively. Moreover, in this prediction method, a new discrimination score for the neural network is used, which shows superiority in terms of discriminative performance (Bendtsen et al. 2004). Updated versions of SignalP are now also available, such as SignalP 4.0 and 4.1. SignalP 4.0 solely works on the neural network-based method and can distinguish between the signal peptide and transmembrane regions (Petersen et al. 2011). It is used in *Nelumbo nucifera* (Gaertn.) (Lum et al. 2013) and *Oryza sativa* (Fan et al. 2011).

10.5.2.5 LOCtree

It is used to predict the protein location in the subcellular compartment by mirroring the mechanism of cellular sorting and using the structural features in the database. This system combines the support vector machine and other prediction systems. It uses an explicit ontology of subcellular localization and the nodes of the tree separate the proteins belonging to different cellular sorting pathways. This helps in achieving more than 74% of accuracy in differentiating the proteins belonging to the five compartments. Another advantage of the hierarchical tree system is that it is more technical in nature and allows the addition of more fine-grained modules at later stages (for example, the split of nuclear into the nuclear matrix and other) (Nair and Rost 2005). LOCtree has been used for Arabidopsis (Ahmad et al. 2019), *Lactuca sativa* Linn. (Yazdi et al. 2019), prunus (Li et al. 2018).

10.5.2.6 PredictProtein (http://www.predictprotein.org)

PredictProtein can be used to predict secondary structure, solvent accessibility, transmembrane helices, globular regions, coiled-coil regions, structural switch regions, B-values, disorder regions, intra-residue contacts, protein–protein and protein–DNA binding sites, subcellular localization, domain boundaries, beta-barrels, cysteine bonds, metal-binding sites, and disulfide bridges. The incorporated databases in the server include UniRef for sequence homology searches by PSI-BLAST, PDB, Pfam, and PROSITE for patterns and profiles. It has been useful in the prediction of proteins in *Lolium rigidum* and *Ocimum basilicum* (Fischer et al. 2001), *Phaseolus vulgaris* (Bashir et al. 2017), Peanuts (Kandula et al. 2014), peach (Guan et al. 2019).

10.5.3 Tools for the Detection of Post-translational Modifications

10.5.3.1 PhosPhAt: A Database of Phosphorylation Sites in *Arabidopsis thaliana* and a Plant-Specific Phosphorylation Site Predictor

PhosPhAt database is an online web tool that can be used for the prediction of phosphorylation sites, however, it holds some information on kinases and their phosphorylation objectives. In 2007, it was originally introduced to predict phosphorylation sites and gathered the available data on phosphorylation sites of each protein detected by mass spectrometry-based in *Arabidopsis thaliana* plant

(Heazlewood et al. 2007). In terms of performance compared with other available prediction tools, although it seemed better than other for serine/threonine predictions, at least in the model *A. thaliana*. It provides the outcomes which include interactive visualization of annotated fragmentation spectra and the capability to export spectra and peptide sequences as text files for use in other applications. With the help of PhosPhAt, phosphorylation sites in each protein can be displayed in tabular form with comprehensive data on each identified phosphopeptide. This database provides a big help to the scientific community to understand the phosphorylation proteins and can be available online without any charges available at the following link http://phosphat.mpimpgolm.mpg.de. Moreover, this tool does not only provide a knowledge based on experimentally identified phosphorylation sites but also gives an effective source for the characterization and annotation of yet unknown phosphoserine locations in model plant *Arabidopsis thaliana* (Heazlewood et al. 2007; Durek et al. 2010).

10.5.3.2 Prediction of Glycosylation Sites

Glycosylation is a main kind of PTM through which a carbohydrate (i.e., a glycosyl donor) is connected to specific functional groups on target proteins (i.e., glycosyl acceptors) and ultimately affects the function of protein via the addition of glycans (Yu et al. 2013). The two most important protein glycosylation forms *N- and O-glycosylation* forms are present in plants (Strasser 2012). There are many tools available that help in detecting the glycans from MS platforms, such as MALDI-TOF and subsequently annotation is carried (Yu et al. 2013).

10.5.3.2.1 Cartoonist

This is an automated tool introduced in 2005 by Goldberg for N-glycan profiling based on MALDI-TOF mass spectra (Goldberg et al. 2005). Furthermore, the Cartoonist tool has some interesting characteristic features like computes a confidence score for each assignment a glycan peak according to match mass and uses a library of biosynthetically plausible cartoons that contain about 2800 cartoons derived from 300 archetypes. The cartoonist tool gives promise N-glycan to a peak based on mass and uses the top 15 intense peaks to calibrate the result (Goldberg et al. 2005).

10.5.3.2.2 GlycoWorkbench

This program was built with the aim of glycan profiling and interpretation, which can support numerous data formats and gives simple spectrum processing tools. It is a very user-friendly tool to annotate the mass spectrum after draw a glycan structure. Moreover, it deals with both MS and tandem MS data, and therefore based on simulated fragmentation mechanisms, we can view and get a putative fragment peak of the glycan in the spectrum. It can be accessed on the following web site http://www.glycoworkbench.org/.

10.5.3.2.3 NetOGlyc-Prediction of Mucin-Type O-Glycosylation Sites Based on Sequence Context and Surface Accessibility

This server is used to identify the O-glycosylated proteins from the sample sets and software access on the following link of the NetOGlyc 4.0 server (http://www.cbs.dtu.dk/services/NetOGlyc-4.0/). In mammalian proteins, the NetOglyc server generates neural network predictions of mucin-type GalNAc O-glycosylation positions. NetOGlyc4.0 is based on a support vector machine to produce a predictor of O-GalNAc glycosylation in the proteome-wide generated data set. However, this vector-based new predictor revealed high overall sensitivity (over 84%) when tested on distinct sets of glycosites. This software is utilizing to describe different types of parameters including transmembrane prediction (TMHMM), surface accessibility (NetSurfP), and protein disorder (DISEMBL) (Steentoft et al. 2013; Chattopadhyay et al. 2010).

10.5.3.2.4 NetNGlyc 1.0

This server was used to the prediction of N-linked glycosylation protein profiling of viruses and access by click on the NetNGlyc 1.0 server link (http://www.cbs.dtu.dk/services/NetNGlyc/). This server predicts N-glycosylation sites in human proteins by using artificial neural networks that observe the sequence context of AsnXaaSer/Thr sequences. However, sequences having N-glycosylation potential >0.5 are considered as cut-off value (Goldberg et al. 2005; Sheta et al. 2016).

Acknowledgments RG was supported by the Ramalingaswami re-entry fellowship (BT/HRD/35/02/2006) from the Department of Biotechnology, Ministry of Science and Technology, Government of India. R.R. was supported by Bilateral State Scholarship 2019/2020 (AK-00210 002/2019) and postdoctoral Bilateral State Scholarship 2020/2021 (BE AK 2020-2021/156635), Tempus Public Foundation.

References

Abdallah C, Dumas-Gaudot E, Renaut J, Sergeant K (2012) Gel-based and gel-free quantitative proteomics approaches at a glance. Int J Plant Genomics 2012:494572

Adams MD, Kelley JM, Gocayne JD, Dubnick M, Polymeropoulos MH, Xiao H, Merril CR, Wu A, Olde B, Moreno RF et al (1991) Complementary DNA sequencing: expressed sequence tags and human genome project. Science 252(5013):1651–1656. https://doi.org/10.1126/science.2047873. PMID: 2047873

Adhikari L, Hussain A, Rasul G (2017) Tapping the potential of neglected and underutilized food crops for sustainable nutrition security in the mountains of Pakistan and Nepal. Sustainability 9(2):291

Afendi FM, Okada T, Yamazaki M, Hirai-Morita A et al (2012) KNApSAcK family databases: integrated metabolite–plant species databases for multifaceted plant research. Plant Cell Physiol 53(2):e1

Afzal M, Alghamdi SS, Migdadi HH, Khan MA, Nurmansyah Mirza SB, El-Harty E (2020) Legume genomics and transcriptomics: from classic breeding to modern technologies. Saudi J Biol Sci 27(1):543–555. https://doi.org/10.1016/j.sjbs.2019.11.018

Ahmad N, Jianyu L, Xu T, Noman M, Jameel A, Na Y, Yuanyuan D, Nan W, Xiaowei L, Fawei W, Xiuming L (2019) Overexpression of a novel cytochrome P450 promotes flavonoid biosynthesis and osmotic stress tolerance in transgenic Arabidopsis. Genes 10(10):756

Akiyama K, Chikayama E, Yuasa H, Shimada Y, Tohge T, Shinozaki K, Hirai MY, Sakurai T, Kikuchi J, Saito K (2008) PRIMe: a web site that assembles tools for metabolomics and transcriptomics. In Silico Biol 8(3–4):339–345

Alagna F, D'Agostino N, Torchia L, Servili M, Rao R, Pietrella M, Giuliano G, Chiusano ML, Baldoni L, Perrotta G (2009) Comparative 454 pyrosequencing of transcripts from two olive genotypes during fruit development. BMC Genomics 10:399

Albert VA, Barbazuk WB, Der JP, Leebens-Mack J, Ma H, Palmer JD et al (2013) The Amborella genome and the evolution of flowering plants. Science 342:1241089. https://doi.org/10.1126/science.1241089

Al-Dous EK, George B, Al-Mahmoud ME, Al-Jaber MY, Wang H, Salameh YM et al (2011) De novo genome sequencing and comparative genomics of date palm (*Phoenix dactylifera*). Nat Biotechnol 29:521–527. https://doi.org/10.1038/nbt.1860

Allen DR, McWhinney BC (2019) Quadrupole time-of-flight mass spectrometry: a paradigm shift in toxicology screening applications. Clin Biochem Rev 40(3):135

Ambrosino L, Colantuono C, Diretto G, Fiore A, Chiusano ML (2020) Bioinformatics resources for plant abiotic stress responses: state of the art and opportunities in the fast evolving -omics era. Plants (Basel, Switzerland) 9(5):591. https://doi.org/10.3390/plants9050591

Arabidopsis Genome Initiative (2000) Analysis of the genome sequence of the flowering plant *Arabidopsis thaliana*. Nature 408:796–815

Arora RK (2014) Diversity in underutilized plant species: an Asia-Pacific perspective. Bioversity International, New Delhi

Aversano R, Contaldi F, Ercolano MR, Grosso V, Iorizzo M, Tatino F et al (2015) The Solanum commersonii genome sequence provides insights into adaptation to stress conditions and genome evolution of wild potato relatives. Plant Cell 27:954–968. https://doi.org/10.1105/tpc.114.135954

Aya K, Kobayashi M, Tanaka J, Ohyanagi H, Suzuki T, Yano K, Takano T, Yano K, Matsuoka M (2015) De novo transcriptome assembly of a fern, *Lygodium japonicum*, and a web resource database, Ljtrans DB. Plant Cell Physiol 56(1):e5. https://doi.org/10.1093/pcp/pcu184

Bailey M (2016) Mainstreaming agrobiodiversity in sustainable food systems: scientific foundations for an agrobiodiversity index-summary. Bioversity International, Rome

Bailey RL, Denby N, Haycock B, Sherif K, Steinbaum S, von Schacky C (2015) Perceptions of a healthy diet: insights from a 3-country survey. Nutr Today 50(6):282–287

Bais P, Moon-Quanbeck SM, Nikolau BJ, Dickerson JA (2012) Plantmetabolomics. org: mass spectrometry-based Arabidopsis metabolomics—database and tools update. Nucleic Acids Res 40(D1):D1216–D1220

Barman P, Choudhary AK, Geeta R (2017) A modified protocol yields high-quality RNA from highly mucilaginous *Dioscorea* tubers. 3 Biotech 7(2):150

Barrios A, Caminero C, García P, Krezdorn N, Hoffmeier K, Winter P, de la Vega MP (2017) Deep Super-SAGE transcriptomic analysis of cold acclimation in lentil (*Lens culinaris* Medik.). BMC Plant Biol 17:111

Bashir Z, Rizwan M, Mushtaq K, Munir A, Ali I (2017) In silico structural and functional prediction of Phaseolus vulgaris hypothetical protein PHA VU_004G136400g. J Proteomics Bioinform 10:207–206

Ben Halima N (2018) Prediction and analysis of GH14 family β-amylases in oat seedling extract: structure and function insights using *in silico* approaches. Int J Biol Macromol 125:361–369. https://doi.org/10.1016/j.ijbiomac.2018.12.065

Bendtsen JD, Nielsen H, Von Heijne G, Brunak S (2004) Improved prediction of signal peptides: SignalP 3.0. J Mol Biol 340(4):783–795

Berry E, Choudhary AK, Mishra G, Tandon R, Geeta R (2020) *Justicia adhatoda* reveals two morphotypes with possible functional significance. J Plant Res 133(6):783–805. https://doi.org/10.1007/s10265-020-01224-w

Blasbalg TL, Wispelwey B, Deckelbaum RJ (2011) Econutrition and utilization of food-based approaches for nutritional health. Food Nutr Bull 32(1 Suppl):S4–S13

Boguski MS, Lowe TM, Tolstoshev CM (1993) dbEST–database for "expressed sequence tags". Nat Genet 4:332–333

Bottcher C, von Roepenack-Lahaye E, Schmidt J, Schmotz C, Neumann S, Scheel D, Clemens S (2008) Metabolome analysis of biosynthetic mutants reveals a diversity of metabolic changes and allows identification of a large number of new compounds in *Arabidopsis*. Plant Physiol 147:2107–2120. https://doi.org/10.1104/pp.108.117754

Bredeson J, Lyons J, Prochnik S, Wu GA et al (2016) Sequencing wild and cultivated cassava and related species reveals extensive interspecific hybridization and genetic diversity. Nat Biotechnol 34:562–570. https://doi.org/10.1038/nbt.3535

Broz AK, Broeckling CD, De-la-Peña C, Lewis MR, Greene E, Callaway RM, Sumner LW, Vivanco JM (2010) Plant neighbor identity influences plant biochemistry and physiology related to defense. BMC Plant Biol 10(1):1–14

Burset M, Seledtsov IA, Solovyev VV (2000) Analysis of canonical and non-canonical splice sites in mammalian genomes. Nucleic Acids Res 28:4364–4375

Bushart TJ, Cannon AE, Ul Haque A, San Miguel P, Mostajeran K, Clark GB et al (2013) RNA-seq analysis identifies potential modulators of gravity response in spores of Ceratopteris (Parkeriaceae): evidence for modulation by calcium pumps and apyrase activity. Am J Bot 100:161–174

Cao J, Li X (2015) Identification and phylogenetic analysis of late embryogenesis abundant proteins family in tomato (Solanum lycopersicum). Planta 241(3):757–772

Carroll AJ, Badger MR, Millar AH (2010) The MetabolomeExpress Project: enabling web-based processing, analysis and transparent dissemination of GC/MS metabolomics datasets. BMC Bioinform 11(1):376

Casneuf T, Van de Peer Y, Huber W (2007) In situ analysis of cross-hybridisation on microarrays and the inference of expression correlation. BMC Bioinformatics 8:461. https://doi.org/10.1186/1471-2105-8-461. PMID: 18039370; PMCID: PMC2213692

Castañeda-Álvarez NP, Khoury CK, Achicanoy HA, Bernau V et al (2016) Global conservation priorities for crop wild relatives. Nat Plants 2(4):1–6

Chandra MS, Naresh RK, Thenua OV, Singh R, Geethanjali D (2020) Improving resource conservation, productivity and profitability of neglected and underutilized crops in the breadbasket of India: a review

Chattopadhyay S, Bagchi P, Dutta D, Mukherjee A, Kobayashi N, Chawlasarkar M (2010) Computational identification of posttranslational modification sites and functional families reveal possible moonlighting role of rotaviral proteins. Bioinformation 4(10):448

Chawla G, Ranjan C (2016) Principle, instrumentation, and applications of UPLC: a novel technique of liquid chromatography. Open Chem J 3:1–16

Chen CY (2011) TCM Database@ Taiwan: the world's largest traditional Chinese medicine database for drug screening in silico. PLoS One 6(1):e15939

Chen J, Huang Q, Gao D, Wang J, Lang Y, Liu T et al (2013) Whole-genome sequencing of *Oryza brachyantha* reveals mechanisms underlying Oryza genome evolution. Nat Commun 4:1595. https://doi.org/10.1038/ncomms2596

Chen H, Chen X, Chai X, Qiu Y, Gong C, Zhang Z, Wang T, Zhang Y, Li J, Wang A (2015) Effects of low temperature on mRNA and small RNA transcriptomes in *Solanum lycopersicoides* leaf revealed by RNA-Seq. Biochem Biophys Res Commun 464:768–773

Chiara M, Horner DS, Spada A (2013) De novo assembly of the transcriptome of the non-model plant *Streptocarpus rexii* employing a novel heuristic to recover locus-specific transcript clusters. PLoS One 8(12):e80961

Chibarabada TP, Mabhaudhi T, Modi AT (2017) Expounding the value of grain legumes in the semi- and arid tropics. Sustainability 9:60. https://doi.org/10.3390/su9010060

Chimonyo VGP, Modi AT, Mabhaudhi T (2016) Water use and productivity of a sorghum-cowpea-bottle gourd intercrop system. Agric Water Manag 165:82–96. https://doi.org/10.1016/j.agwat.2015.11.014

Chivenge P, Mabhaudhi T, Modi AT, Mafongoya P (2015) The potential role of neglected and underutilised crop species as future crops under water scarce conditions in Sub-Saharan Africa. Int J Env Res Public Health 12(6):5685–5711

Choksawangkarn W, Edwards N, Wang Y et al (2012) Comparative study of workflows optimized for in-gel, in-solution, and on-filter proteolysis in the analysis of plasma membrane proteins. J Proteome Res 11:3030–3034

Choudhary AK (2018) Fatty acid profiling in genus Leucas R.Br. (Lamiaceae), characterization of fatty acid desaturases and comparative transcriptome analysis to study unusual fatty acid biosynthesis in *L. cephalotes* (Roth) Spreng. Dissertation, University of Delhi. http://crl.du.ac.in/Doc.Bib/2018/Botany.pdf

Choudhary AK, Sunojkumar P, Mishra G (2017) Fatty acid profiling and multivariate analysis in the genus *Leucas* reveals its nutritional, pharmaceutical and chemotaxonomic significance. Phytochemistry 143:72–80

Churko JM, Mantalas GL, Snyder MP, Wu JC (2013) Overview of high throughput sequencing technologies to elucidate molecular pathways in cardiovascular diseases. Circul Res 112 (12):1613–1623. https://doi.org/10.1161/CIRCRESAHA.113.300939

Coemans B, Matsumura H, Terauchi R, Remy S, Swennen R, Sagi L (2005) SuperSAGE combined with PCR walking allows global gene expression profiling of banana (*Musa acuminata*), a non-model organism. Theor Appl Genet 111:1118–1126

Collakova E, Yen JY, Senger RS (2012) Are we ready for genome-scale modeling in plants? Plant Sci 191:53–70

Collen J, Porcel B, Carre W, Ball SG, Chaparro C, Tonon T, Barbeyron T, Michel G, Noel B, Valentin K, Elias M, Artiguenave F, Arun A, Aury J-M, Barbosa-Neto JF, Bothwell JH, Bouget F-Y, Brillet L, Cabello-Hurtado F, Capella-Gutierrez S, Charrier B, Cladiere L, Cock JM, Coelho SM, Colleoni C, Czjzek M, Da Silva C, Delage L, Denoeud F, Deschamps P, Dittami SM, Gabaldon T, Gachon CMM, Groisillier A, Herve C, Jabbari K, Katinka M, Kloareg B, Kowalczyk N, Labadie K, Leblanc C, Lopez PJ, McLachlan DH, Meslet-Cladiere L, Moustafa A, Nehr Z, Nyvall Collen P, Panaud O, Partensky F, Poulain J, Rensing SA, Rousvoal S, Samson G, Symeonidi A, Weissenbach J, Zambounis A, Wincker P, Boyen C (2013) Genome structure and metabolic features in the red seaweed Chondrus crispus shed light on evolution of the Archaeplastida. Proc Natl Acad Sci 110(13):5247–5252

Collins WW, Hawtin GC (1999) Conserving and using crop plant biodiversity in agroecosystems. In: Collins WW, Qualset CO (eds) Biodiversity in agroecosystems. CRC Press, Boca Raton, pp 267–281

Crick F (1970) Central dogma of molecular biology. Nature 227(5258):561–563. https://doi.org/10.1038/227561a0

Daccord N, Celton J, Linsmith G et al (2017) High-quality de novo assembly of the apple genome and methylome dynamics of early fruit development. Nat Genet 49:1099–1106. https://doi.org/10.1038/ng.3886

Damerval C, De Vienne D, Zivy M, Thiellement H (1986) Technical improvements in two-dimensional electrophoresis increase the level of genetic variation detected in wheat-seedling proteins. Electrophoresis 7:52–54

Davari SA, Gokhale NB, Palsande VN, Kasture MC (2018) Wal (*Lablab purpureus* L.): an unexploited potential food legumes. Int J Chem Stud 6(2):946–949

Davey MW, Graham NS, Vanholme B et al (2009) Heterologous oligonucleotide microarrays for transcriptomics in a non-model species; a proof-of-concept study of drought stress in Musa. BMC Genomics 10:436. https://doi.org/10.1186/1471-2164-10-436

De Vos RC, Moco S, Lommen A, Keurentjes JJ, Bino RJ, Hall RD (2007) Untargeted large-scale plant metabolomics using liquid chromatography coupled to mass spectrometry. Nat Protoc 2:778–791. https://doi.org/10.1038/nprot.2007.95

Deborde C, Jacob D (2014) MeRy-B, a metabolomic database and knowledge base for exploring plant primary metabolism. Methods Mol Biol 1083:3–16

Dehghan Nayeri F, Yarizade K (2014) Bioinformatics study of delta-12 fatty acid desaturase 2 (FAD2) gene in oilseeds. Mol Biol Rep 41:5077–5087. https://doi.org/10.1007/s11033-014-3373-5

Deng F, Jia X, Lai S, Liu Y, Chen S (2015) Transcript assembly and quality assessment. Sheng Wu Gong Cheng Xue Bao Chinese J Biotech 31(9):1271–1278. PMID: 26955705

Der JP, Barker MS, Wickett NJ, dePamphilis CW, Wolf PG (2011) *De novo* characterization of the gametophyte transcriptome in bracken fern, *Pteridium aquilinum*. BMC Genomics 12:99. https://doi.org/10.1186/1471-2164-12-99

Dixon RA, Strack D (2003) Phytochemistry meets genome analysis, and beyond. Phytochemistry 62:815–816

Dohm JC, Minoche AE, Holtgräwe D, Capella-Gutiérrez S, Zakrzewski F, Tafer H et al (2014) The genome of the recently domesticated crop plant sugar beet (*Beta vulgaris*). Nature 505:546–549. https://doi.org/10.1038/nature12817

Dreher K (2014) Putting the Plant Metabolic Network pathway databases to work: going offline to gain new capabilities. Methods Mol Biol 1083:151–171

Dulloo E, Hunter D, Leaman DL (2014) Plant diversity in addressing food, nutrition and medicinal needs. In: Garib-Fakim A (ed) Novel plant bioresources: application in food, medicine and cosmetics. Chichester, Wiley-Blackwell, pp 1–21

Duraipandiyan V, Al-Dhabi NA, Balachandran C, Ignacimuthu S, Sankar C, Balakrishna K (2015) Antimicrobial, antioxidant, and cytotoxic properties of vasicine acetate synthesized from vasicine isolated from *Adhatoda vasica* L. BioMed Res Int 1:2015. https://doi.org/10.1155/2015/727304

Durek P, Schmidt R, Heazlewood JL, Jones A, MacLean D, Nagel A, Kersten B, Schulze WX (2010) PhosPhAt: the *Arabidopsis thaliana* phosphorylation site database. An update. Nucleic Acids Res 38:D828–D834

Dutta A, Shetty P, Bhat S, Ramachandra Y, Hegde S (2012) A mass spectrometric study for comparative analysis and evaluation of metabolite recovery from plasma by various solvent systems. J Biomol Tech 23:128–135. https://doi.org/10.7171/jbt.12-2304-001

Ekblom R, Galindo J (2011) Applications of next generation sequencing in molecular ecology of non-model organisms. Heredity 107:1–15. https://doi.org/10.1038/hdy.2010.152

Exner V, Hirsch-Hoffmann M, Gruissem W, Hennig L (2008) PlantDB–a versatile database for managing plant research. Plant Methods 4(1):1

Fàbregas N, Lozano-Elena F, Blasco-Escámez D, Tohge T, Martínez-Andújar C, Albacete A, Osorio S, Bustamante M, Riechmann JL, Nomura T, Yokota T (2018) Overexpression of the vascular brassinosteroid receptor BRL3 confers drought resistance without penalizing plant growth. Nat Commun 9(1):1–13

Facchini PJ, Bohlmann J, Covello PS, De Luca et al (2012) Synthetic biosystems for the production of high-value plant metabolites. Trends Biotechnol 30:127–131

Fan J, Hu J, Ye C (2011) May bioinformatics approaches in the study of a protein similar to anthocyanin regulatory B protein. In: 2011 5th international conference on bioinformatics and biomedical engineering. IEEE, pp 1–4

Fanzo J, Hunter D, Borelli T, Mattei F (eds) (2013) Diversifying food and diets: using agricultural biodiversity to improve nutrition and health. Routledge, London

FAO, IFAD, UNICEF, WFP, WHO (2019) The state of food security and nutrition in the world: safeguarding against economic slowdowns and downturns. FAO, Rome

Farag MA, Huhman DV, Dixon RA, Sumner LW (2008) Metabolomics reveals novel pathways and differential mechanistic and elicitor-specific responses in phenylpropanoid and isoflavonoid biosynthesis in *Medicago truncatula* cell cultures. Plant Physiol 146:387–402. https://doi.org/10.1104/pp.107.108431

Fernandez P, Soria M, Blesa D et al (2012) Development, characterization and experimental validation of a cultivated sunflower (*Helianthus annuus* L.) gene expression oligonucleotide microarray. PLoS One 7(10):e45899. https://doi.org/10.1371/journal.pone.0045899

Fernie AR (2007) The future of metabolic phytochemistry: larger numbers of metabolites, higher resolution, greater understanding. Phytochemistry 68:2861–2880

Fernie AR, Trethewey RN, Krotzky AJ, Willmitzer L (2004) Metabolite profiling: from diagnostics to systems biology. Nat Rev Mol Cell Biol 5:763–769. https://doi.org/10.1038/nrm1451

Ferrer I, Thurman EM (2003) Liquid chromatography/time-of-flight/mass spectrometry (LC/ToF/MS) for the analysis of emerging contaminants. Trends Anal Chem 22:750–756

Fiehn O, Kopka J, Dormann P, Altmann T, Trethewey RN, Willmitzer L (2000) Metabolite profiling for plant functional genomics. Nat Biotechnol 18:1157–1161. https://doi.org/10.1038/81137

Fischer TC, Klattig J, Gierl A (2001) A general cloning strategy for divergent plant cytochrome P450 genes and its application in Lolium rigidum and Ocimum basilicum. Theor Appl Genet 103(6–7):1014–1021

Fukushima A, Kusano M, Mejia RF, Iwasa M, Kobayashi M, Hayashi N et al (2014) Metabolomic characterization of knockout mutants in Arabidopsis: development of a metabolite profiling database for knockout mutants in Arabidopsis. Plant Physiol 165:948–961

Gardy JL, Spencer C, Wang K, Ester M, Tusnady GE, Simon I, Hua S, deFays K, Lambert C, Nakai K (2003) PSORT-B: improving protein subcellular localization prediction for Gram-negative bacteria. Nucleic Acids Res 31:3613–3617

Ghatak A, Chaturvedi P, Paul P et al (2017) Proteomics survey of Solanaceae family: Current status and challenges ahead. J Proteomics 169:41–57. https://doi.org/10.1016/j.jprot.2017.05.016

Gogna M, Choudhary A, Mishra G, Kapoor R, Bhatla SC (2020) Changes in lipid composition in response to salt stress and its possible interaction with intracellular Na^+-K^+ ratio in sunflower (*Helianthus annuus* L.). Environ Exp Bot 178:104147

Goldberg D, Sutton-Smith M, Paulson J, Dell A (2005) Automatic annotation of matrix-assisted laser desorption/ionization N-glycan spectra. Proteomics 5(4):865–875

Govindarajan R, Duraiyan J, Kaliyappan K, Palanisamy M (2012) Microarray and its applications. J Pharm Bioallied Sci 4(Suppl 2):S310–S312. https://doi.org/10.4103/0975-7406.100283

Grennan AK (2009) MoTo DB: a metabolic database for tomato. Plant physiol 151(4):1701–1702

Guan D, Hu X, Diao D, Wang F, Liu Y (2019) Genome-wide analysis and identification of the Aux/IAA gene family in peach. Int J Mol Sci 20(19):4703

Guo S, Zhang J, Sun H, Salse J, Lucas WJ, Zhang H et al (2013) The draft genome of watermelon (*Citrullus lanatus*) and resequencing of 20 diverse accessions. Nat Genet 45:51–58. https://doi.org/10.1038/ng.2470

Gupta R, Deswal R (2012) Low temperature stress modulated secretome analysis and purification of antifreeze protein from *Hippophae rhamnoides*, a Himalayan wonder plant. J Proteome Res 11:2684–2696

Gupta OP, Sharma ML, Ghatak BJ, Atal CK (1977) Potent uterine activity of alkaloid vasicine. Indian J Med Res 66(5):865

Gupta R, Wang Y, Agrawal GK et al (2015) Time to dig deep into the plant proteome: a hunt for low-abundance proteins. Front Plant Sci 6:1–3. https://doi.org/10.3389/fpls.2015.00022

Gupta R, Lee SJ, Min CW et al (2016) Coupling of gel-based 2-DE and 1-DE shotgun proteomics approaches to dig deep into the leaf senescence proteome of *Glycine max*. J Proteomics 148:65–74. https://doi.org/10.1016/j.jprot.2016.07.025

Gupta R, Min CW, Kim SW, Yoo JS, Moon AR, Shin AY, Kwon SY, Kim ST (2020) A TMT-based quantitative proteome analysis to elucidate the TSWV induced signaling cascade in susceptible and resistant cultivars of Solanum lycopersicum. Plants 9(3):290

Haas BJ, Zody MC (2010) Advancing RNA-seq analysis. Nat Biotechnol 28:421–423

Hadebe ST, Modi AT, Mabhaudhi T (2017) Drought tolerance and water use of cereal crops: a focus on sorghum as a food security crop in sub-Saharan Africa. J Agro Crop Sci 203:177–191. https://doi.org/10.1111/jac.12191

Hartmann T (2007) From waste products to ecochemicals: fifty years research of plant secondary metabolism. Phytochemistry 68:2831–2846

Heazlewood JL, Durek P, Hummel J, Selbig J, Weckwerth W, Walther D, Schulze WX (2007) PhosPhAt: a database of phosphorylation sites in *Arabidopsis thaliana* and a plant-specific phosphorylation site predictor. Nucleic Acids Res 36:D1015–D1021

Hefferon K (2019) Biotechnological approaches for generating zinc-enriched crops to combat malnutrition. Nutrients 11(2):253

Hernández-de-Diego R, Tarazona S, Martínez-Mira C, Balzano-Nogueira L, Furió-Tarí P, Pappas GJ Jr, Conesa A (2018) PaintOmics 3: a web resource for the pathway analysis and visualization of multi-omics data. Nucleic Acids Res 46(W1):W503–W509

Hirakawa H, Shirasawa K, Miyatake K, Nunome T, Negoro S, Ohyama A et al (2014) Draft genome sequence of eggplant (*Solanum melongena* L.): the representative Solanum species indigenous to the old world. DNA Res 21:649–660. https://doi.org/10.1093/dnares/dsu027

Horai H, Arita M, Kanaya S, Nihei Y, Ikeda T et al (2010) MassBank: a public repository for sharing mass spectral data for life sciences. J Mass Spectrom 45(7):703–714

Huang S, Ding J, Deng D, Tang W, Sun H, Liu D et al (2013) Draft genome of the kiwifruit *Actinidia chinensis*. Nat Commun 4:2640. https://doi.org/10.1038/ncomms3640

Hurkman WJ, Tanaka CK (1986) Solubilization of plant membrane proteins for analysis by two-dimensional gel electrophoresis. Plant Physiol 81:802–806

IPGRI (2002) Neglected and underutilized plant species: strategic action plan of the International Plant Genetic Resources Institute. International Plant Genetic Resources Institute, Rome

Islam S, Kjallquist U, Moliner A, Zajac P, Fan J-B, Lonnerberg P, Linnarsson S (2011) Characterization of the single-cell transcriptional landscape by highly multiplex RNA-seq. Genome Res 21 (7):1160–1167

Islam S, Kjallquist U, Moliner A, Zajac P, Fan JB, Lonnerberg P, Linnarsson S (2012) Highly multiplexed and strand-specific single-cell RNA 5′ end sequencing. Nat Protoc 7:813–828

Jacobs A, Lunde C, Bacic A, Tester M, Roessner U (2007) The impact of constitutive heterologous expression of moss Naþ transporter on the metabolomes of rice and barley. Metabolomics 3:307–317

Jarvis D, Ho Y, Lightfoot D et al (2017) The genome of *Chenopodium quinoa*. Nature 542:307–312. https://doi.org/10.1038/nature21370

Jayaraman A et al (2006) Computer simulation study of molecular recognition in model DNA microarrays. Biophys J 91:2227–2236

Jorge TF, Rodrigues JA, Caldana C, Schmidt R, van Dongen JT, Thomas-Oates J, António C (2016) Mass spectrometry-based plant metabolomics: metabolite responses to abiotic stress. Mass Spectrom Rev 35(5):620–649

Jung KH, An G (2012) Application of MapMan and RiceNet drives systematic analyses of the early heat stress transcriptome in rice seedlings. J Plant Biol 55(6):436–449

Junttila S, Rudd S (2012) Characterization of a transcriptome from a non-model organism, *Cladonia rangiferina*, the grey reindeer lichen, using high-throughput next generation sequencing and EST sequence data. BMC Genomics 13:575. https://doi.org/10.1186/1471-2164-13-575

Kagale S, Koh C, Nixon J, Bollina V, Clarke WE, Tuteja R et al (2014) The emerging biofuel crop Camelina sativa retains a highly undifferentiated hexaploid genome structure. Nat Commun 5:3706. https://doi.org/10.1038/ncomms4706

Kandula V, Gottschalk VA, Katam R, Anupalli RR (2014) Peanut bioinformatics: tools and applications for developing more effective immunotherapies for peanut allergy and improving food safety. In: Agricultural bioinformatics. Springer, New Delhi, pp 129–137

Kawasaki ES (2004) Microarrays and the gene expression profile of a single cell. Ann N Y Acad Sci 1020(1):92–100

Kermali S, Anishetty N, Cooper H (1997) Promoting development and commercialization of underutilized crops and species in the FAO global plan of action. ICUC, Southampton

Kilonzi SM, Makokha AO, Kenji GM (2017) Physical characteristics, proximate composition and anti-nutritional factors in grains of lablab bean (*Lablab purpureus*) genotypes from Kenya. J Appl Biosci 114(1):11289–11298

Kim YC, Wang SM (2009) Decoding neuron transcriptome by SAGE. In: Encyclopedia of neuroscience. Elsevier BV, pp 357–363. https://doi.org/10.1016/B978-008045046-9.00879-2

Kim SW, Lee SH, Min CW et al (2017) Ginseng (Panax sp.) proteomics: an update. Appl Biol Chem 60:311–320. https://doi.org/10.1007/s13765-017-0283-y

Kim SW, Gupta R, Min CW et al (2019) Label-free quantitative proteomic analysis of *Panax ginseng* leaves upon exposure to heat stress. J Ginseng Res 43:143–153

Kreps JA, Wu YJ, Chang HS, Zhu T, Wang X, Harper JF (2002) Transcriptome changes for Arabidopsis in response to salt, osmotic, and cold stress. Plant Physiol 130(4):2129–2141. https://doi.org/10.1104/pp.008532

Krogh A, Larsson B, Von Heijne G, Sonnhammer EL (2001) Predicting transmembrane protein topology with a hidden Markov model: application to complete genomes. J Mol Biol 305 (3):567–580

Kusano M, Tohge T, Fukushima A, Kobayashi M, Hayashi N et al (2011) Metabolomics reveals comprehensive reprogramming involving two independent metabolic responses of Arabidopsis to ultraviolet-B light. Plant J 67(2):354–369. https://doi.org/10.1111/j.1365-313X.2011.04599.x

Lei Z, Huhman DV, Sumner LW (2011) Mass spectrometry strategies in metabolomics. J Biol Chem 286:25435–25442

Leushkin EV, Sutormin RA, Nabieva ER, Penin AA, Kondrashov AS, Logacheva MD (2013) The miniature genome of a carnivorous plant *Genlisea aurea* contains a low number of genes and short non-coding sequences. BMC Genomics 14:476. https://doi.org/10.1186/1471-2164-14-476

Li FL, Wu XZ, Tsang E, Cutler AJ (2005) Transcriptional profiling of imbibed Brassica napus seed. Genomics 86(6):718–730. https://doi.org/10.1016/j.ygeno.2005.07.006

Li X, Korir NK, Liu L, Shangguan L, Wang Y, Han J, Chen M, Fang J (2012) Microarray analysis of differentially expressed genes engaged in fruit development between *Prunus mume* and *Prunus armeniaca*. J Plant Physiol 169(17):1776–1788. https://doi.org/10.1016/j.jplph.2012.07.016

Li F, Fan G, Wang K, Sun F, Yuan Y, Song G et al (2014) Genome sequence of the cultivated cotton *Gossypium arboreum*. Nat Genet 46:567–572. https://doi.org/10.1038/ng.2987

Li CX, Xu ZG, Dong RQ, Chang SX, Wang LZ, Khalil-Ur-Rehman M, Tao JM (2017a) An RNA-Seq analysis of grape plantlets grown in vitro reveals different responses to blue, green, red led light, and white fluorescent light. Front Plant Sci 8:78

Li Y, Huang J, Song X, Zhang Z, Jiang Y, Zhu Y, Zhao H, Ni D (2017b) An RNA-Seq transcriptome analysis revealing novel insights into aluminum tolerance and accumulation in tea plant. Planta 246:1–13

Li HF, Dong QL, Li GX, Ran K (2018) Identification and expression analysis of 11 MADS-box genes in peach (Prunus persica var. nectarina 'Luxing'). J Hortic Sci Biotechnol 93(3):232–243

Lightfoot DJ, Jarvis DE, Ramaraj T et al (2017) Single-molecule sequencing and Hi-C-based proximity-guided assembly of amaranth (*Amaranthus hypochondriacus*) chromosomes provide insights into genome evolution. BMC Biol 15:74. https://doi.org/10.1186/s12915-017-0412-4

Ling HQ, Zhao S, Liu D, Wang J, Sun H, Zhang C et al (2013a) Draft genome of the wheat A-genome progenitor *Triticum urartu*. Nature 496:87–90. https://doi.org/10.1038/nature11997

Ling MH, Rabara RC, Tripathi P, Rushton PJ, Ge X (2013b) Extending MapMan ontology to tobacco for visualization of gene expression. Dataset Pap Biol 2013:706465

Liu MJ, Zhao J, Cai QL, Liu GC, Wang JR, Zhao ZH et al (2014) The complex jujube genome provides insights into fruit tree biology. Nat Commun 5:5315. https://doi.org/10.1038/ncomms6315

Liu J, Wang Y, Li Q (2017) Analysis of differentially expressed genes and adaptive mechanisms of *Prunus triloba* Lindl. under alkaline stress. Hereditas 154:10

Loy A, Bodrossy L (2006) Highly parallel microbial diagnostics using oligonucleotide microarrays. Clin Chim Acta 363:106–119

Lum G, VanBuren R, Ming R, Min XJ (2013) Secretome prediction and analysis in sacred lotus (Nelumbo nucifera Gaertn.). Trop Plant Biol 6(2-3):131–137

Luo W, Pant G, Bhavnasi YK, Blanchard SG Jr, Brouwer C (2017) Pathview web: user friendly pathway visualization and data integration. Nucleic Acids Res 45(W1):W501–W508

Mabhaudhi T, Chimonyo V, Modi A (2017) Status of underutilised crops in South Africa: opportunities for developing research capacity. Sustainability 9:1569. https://doi.org/10.3390/su9091569

Mann IK, Wegrzyn JL, Rajora OP (2013) Generation, functional annotation and comparative analysis of black spruce (Picea mariana) ESTs: an important conifer genomic resource. BMC Genomics 14(1):702

Manzoni C, Kia DA, Vandrovcova J, Hardy J, Wood NW, Lewis PA, Ferrari R (2018) Genome, transcriptome and proteome: the rise of omics data and their integration in biomedical sciences. Brief Bioinformatics 19(2):286–302. https://doi.org/10.1093/bib/bbw114

Maroli AS, Nandula VK, Duke SO, Gerard P, Tharayil N (2018) Comparative metabolomic analyses of Ipomoea lacunosa biotypes with contrasting glyphosate tolerance captures herbicide-induced differential perturbations in cellular physiology. J Agric Food Chem 66 (8):2027–2039

Martin JA, Wang Z (2011) Next-generation transcriptome assembly. Nat Rev Genet 12 (10):671–682. https://doi.org/10.1038/nrg3068

Martin G, Baurens F, Droc G et al (2016) Improvement of the banana "*Musa acuminata*" reference sequence using NGS data and semi-automated bioinformatics methods. BMC Genomics 17:243. https://doi.org/10.1186/s12864-016-2579-4

Mascher M, Gundlach H, Himmelbach A et al (2017) A chromosome conformation capture ordered sequence of the barley genome. Nature 544:427–433. https://doi.org/10.1038/nature22043

Matsuda F, Yonekura-Sakakibara K, Niida R, Kuromori T, Shinozaki K, Saito K (2009) MS/MS spectral tag-based annotation of nontargeted profile of plant secondary metabolites. Plant J 57:555–577. https://doi.org/10.1111/j.1365-313X.2008.03705.x

Matsuda F, Hirai MY, Sasaki E, Akiyama K, Yonekura-Sakakibara K, Provart NJ, Sakurai T, Shimada Y, Saito K (2010) AtMetExpress development: a phytochemical atlas of Arabidopsis development. Plant Physiol 152:566–578

Mattick JS, Makunin IV (2006) Non-coding RNA. Hum Mol Genet 15(Suppl 1):R17–R29. https://doi.org/10.1093/hmg/ddl046

Mayes S, Massawe FJ, Alderson PG, Roberts JA, Azam-Ali SN, Hermann M (2012) The potential for underutilized crops to improve security of food production. J Exp Bot 63(3):1075–1079

Medina-Puche L, Blanco-Portales R, Molina-Hidalgo FJ et al (2016) Extensive transcriptomic studies on the roles played by abscisic acid and auxins in the development and ripening of strawberry fruits. Funct Integr Genomics 16(6):671–692. https://doi.org/10.1007/s10142-016-0510-3

Mejía-Guerra MK, Li W, Galeano NF, Vidal M, Gray J, Doseff AI, Grotewold E (2015) Core promoter plasticity between maize tissues and genotypes contrasts with predominance of sharp transcription initiation sites. Plant Cell 27:3309–3320

Meng Q, Gupta R, Min CW et al (2019) Proteomics of rice—*Magnaporthe oryzae* interaction: what have we learned so far? Front Plant Sci 10:1383

Miller MB, Tang YW (2009) Basic concepts of microarrays and potential applications in clinical microbiology. Clin Microbiol Rev 22(4):611–633. https://doi.org/10.1128/CMR.00019-09

Min CW, Lee SH, Cheon YE et al (2017) Gel-based and gel-free proteome data associated with controlled deterioration treatment of Glycine max seeds. Data Brief 15:449–453. https://doi.org/10.1016/j.dib.2017.09.056

Min CW, Gupta R, Agrawal GK et al (2019a) Concepts and strategies of soybean seed proteomics using the shotgun proteomics approach. Expert Rev Proteomics 16:795–804

Min CW, Jung WY, Park HJ et al (2019b) Label-free quantitative proteomic analysis determines changes in amino acid and carbohydrate metabolism in three cultivars of Jerusalem artichoke tubers. Plant Biotechnol Rep 13:111–122

Ming R, VanBuren R, Liu Y, Yang M, Han Y, Li LT et al (2013) Genome of the long-living sacred lotus (*Nelumbo nucifera* Gaertn.). Genome Biol 14:R41. https://doi.org/10.1186/gb-2013-14-5-r41

Missoum A (2018) DNA microarray and bioinformatics technologies: a mini-review. Proc Nat Res Soc 2. https://doi.org/10.11605/j.pnrs.201802010

Moliterni VMC, Paris R, Onofri C, Orrù L, Cattivelli L, Pacifico D, Avanzato C, Ferrarini A, Delledonne M, Mandolino G (2015) Early transcriptional changes in Beta vulgaris in response to low temperature. Planta 242:187–201

Moreau H, Verhelst B, Couloux A, Derelle E, Rombauts S, Grimsley N, Van Bel M, Poulain J, Katinka M, Hohmann-Marriott MF, Piganeau G, Rouzé P, Da Silva C, Wincker P, Van de Peer Y, Vandepoele K (2012) Gene functionalities and genome structure in Bathycoccus prasinos reflect cellular specializations at the base of the green lineage. Genome Biol 13(8):R74

Musungu B, Bhatnagar D, Quiniou S, Brown RL, Payne GA, O'Brian G, Fakhoury AM, Geisler M (2020) Use of dual RNA-seq for systems biology analysis of Zea mays and Aspergillus flavus interaction. Front Microbiol 11:853

Nair R, Rost B (2005) Mimicking cellular sorting improves prediction of subcellular localization. J Mol Biol 348(1):85–100

Naoumkina M, Farag MA, Sumner LW, Tang Y, Liu CJ, Dixon RA (2007) Different mechanisms for phytoalexin induction by pathogen and wound signals in *Medicago truncatula*. Proc Natl Acad Sci U S A 104:17909–17915. https://doi.org/10.1073/pnas.0708697104

Natsume S, Takagi H, Shiraishi A, Murata J, Toyonaga H, Patzak J et al (2014) The draft genome of hop (*Humulus lupulus*), an essence for brewing. Plant Cell Physiol 56:428–441. https://doi.org/10.1093/pcp/pcu169

Ntie-Kang F, Zofou D, Babiaka SB, Meudom R et al (2013) AfroDb: a select highly potent and diverse natural product library from African medicinal plants. PLoS One 8:e78085. https://doi.org/10.1371/journal.pone.0078085

Okazaki Y, Kamide Y, Hirai MY, Saito K (2011) Plant lipidomics based on hydrophilic interaction chromatography coupled to ion trap time-of-flight mass spectrometry. Metabolomics 9 (1):121–131. https://doi.org/10.1007/s11306-011-0318-z

Padulosi S, Heywood V, Hunter D, Jarvis A (2011) Underutilized species and climate change: current status and outlook. In: Yadav SS, Redden RJ, Hatfield JL, Lotze-Campen H, Hall AE (eds) Crop adaptation to climate change, 1st edn. Blackwell Publishing, pp 507–521

Pai TW, Li KH, Yang CH, Hu CH, Lin HJ, Wang WD, Chen YR (2018) Multiple model species selection for transcriptomics analysis of non-model organisms. BMC Bioinformatics 19(Suppl 9):284. https://doi.org/10.1186/s12859-018-2278-z

Peltier JB, Ytterberg AJ, Sun Q, van Wijk KJ (2004) New functions of the thylakoid membrane proteome of *Arabidopsis thaliana* revealed by a simple, fast, and versatile fractionation strategy. J Biol Chem 279(47):49367–49383

Peng Y, Lai Z, Lane T, Nageswara-Rao M, Okada M, Jasieniuk M et al (2014) De novo genome assembly of the economically important weed horseweed using integrated data from multiple sequencing platforms. Plant Physiol 166:1241–1254. https://doi.org/10.1104/pp.114.247668

Petersen TN, Brunak S, Von Heijne G, Nielsen H (2011) SignalP 4.0: discriminating signal peptides from transmembrane regions. Nat Methods 8(10):785–786

Pevzner PA, Tang H, Waterman MS (2001) An Eulerian path approach to DNA fragment assembly. Proc Natl Acad Sci U S A 98:9748–9753

Polashock J, Zelzion E, Fajardo D, Zalapa J, Georgi L, Bhattacharya D et al (2014) The American cranberry: first insights into the whole genome of a species adapted to bog habitat. BMC Plant Biol 14:165. https://doi.org/10.1186/1471-2229-14-165

Poon TC, Johnson PJ (2001) Proteome analysis and its impact on the discovery of serological tumor markers. Clin Chim Acta 313(1–2):231–239

Pootakham W, Sonthirod C, Naktang C et al (2017) De novo hybrid assembly of the rubber tree genome reveals evidence of paleotetraploidy in *Hevea* species. Sci Rep 7:41457. https://doi.org/10.1038/srep41457

Price DC, Chan CX, Yoon HS, Yang EC, Qiu H, Weber APM et al (2012) *Cyanophora paradoxa* genome elucidates origin of photosynthesis in algae and plants. Science 335:843–847. https://doi.org/10.1126/science.1213561

Rai A, Saito K, Yamazaki M (2017) Integrated omics analysis of specialized metabolism in medicinal plants. Plant J 90(4):764–787

Ralston-Hooper K, Hopf A, Oh C, Zhang X, Adamec J, Sepulveda MS (2008) Development of GC9GC/TOF–MS metabolomics for use in ecotoxicological studies with invertebrates. Aquat Toxicol 88:48–52. https://doi.org/10.1016/j.aquatox.2008.03.002

Ramya M, Park PH, Chuang YC, Kwon OK, An HR, Park PM, Baek YS, Kang BC, Tsai WC, Chen HH (2019) RNA sequencing analysis of *Cymbidium goeringii* identifies floral scent biosynthesis related genes. BMC Plant Biol 19(1):337. https://doi.org/10.1186/s12870-019-1940-6

Ranasinghe A, Ramanathan R, Jemal M, D'Arienzo CJ, Humphreys WG, Olah TV (2012) Integrated quantitative and qualitative workflow for in vivo bioanalytical support in drug discovery using hybrid Q-TOF-MS. Bioanalysis 4:511–528

Raymond O, Gouzy J, Just J et al (2018) The Rosa genome provides new insights into the domestication of modern roses. Nat Genet 50:772–777. https://doi.org/10.1038/s41588-018-0110-3

Rensing SA, Lang D, Zimmer AD, Terry A, Salamov A, Shapiro H, Nishiyama T et al (2008) The Physcomitrella genome reveals evolutionary insights into the conquest of land by plants. Science 319(5859):64–69. https://doi.org/10.1126/science.1150646

Reyes-Chin-Wo S, Wang Z, Yang X et al (2017) Genome assembly with in vitro proximity ligation data and whole-genome triplication in lettuce. Nat Commun 8:14953. https://doi.org/10.1038/ncomms14953

Rocheta M, Sobral R, Magalhães J, Amorim MI, Ribeiro T, Pinheiro M, Egas C, Morais-Cecílio L, Costa MM (2014) Comparative transcriptomic analysis of male and female flowers of monoecious *Quercus suber*. Front Plant Sci 5:599. https://doi.org/10.3389/fpls.2014.00599

Roessner U, Luedemann A, Brust D, Fiehn O, Linke T, Willmitzer L, Fernie A (2001) Metabolic profiling allows comprehensive phenotyping of genetically or environmentally modified plant systems. Plant Cell 13:11–29. https://doi.org/10.1105/tpc.13.1.11

Roessner U, Patterson JH, Forbes MG, Fincher GB, Langridge P, Bacic A (2006) An investigation of boron toxicity in barley using metabolomics. Plant Physiol 142:1087–1101

Roh SW, Abell GC, Kim KH, Nam YD, Bae JW (2010) Comparing microarrays and next-generation sequencing technologies for microbial ecology research. Trends Biotechnol 28 (6):291–299. https://doi.org/10.1016/j.tibtech.2010.03.001

Romanelli C, Cooper D, Campbell-Lendrum D, Maiero M, Karesh WB, Hunter D, Golden CD (2015) Connecting global priorities: biodiversity and human health: a state of knowledge review. World Health Organistion/Secretariat of the UN Convention on Biological Diversity, Geneva

Russo G, Zegar C, Giordano A (2003) Advantages and limitations of microarray technology in human cancer. Oncogene 22:6497–6507. https://doi.org/10.1038/sj.onc.1206865

Sato S, Nakamura Y, Kaneko T, Asamizu E, Kato T et al (2008) Genome structure of the legume, *Lotus japonicus*. DNA Res 15:227–239

Schauer N, Fernie AR (2006) Plant metabolomics: towards biological function and mechanism. Trends Plant Sci 11:508–516

Schena M, Shalon D, Davis RW, Brown PO (1995) Quantitative monitoring of gene expression patterns with a complementary DNA microarray. Science 270(5235):467–470. https://doi.org/10.1126/science.270.5235.467. PMID: 7569999

Schmutz J, Cannon SB, Schlueter J, Ma J, Mitros T, Nelson W, Hyten DL, Song Q, Thelen JJ, Cheng J, Xu (2010) Genome sequence of the palaeopolyploid soybean. Nature 463 (7278):178–183

Schwacke R, Ponce-Soto GY, Krause K, Bolger AM et al (2019) MapMan4: a refined protein classification and annotation framework applicable to multi-omics data analysis. Mol Plant 12 (6):879–892

Schwarze K, Buchanan J, Taylor JC et al (2018) Are whole-exome and whole-genome sequencing approaches cost-effective? A systematic review of the literature. Genet Med 20:1122–1130. https://doi.org/10.1038/gim.2017.247

Seaver SM, Henry CS, Hanson AD (2012) Frontiers in metabolic reconstruction and modeling of plant genomes. J Exp Bot 63:2247–2258

Shahaf N, Rogachev I, Heinig U, Meir S, Malitsky S, Battat M, Wyner H, Zheng S, Wehrens R, Aharoni A (2016) The WEIZMASS spectral library for high-confidence metabolite identification. Nat Commun 7(1):1–13

Shelef O, Weisberg PJ, Provenza FD (2017) The value of native plants and local production in an era of global agriculture. Front Plant Sci 8:2069

Shellie RA, Welthagen W, Zrostlikova J, Spranger J, Ristow M, Fiehn O, Zimmermann R (2005) Statistical methods for comparing comprehensive two-dimensional gas chromatography-time-offlight mass spectrometry results: metabolomic analysis of mouse tissue extracts. J Chromatogr A 1086:83–90. https://doi.org/10.1016/j.chroma.2005.05.088

Shendure J, Ji H (2008) Next-generation DNA sequencing. Nat Biotechnol 26:1135–1145. https://doi.org/10.1038/nbt1486

Sheta R, Roux-Dalvai F, Woo CM, Fournier F, Bourassa S, Bertozzi CR, Droit A, Bachvarov D (2016) Proteomic dataset for altered glycoprotein expression upon GALNT3 knockdown in ovarian cancer cells. Data Brief 8:342–349

Shulaev V, Sargent DJ, Crowhurst RN, Mockler TC, Folkerts O, Delcher AL et al (2011) The genome of woodland strawberry (*Fragaria vesca*). Nat Genet 43:109–116. https://doi.org/10.1038/ng.740

Singh H, Rai K, Upadhyay S et al (2015) Transcriptome sequencing of a thalloid bryophyte; *Dumortiera hirsuta* (Sw) Nees: assembly, annotation and marker discovery. Sci Rep 5:15350. https://doi.org/10.1038/srep15350

Singh D, Singh CK, Taunk J, Tomar RSS, Chaturvedi AK, Gaikwad K, Pal M (2017) Transcriptome analysis of lentil (*Lens culinaris* Medikus) in response to seedling drought stress. BMC Genomics 18:206

Smith H, Dal Grande F, Muggia L, Keuler R, Divakar PK et al (2020) Metagenomic data reveal diverse fungal and algal communities associated with the lichen symbiosis. https://doi.org/10.1101/2020.03.04.966853. PPR:PPR116189

Somayajula D, Desai N (2019) Optimization of protein extraction and proteomic studies in *Cenchrus polystachion* (L.) Schult. Heliyon 5:e02968

Steentoft C, Vakhrushev SY, Joshi HJ, Kong Y et al (2013) Precision mapping of the human O-GalNAc glycoproteome through SimpleCell technology. EMBO J 32(10):1478–1488

Stegle O, Teichmann SA, Marioni JC (2015) Computational and analytical challenges in single-cell transcriptomics. Nat Rev Genet 16(3):133–145. https://doi.org/10.1038/nrg3833

Stein SE (1999) An integrated method for spectrum extraction and compound identification from gas chromatography/mass spectrometry data. J Am Socr Mass Spectr 10(8):770–781

Strasser R (2012) Challenges in O-glycan engineering of plants. Front Plant Sci 3:218

Thakur M (2014) Underutilized food crops: treasure for the future India. Food Sci Res J 5 (2):174–183

Thimm O, Bläsing O, Gibon Y, Nagel A, Meyer S, Krüger P et al (2004) MAPMAN: a user-driven tool to display genomics data sets onto diagrams of metabolic pathways and other biological processes. Plant J 6:914–939

Tokimatsu T, Sakurai N, Suzuki H, Ohta H, Nishitani K et al (2005) KaPPA-view. A web-based analysis tool for integration of transcript and metabolite data on plant metabolic pathway maps. Plant Physol 138(3):1289–1300

Toledo Á, Burlingame B (2006) Biodiversity and nutrition: a common path toward global food security and sustainable development. J Food Compos Anal 19(6–7):477–483

Tounsi S, Feki K, Kamoun Y, Saïdi MN, Jemli S, Ghorbel M, Alcon C, Brini F (2019) Highlight on the expression and the function of a novel MnSOD from diploid wheat (T. monococcum) in response to abiotic stress and heavy metal toxicity. Plant Physiol Biochem 142:384–394

Trindade I, Capitao C, Dalmay T, Fevereiro MP, Santos DM (2010) miR398 and miR408 are up-regulated in response to water deficit in *Medicago truncatula*. Planta 231:705–716

Turner MF, Heuberger AL, Kirkwood JS, Collins CC, Wolfrum EJ, Broeckling CD, Prenni JE, Jahn CE (2016) Non-targeted metabolomics in diverse sorghum breeding lines indicates primary and secondary metabolite profiles are associated with plant biomass accumulation and photosynthesis. Front Plant Sci 7:953

Udayakumar M, Chandar DP, Arun N, Mathangi J, Hemavathi K, Seenivasagam R (2012) PMDB: plant of database—a metabolomic approach. Med Chem Res 21:47–52

Unamba CI, Nag A, Sharma RK (2015) Next generation sequencing technologies: the doorway to the unexplored genomics of non-model plants. Front Plant Sci 6:1074. https://doi.org/10.3389/fpls.2015.01074

Urano K, Maruyama K, Ogata Y, Morishita Y, Takeda M, Sakurai N et al (2009) Characterization of the ABA-regulated global responses to dehydration in Arabidopsis to metabolomics. Plant J 57:1065–1078

Utsumi Y, Tanaka M, Morosawa T, Kurotani A, Yoshida T, Mochida K et al (2012) Transcriptome analysis using a high-density oligomicroarray under drought stress in various genotypes of cassava: an important tropical crop. DNA Res 19(4):335–345. https://doi.org/10.1093/dnares/dss016

Van Bakel H, Stout JM, Cote AG, Tallon CM, Sharpe AG, Hughes TR et al (2011) The draft genome and transcriptome of *Cannabis sativa*. Genome Biol 12:R102. https://doi.org/10.1186/gb-2011-12-10-r102

Van Zhong G, Burns JK (2003) Profiling ethylene-regulated gene expression in Arabidopsis thaliana by microarray analysis. Plant Mol Biol 53(1):117–131. https://doi.org/10.1023/B:PLAN.0000009270.81977.ef

Varshney RK, Nayak SN, May GD, Jackson SA (2009) Next-generation sequencing technologies and their implications for crop genetics and breeding. Trends Biotechnol 27(9):522–530

Varshney RK, Chen W, Li Y, Bharti AK, Saxena RK et al (2012) Draft genome sequence of pigeonpea (*Cajanus cajan*), an orphan legume crop of resource-poor farmers. Nat Biotechnol 30:83–89

Varshney RK, Gaur PM, Chamarthi SK, Krishnamurthy L et al (2013a) Fast-track introgression of 'QTL-hotspot' for root traits and other drought tolerance traits in JG 11, an elite and leading variety of chickpea. Plant Genome 6:1–9

Varshney RK, Murali Mohan S, Gaur PM et al (2013b) Achievements and prospects of genomics-assisted breeding in three legume crops of the semi-arid tropics. Biotechnol Adv 31:1120–1134

Varshney RK, Song C, Saxena RK, Azam S et al (2013c) Draft genome sequence of chickpea (*Cicer arietinum*) provides a resource for trait improvement. Nat Biotechnol 31:240–246

Vatansever R, Özyiğit II, Filiz E (2016) Comparative and phylogenetic analysis of zinc transporter genes/proteins in plants. Turk J Biol 40(3):600–611

Verdonk JC, de Vos R, Verhoeven HA, Haring MA, van Tunen AJ, Schuurink RC (2003) Regulation of floral scent production in petunia revealed by targeted metabolomics. Phytochemistry 62:997–1008

Vital CE, Giordano A, de Almeida Soares E, Rhys Williams TC et al (2017) An integrative overview of the molecular and physiological responses of sugarcane under drought conditions. Plant Mol Biol 94(6):577–594

Wang W, Scali M, Vignani R et al (2003) Protein extraction for two-dimensional electrophoresis from olive leaf, a plant tissue containing high levels of interfering compounds. Electrophoresis 24:2369–2375

Wang W, Tai F, Chen S (2008) Optimizing protein extraction from plant tissues for enhanced proteomics analysis. J Sep Sci 31:2032–2039

Wang Z, Gerstein M, Snyder M (2009) RNA-Seq: a revolutionary tool for transcriptomics. Nat Rev Genet 10(1):57–63. https://doi.org/10.1038/nrg2484

Wang K, Wang Z, Li F, Ye W, Wang J, Song G et al (2012a) The draft genome of a diploid cotton *Gossypium raimondii*. Nat Genet 44:1098–1103. https://doi.org/10.1038/ng.2371

Wang Z, Hobson N, Galindo L, Zhu S, Shi D, McDill J et al (2012b) The genome of flax (*Linum usitatissimum*) assembled de novo from short shotgun sequence reads. Plant J 72:461–473. https://doi.org/10.1111/j.1365-313X.2012.05093.x

Wang X, Kayesh E, Han J, Liu C, Wang C, Song C, Ge A, Fang J (2014a) Microarray analysis of differentially expressed genes engaged in fruit development between table and wine grape. Mol Biol Rep 41(7):4397–4412. https://doi.org/10.1007/s11033-014-3311-6

Wang M, Yu Y, Haberer G, Marri PR, Fan C, Goicoechea JL et al (2014b) The genome sequence of African rice (*Oryza glaberrima*) and evidence for independent domestication. Nat Genet 46:982–988. https://doi.org/10.1038/ng.3044

Wang W, Haberer G, Gundlach H, Gläßer C, Nussbaumer T, Luo MC et al (2014c) The *Spirodela polyrhiza* genome reveals insights into its neotenous reduction fast growth and aquatic lifestyle. Nat Commun 5:3311. https://doi.org/10.1038/ncomms4311

Wegrzyn JL, Liechty JD, Stevens KA, Wu L-S, Loopstra CA, Vasquez-Gross HA, Dougherty WM, Lin BY, Zieve JJ, Martínez-García PJ, Holt C, Yandell M, Zimin AV, Yorke JA, Crepeau MW, Puiu D, Salzberg SL, de Jong PJ, Mockaitis K, Main D, Langley CH, Neale DB (2014) Unique features of the Loblolly Pine (L.) megagenome revealed through sequence annotation. Genetics 196(3):891–909

Wessel DM, Flügge UI (1984) A method for the quantitative recovery of protein in dilute solution in the presence of detergents and lipids. Anal Biochem 138:141–143

Wiśniewski JR, Zougman A, Nagaraj N, Mann M (2009) Universal sample preparation method for proteome analysis. Nat Methods 6:359–362. https://doi.org/10.1038/nmeth.1322

World Health Organization (2009) Global prevalence of vitamin A deficiency in populations at risk 1995-2005: WHO global database on vitamin A deficiency

World Health Organization (2018) The state of food security and nutrition in the world: building climate resilience for food security and nutrition. Food & Agriculture Org. Licence: CC BY-NC-SA 3.0 IGO

Wren JD (2002) Cross-hybridization on PCR-spotted microarrays. IEEE Eng Med Biol Mag 21:71–75

Wu J, Wang Z, Shi Z, Zhang S, Ming R, Zhu S et al (2013) The genome of the pear (*Pyrus bretschneideri* Rehd.). Genome Res 23:396–408. https://doi.org/10.1101/gr.144311.112

Wu X, Gong F, Wang W (2014) Protein extraction from plant tissues for 2DE and its application in proteomic analysis. Proteomics 14:645–658

Xia Z, Xu H, Zhai J, Li D, Luo H, He C, Huang X (2011) RNA-Seq analysis and de novo transcriptome assembly of *Hevea brasiliensis*. Plant Mol Biol 77(3):299

Xie C, Wang D, Yang X (2009) Protein extraction methods compatible with proteomic analysis for the cotton seedling. Crop Sci 49:395–402

Xie C, Zhong D, Yu K, Chen X (2012) Recent advances in metabolite identification and quantitative bioanalysis by LC-Q-TOF MS. Bioanalysis 4:937–959

Xing T, Higgins VJ, Blumwald E (1996) Regulation of plant defense response to fungal pathogens: two types of protein kinases in the reversible phosphorylation of the host plasma membrane H^+-ATPase. Plant Cell 8(3):555–564

Yang H, Tao Y, Zheng Z, Zhang Q, Zhou G, Sweetingham MW et al (2013) Draft genome sequence, and a sequence-defined genetic linkage map of the legume crop species *Lupinus angustifolius* L. PLoS One 8:e64799. https://doi.org/10.1371/journal.pone.0064799

Yazdi M, Kolahi M, Kazem EM, Barnaby AG (2019) Study of the contamination rate and change in growth features of lettuce (Lactuca sativa Linn.) in response to cadmium and a survey of its phytochelatin synthase gene. Ecotoxicol Environ Saf 180:295–308

Young ND, Debellé F, Oldroyd GE, Geurts R, Cannon SB, Udvardi MK et al (2011) The Medicago genome provides insight into the evolution of rhizobial symbioses. Nature 480(7378):520–524

Yu CS, Lin CJ, Hwang JK (2004) Predicting subcellular localization of proteins for Gram-negative bacteria by support vector machines based on n-peptide compositions. Protein Sci 13 (5):1402–1406

Yu CS, Chen YC, Lu CH, Hwang JK (2006) Prediction of protein subcellular localization. Proteins 64(3):643–651

Yu CY, Mayampurath A, Tang H (2013) Software tools for glycan profiling. In: Mass spectrometry of glycoproteins. Humana Press, Totowa, pp 269–276

Zenoni S, D'Agostino N, Tornielli GB, Quattrocchio F et al (2011) Revealing impaired pathways in the an11 mutant by high-throughput characterization of *Petunia axillaris* and *Petunia inflata* transcriptomes. Plant J 68:11–27

Zhang Q, Chen W, Sun L, Zhao F, Huang B, Yang W et al (2012) The genome of *Prunus mume*. Nat Commun 3:1318. https://doi.org/10.1038/ncomms2290

Zhang J, Zhang X, Tang H et al (2018) Allele-defined genome of the autopolyploid sugarcane, *Saccharum spontaneum* L. Nat Genet 50:1565–1573. https://doi.org/10.1038/s41588-018-0237-2

Zhao CR, Sawaki Y, Sakurai N, Shibata D, Koyama H (2010) Transcriptomic profiling of major carbon and amino acid metabolism in the roots of Arabidopsis thaliana treated with various rhizotoxic ions. Soil Sci Plant Nutr 56(1):150–162

Zheng C, Zhao L, Wang Y, Shen J, Zhang Y, Jia S, Li Y, Ding Z (2015) Integrated RNA-seq and sRNA-seq analysis identifies chilling and freezing responsive key molecular players and pathways in tea plant (*Camellia sinensis*). PLoS One 10(4):e0125031. https://doi.org/10.1371/journal.pone.0125031

Understanding the Barriers of Underutilization and Applicability of Biotechnological Tools for Augmenting Spice Crop Production

11

Arjun Sharma, Shreekar Pant, Zishan Ahmad Wani, and Reetika Mahajan

Abstract

Since times immemorial, spices play an imperative part in the lifestyle of local inhabitants throughout the world. Spices have provided frequent role all the way through history, including as coloring and flavoring agents, preservatives, food additives, and medicine. Spices are also familiar for ample range of therapeutic potentials, like antiseptics, antibiotics, anti-oxygenic utilized by human race throughout the world. Many spices, such as Cardamom, Pepper, Turmeric, Clove, Coriander, Mustard, Ginger, Onion, Garlic, etc. are used in daily food preparations. All of these daily used spices have medicinal properties, such as purgative, laxative, expectorant, carminative, diuretic, etc. However, since last few decades, modern agricultural practices have resulted in decline in production of large varieties of spice crops. Apart from underutilization, few spice crops have been overexploited without employing any conservation strategies. Various environmental factors like local cultivation practices and genetic erosion have threatened diversity of spice crops. In this chapter, we first describe the importance of spice crops and the backdrops of underutilizations. We then explore the utility of biotechnological tools for improving production and cultivation of spice crops.

A. Sharma (✉)
Department of Biotechnology, BGSB University, Rajouri, Jammu and Kashmir, India

S. Pant · Z. A. Wani
Department of Botany, BGSB University, Rajouri, Jammu and Kashmir, India

R. Mahajan
Division of Plant Biotechnology, SKUAST-K, Shalimar, Srinagar, Jammu and Kashmir, India

S. M. Zargar et al. (eds.), *Neglected and Underutilized Crops - Towards Nutritional Security and Sustainability*, https://doi.org/10.1007/978-981-16-3876-3_11

Keywords

Spice crops · Medicinal properties · Underutilized crops · Genetic diversity · Biotechnological tools

11.1 Introduction

India is recognized as a "mega-biodiversity" country and its recognition partly derives from the Himalaya distinguished as a global biodiversity hotspot (Rao and Pant 2001). The Himalayan agro-ecosystem is the planet's highest and home to the world's unique crop biodiversity. Spices constitute an imperative assemblage of agricultural crops, which are vital in the gastronomic art and are in use since times immemorial (Bhardwaj et al. 2011). Spices have distinctive feature in fast food recipes, particularly in bakery foodstuffs and confectionaries. Spices account for the major category in Horticultural commodities and are viewed as absolutely, necessary for flavoring foods. Some spices are used in medication; aroma, beauty aids, and several other industries. In addition, few spices are routinely employed as coloring agents, preservatives, antioxidants, antiseptics, and antibiotics. They also play a key function in the national economy of India and in miscellaneous nations of the world, involved in import-export of spices (Edison et al. 1991; Kumar et al. 2010; Chadha and Retrinam 2010).

India is recognized as "land of spices," contributing about 20–25% of the trade in spices globally. India is bestowed with diverse agro-climatic zones, encompassing tropical, sub-tropical, and temperate zones, is heaven to harbor more spice crops. Most of the spices are utilized in the form raw plant parts or in powdered form. Essential oils and oleoresins of spices are extensively exploited in seasoning of foodstuffs and providing whiff, flavor and savor to the food products. Spices having utmost content of essential oil, oleoresin, and active biochemicals possess good export demand (Kumar et al. 2010).

Use of spices in Ayurvedic medicines is unbelievable. Indian spices, such as pepper, cardamom, chilli, ginger, turmeric, coriander, cumin, fennel, fenugreek, celery, saffron, tamarind, and garlic have medicinal properties, however cloves flowers (*Syzygium aromaticum*) have analgesic property and essential oil can be used as painkiller in dentistry (Duke et al. 2003). Cardamom (*Amomum subulatum*) stimulates appetite in people suffering with anorexia nervosa and its seeds have aphrodisiac property (Duke et al. 2003). The rhizome of ginger (*Zingiber officinale*) has gastrointestinal properties and also provides relief against dyspepsia, flatulence, abdominal discomfort, and nausea (Kikuzaki et al. 1991; Phillips et al. 1993). Spices also provide considerable amount of minerals and micro-nutrients, such as iron, calcium, magnesium, etc.

Spices crops are facing so many challenges in the present time. Majority in the loss of spice crops are due to biological stresses particularly pathogens, meager disease management system, and lack of robust host resistance (Sarma et al. 2014). Major threat in these crops including foot rot disease in black pepper, *Fusarium* yellows and bacterial wilt in ginger, turmeric and root rots, wilts in cumin, coriander, and fenugreek (Sarma 2010). Integrated pest management (IPM), such as microbial biocontrol technology will be an ideal option. This technology will help in reducing pesticide accumulation in the ecosystem and has been attained a substantial consideration. Biological control has been found to be an effectual method in lessening the crop losses in soil-borne diseases not only in case of spice crops but other crops also (Cook 1985; Sarma et al. 2014). Hyper parasitic fungi including *Trichoderma harzianum, T. viride, Pochonia chlamydosporia, Paecilomyces lilacinus,* and plant growth-promoting Rhizobacteria, such as *Pseudomonas fluorescens, Pseudomonas cepacia, Bacillus subtilis,* and Arbuscular mycorrhiza have been found to be effective in reducing crop losses caused by soil-borne plant pathogens in the spice crops (Fridlender et al. 1999). By adopting solid and liquid fermentation technologies, large-scale production of these microbes can be achieved (Chet 1987; Batta 2004). In seed spices, treatment of seeds is important where seed contamination is the chief cause of infectivity (Singh et al. 1972; Sarma and Anandaraj 1998).

Through intensive investigations, it has been established that *Trichoderma* is an effective biocontrol agent to suppress the activity of several soil-borne plant pathogens (Papavizas 1985). These biocontrol agents have been received a huge attention in the management of spice crop diseases in black pepper, small cardamom, ginger, turmeric, cumin, coriander, and fenugreek (Sharma and Jain 1979; Suseela Bhai et al. 1993; Usman et al. 1996; Suseela Bhai 1998; Vyas and Mathur 2002; Saju 2004; Sarma and Saju 2004; Haggag and Abo Sedera 2005; Sarma 2010; Sarma et al. 2013, Kakani et al. 2009). *Trichoderma* consists of lytic enzyme like cellulases, glucanases, and chitinase, and antibiotics that are responsible for the rupture of cell wall of pathogen in spice crops (Suseela Bhai 1998). Root rot suppression in cardamom has been confirmed with soil application of *T. harzianum* (Sarma 2006). Similarly, rhizome rot of ginger caused by *Pythium aphanidermatum* has been suppressed by the application of *T. harzianum*, both as seed treatment and soil application coupled with neem cakes (Usman et al. 1996). In fenugreek, root rot is caused by *Rhizoctonia solani.* It has been lucratively restricted with the application of *T. viride* by treatment in soil as well as seed coating (Kakani et al. 2009).

Insufficient accessibility of high yielding varieties, absence of pest and disease resistant genotypes and dearth of variability affects the productivity of many spice crops. Biotic stresses have serious effects on spice crops (Kumar et al. 2015). Due to long pre bearing age, crop improvement in majority of spice crops is a tedious and long-term process (Ravindran et al. 1996). Some of the constraints for spice production including soil-borne nature of the causal organisms, meager flexibility for effectual disease management, and lack of robust host resistance (Sarma et al. 2014).

Spices growing in India are prominent for their admirable fragrance, flavor, and tanginess, not common in spices growing in any other country (Rathore and Shekhawat 2008). Furthermore, spices are rich in certain bioactive compounds like linalool, allicin, betalains, carvone, and limonene (Anupam et al. 2018). So many spices like Ajmoda, Betel leaf, Black mustard, Green mint, etc. which not only need commercial exploitation, but also are required to make familiar for their specific use, as they have lot of medicinal properties. Due to the presence of bioactive compounds, spices are indispensible for human health. In traditional health care systems of India, spices are being used from ancient times for curing day-to-day health problems like headache, stomachache, fever, cough, cold, skin diseases, and hair care. Unfortunately, only few spices growing in Indian Himalayan Region are being exploited for their commercial values. This is the reason that spices export of India is notable although not so extravagant taking into account the historical significance of India as a land of spices (IISR 2013). So, there is a need to exploit the underutilized spices of the Indian Himalayan Region and promote the cultivation and marketing awareness among the masses. Proceeding parts of this chapter provides up-to-date documentation of spices in India, along with their phytochemistry, biological activities, production, marketing strategies, and constraints in production. Furthermore, the role of biotechnological research in promoting and enhancing the cultivation of spices has been discussed.

11.1.1 Phytochemicals and Their Biological Activities

Spices are also used for making processed meat, soups, sauces, pickles, dairy products, baked foods, confectionary, snacks, beverages, cosmetics, perfumes, hygiene products, and aerosols. Besides these, spices are a rich source of phytochemicals and thus play an imperative role in human health care systems (Table 11.1). Phytochemicals are naturally occurring biologically active compounds, which afford health benefits for humans supplementary than those endorsed to macro- and micro-nutrients (Hasler and Blumberg 1999). Various bioactive compounds present in the spices growing all over India and rest of the countries with potential beneficial effects (Table 11.2). Chemical structure of some of the bioactive compounds is given in Fig. 11.1. Owing to the presence of these bioactive chemicals, spices have a high medicinal potential (Dubey 2017). Spices are a rich source of bioactive compounds that show many biological activities, viz., Innocuous, carminative, diaphoretic, emmenagogue, expectorant, febrifuge, hypotensive, stimulant, vermifuge, cardio-protective, anti-bacterial, antifungal, antiviral, antitumor, anti-inflammatory, anti-ulcer, anti-asthmatic, anti-diabetic, digestive, hepato-protective, laxative, diuretic, anti-depressant, antioxidant, anti-secretory, allergenic, aperients, rubefacient, anthelminthic, neuroprotective, antiscorbutic, anti-pyretic, anti-leishmanial, analgesic, anti-malarial, anti-obesity, ophthalmic, and sedative.

11.1.2 Production and Global Marketing Scenario

Spices are high worth and low volume products of trade on the global market. All over the globe, the fast growing nutraceutical industry relies chiefly on spices for

Table 11.1 List of the spices along with their habit, biological activities, and active compounds

S. No.	Taxa	Family	Common name	Habit	Part (s) used	Biological activities	Active compounds
1	*Acorus calamus* L.	Acoraceae	Sweet Flag	H	Rhizomes	Antiseptic, aphrodisiac, carminative, diaphoretic, emmenagogue, expectorant, febrifuge, hallucinogenic, hypotensive, sedative, stimulant, stomachic, and vermifuge	Linalool, Chalocone, Eugenol, 1,8-cineole, Phenyllindane, Phenyl propane, Acoradin, Asaraldehyde
2	*Allium griffithianum* Boiss.	Alliaceae	–	H	Seeds	Anti-bacterial	Allicin
3	*Allium humile* Kunth.	Alliaceae	Small alpine onion	H	Seeds	Cardio-protective, anti-bacterial, anti-inflammatory, anti-asthmatic, anti-diabetic	Allicin
4	*Allium roylei* Stearn in Herb.	Alliaceae	–	H	Leaves	Antioxidant, anti-bacterial, antifungal	Allicin
5	*Allium straycheyi* Baker	Alliaceae	Jambu	H	Leaves	Hepato-protective, antioxidant, anti-inflammatory	Allicin, dimethyl trisulfide, dimethyl disulfide, 2,4-dimethylthiophene, 1,2-bis (methylthio) ethene, 16-hentriacontanone, 4-nitrophenyl ester, hexadecanoic acid, Heneicosane, isobutyl octadecyl ester oxalic acid
6	*Alpine galanga* Wild.	Zingiberaceae	Greater Galangal	H	Rhizomes	Aphrodisiac, aromatic, bitter, digestive, expectorant, antifungal, anti-bacterial, antiviral, anti-diabetic	Galangin, Myrcene, Ocimene, Pinene, Borneol, Curcuminoid

(continued)

Table 11.1 (continued)

S. No.	Taxa	Family	Common name	Habit	Part (s) used	Biological activities	Active compounds
7	*Amaranthus cruentus* L.	Amaranthaceae	Red amaranth	H	Leaves, seeds	Laxative, expectorant, anti-tumor, diuretic	Palmitic acid, Stearic acid, Oleic acid, linoleic acid, Docosaenoic acid, flavonoids, Iridoïds, Cardenolids, Carotenoids, Betalains
8	*Amaranthus spinosus* L.	Amaranthaceae	Spiny amaranth	H	Leaves	Hepato-protective, diuretic, anti-depressant, anti-diabetic, anti-cholesterolemic, anti-malarial anti-allergenic	Hexatriacontane, amaranthine, isoamaranthine, quercetin, Rutin, α-spinasterol, β-sitosterolglucoside, β-D-ribofuranosyl adenine, α-xylofuranosyluracil
9	*Anethum graveolens* L.	Apiaceae	Dill	H	Seeds, leaves	Carminative, stomachic, diuretic, anti-ulcer, anti-secretory, antioxidant	Linolenic acid, Palmitic acid, D-Carvone, Dilapiol, Anethofuran, Limonene, Pinene, Diterpene, Dihydrocarvone, Cineole, Myrcene, paramyrcene, Apiol, Myristicin, quercetin isoharmentin
10	*Angelica glauca* Edgew.	Apiaceae	Angelica	H	Fruits, roots	Stomachic, anti-allergenic, anti-asthmatic	Furocoumarin, Archangelin, Oxypeucedanin
11	*Apium graveolens* L.	Apiaceae	Celery	H	Leaves, fruits, roots	Anticoagulant, cardio-protective, antioxidant, diuretic, hepato-protective	Limonene, β-Selinine, Apiole, Sedanolide, Sphondin, 3-Butylphthalide, Psoralen, Bergapten, Xanthotoxin, apiumetin, apiumoside, apigravrin

12	*Armoracia rusticana* Gaertn.	Brassicaceae	Horse reddish	H	Roots	Antiseptic, aperient, digestive, diuretic, expectorant, rubefacient, and stimulant	Isothiocyanate
13	*Artemisia absinthium* L.	Asteraceae	Worm wood	H	Leaves	Anthelminthic, antifungal, neuroprotective, hepato-protective, anti-malarial, antioxidant, and anti-depressant	Z-epoxyocimene, chrysanthenol, chrysanthenyl acetate, artabsin, absinthin, β-pinene, β-thujone, α-thujone, sabinyl acetate, Lavandulol, cis-Dihydrocarvone, Terpinolene, Limonene
14	*Artemisia dracunculus* L.	Asteraceae	Tarragon	H	Leaves	Anti-inflammatory, hepato-protective, anti-hyperglycemic, antiscorbutic, diuretic, emmenagogue, hypnotic, and stomachic	A-pinene, Sabinene, β-pinene, β-myrcene, α-phellandrene, Limonene, *p*-cymene, β-ocimene, Linalool, Borneol, Thymol, Bornyl acetate, Anethole
15	*Artemisia scoparia* Waldst. & Kit.	Asteraceae			Seeds; leaves	Antioxidant, antimicrobial, anthelminthic, anti-nociceptive, anti-inflammatory, ant-pyretic activity, anti-leishmanial, analgesic, anti-malarial, anti-obesity	Quarcetin-3-glycogalactoside, Kaempferol-3-glycigalactoside, quercetin-3,7-rutinosodigalactoside, querccetin-3, 7-rutinosoga-lactoside, β-thujone, α-thujone, 1,8-cinelol, scoparon
16	*Brassica juncea* L. Czern.	Brassicaceae	Indian Mustard	H	Seeds	Anodyne, aperitif, diuretic, emetic, rubefacient, and stimulant	Sinigrin, Allyl-3-butenyl Isothiocynate, Carotene
17	*Brassica napus* L.	Brassicaceae	Rape	H	Seeds	Anti-inflammatory, hepato-protective, anti-scurvy, diuretic	Oleic acid, linoleic acid, α-linolenic acid, erucic acid, progoitrin, gluconapin, *p*-hydroxybenzoic, vanillic acid, gentisic acid, protocatechuic

(continued)

Table 11.1 (continued)

S. No.	Taxa	Family	Common name	Habit	Part (s) used	Biological activities	Active compounds
							acid, syringic acid, *p*-coumaric acid, ferulic acid
18	*Cannabis sativa* L.	Cannabaceae	Hemp	H	Seeds	Anodyne, anthelmintic, antiemetic, anti-inflammatory, antiperiodic, antispasmodic, cholagogue, diuretic, emollient, hypnotic, hypotensive, laxative, narcotic, ophthalmic, and sedative	Cannabidiol, cannabichromene, cannabigerol
19	*Capparis spinosa* L.	Capparidaceae	Caper Bush	S	Flowers	Anti-diabetic, anti-obesity, anti-hypertensive, antimicrobial, anti-inflammatory, anti-hepatotoxic	Rutin, Carpic acid, Glucocapparin, Glucocleomin, Isothiocynate, Thiocyanate
20	*Celosia argentea* L.	Amaranthaceae	Silver cock's comb	H	Flower, leaves	Astringent, hemostatic, ophthalmic, parasiticidal, poultice, hypotensive, ophthalmic, anti-diarrhea, anti-tumor	Saponins, Cycpeptides, Lutin, Epigallocatechin, Gallic acid, Rosmarinic acid, Quercetin, Ascorbic acid
21	*Chenopodium album* L.	Amaranthaceae	Fat Hen	H	Leaves	Anthelmintic, antiphlogistic, anti-rheumatic, mildly laxative, odontalgic	Caempferol, quercetin, isoramnetin, apocarotenoids, flavonoids, Kaempferol, kaempferol 3-*O*-β-glucoside, kaempferol 3-*O*-β-diglucoside, kaempferol-3-*O*-arabinoglucoside
22	*Cleome viscosa* L.	Cleomaceae	Tick weed	H	Seeds	Anthelmintic, carminative, rubefacient, stimulant, and vesicant	Diterpene, Cleomeolide, Cleomaldeic acid, Cleomiscosin

23	*Garcinia cambogia* L.		Gamboges	T	Fruits	Hepato-protective, anti-inflammatory	Hydroxycitric acid, ascorbic acid
24	*Hyssopus officinalis* L.	Lamiaceae	Hyssop	T	Leaves	Stimulant, carminative, sedative, antispasmodic, diuretic, stomachic, and emmenagogue	Quercetin, Apigenin, Luteolin, Acacetin, Diosmin
25	*Illicium verum* Hook.f.	Magnoliaceae	Star Anise	T	Fruits	Anti-bacterial, antifungal, carminative, diuretic, odontalgic, stimulant, and stomachic	A-pinene, α-phellandrene, *p*-cymene, Limonene, Linalool, α-terpineol, Estragole, Chavicol, cis-anethole, Eugenol, methylchavicol, safrole
26	*Impatiens balsamina* L.	Balsaminaceae	Rose Balsam	H	Leaves	Cathartic, diuretic, anti-inflammatory, and emetic	Alkaloids, glycosides, phenols, flavonoids, saponins, steroids, tannins, triterpenes
27	*Levisticum officinale* Koch.	Apiaceae	Lovage	H	Leaves, seeds	Antispasmodic, aromatic, carminative, diaphoretic, digestive, diuretic, mildly expectorant, and stimulant	A-pinene, Sabinene, β-pinene, myrcene, α-phellandrene, Terpinolene, Tricosane, Globulol, Isomenthol, Zingiberene, Elemol
28	*Mangifera indica* L.	Anacardiaceae	Mango	T	Unripe fruit	Antiseptic, astringent, diaphoretic, stomachic, vermifuge, tonic, laxative, and diuretic	Protocatechic acid, Catechin, Mangiferin, Alanine, Glycine, Kinic acid, Shikimic acid
29	*Majorana hortensis* Mon.	Lamiaceae	Sweet Marjoram	H	Leaf	Antioxidant, anti-tumor, synergistic	β-pinene, Sabinene, β-Myrcene, Limonene, 3-octamol, Linalool
30	*Myristica fragans* Houtt.	Myristicaceae	Nutmeg	T	Seed	Astringent, anti-inflammatory, digestive, stimulant, carminative, and aphrodisiac	Maslignan, α-pinene, Camphene, Sabinene, Limonene, Linalool, Terpinolene, 4-terpinol, Safrole, Elimicin, Myristic acid, Etearic acid, Ethyl palmitate

(continued)

Table 11.1 (continued)

S. No.	Taxa	Family	Common name	Habit	Part (s) used	Biological activities	Active compounds
31	*Ocimum basilicum* L.	Lamiaceae	Sweet basil	H	Leaf	Antispasmodic, aromatic, carminative, digestive, galactagogue, stomachic, and tonic	Linalool, Methyl Chavicol
32	*Ocimum sanctum* L.	Lamiaceae	Holy Basil	H	Leaf	Antifertility, anticancer, anti-diabetic, antifungal, antimicrobial, hepato-protective, cardio-protective, antiemetic, antispasmodic, analgesic, adaptogenic, and diaphoretic	1,8-Cineole, β-ocimene, β-caryophyllene, α-Humulene, Luteolin, Apigenin, Molludistin
33	*Origanum vulgare* L.	Lamiaceae	Oregano	H	Leaf, flower	Antiseptic, antispasmodic, carminative, cholagogue, diaphoretic, emmenagogue, expectorant, stimulant, stomachic, and mildly tonic	Palmitic acid, Stearic acid, Oleic acid, Linolenic acid, Apeginin
34	*Papaver somniferum* L.	Papaveraceae	Opium Poppy	H	Seed	Anodyne, antitussive, astringent, diaphoretic, emmenagogue, hypnotic, narcotic, and sedative	Morphine, Codeine, Papaverine, Narcotine
35	*Perilla frutescens* (L.) Britton	Lamiaceae	Beefsteak	H	Seeds	Antioxidant, anti-tumor, anti-bacterial, anti-depressant, anti-inflammatory	Luteolin, Chrysoeriol, Apigenin, rosmarinic acid
36	*Petroselinum crispum* (Mill.) Fuss.	Apiaceae	Parsley	H	Leaves	Antimicrobial, antioxidant, anti-diabetic, cardio-protective, analgesic, spasmolytic, anticancer	A-pinene, Sabinene, β-pinene, β-myrcene, Terpinolene, *p*-cymene
37	*Pimpinella anisum* L.	Apiaceae	Aniseed	H	Leaves, seeds	Antiseptic, antispasmodic, aromatic, carminative, digestive,	*Trans*-anethole, methylchavicol, anisaldehyde, estragole,

						expectorant, pectoral, stimulant, stomachic, and tonic	coumarins, scopoletin, umbelliferone, estrols, terpene hydrocarbons, polyenes, and polyacetylenes
38	*Punica granatum* L.	Punicaceae	Pomegranate	S	Seed, bark	Anti-bacterial, antifungal, astringent, demulcent, stomachic, antioxidant	Castalagin, Casuariin, Casuarinin, Corilagin, Granatin A, Granatin B
39	*Rumex hastatus* Don.	Polygonaceae	Arrowleaf Dock	H	Leaves, roots	Antioxidant, laxative, tonic, anti-rheumatic	Rutin, Vitexin, Luteolin, Luteolin-7-glucoside
40	*Salvia officinalis* L.	Lamiaceae	Sage	S	Leaves, flowers	Anticancer, anti-inflammatory, anti-nociceptive, antioxidant, antimicrobial, antimutagenic, antidementia, hypoglycemic, and hypolipidemic	Rosmarinic acid, ellagic acid, caryophyllene, cineole, elemene, humulene, ledene, pinene, thujone, epicatecin, epigallocatechin gallate, quercetin, rutin, luteolin-7-glucoside
41	*Satureja hortensis* L.	Lamiaceae	Summer savory	H	Leaves	Antioxidant, antimicrobial, sedative, anti-inflammatory	Thymol, γ-terpinene, carvacrol, *p*-cymene, rosmarinic acid, caffeic acid, naringenin, Apigenin
42	*Thymus linearis* Benth. ex Benth.	Lamiaceae	Himalayan Thyme	H	Seeds	Antiseptic, analgesic, anti-pyretic, expectorant, antispasmodic, anti-inflammatory, and carminative	Thymol, *p*-cymene, γ-terpinene, carvacrol, *p*-cymen-8-ol, borneol, terpinen-4-ol, thymol methyl ether
43	*Zanthoxylum armatum* DC.	Rutaceae	Winged Prickly Ash	S	Fruit	Stomachic, vermifuge, carminative, stimulant	α-Fenchol, α-Terpinene, α-Thujene, α-Thujone, α-Pinene, Geraniol, γ-Terpinene, Limonene, Linalool, Linanyl acetate, Myrcene, Phellandrene, Sabinene, Nevadensin, β-Sitosterol-β-D-glucoside

Table 11.2 Bioactive compound and potential benefits of spices

Spices	Bioactive compounds	Potential beneficial effects	References
Cinnamon	Procyanidin	Antioxidant	Gruenwald et al. (2010)
Ginger	Zerumbone	Anti-inflammatory	Bordia et al. (1997)
Garlic	Allicin	Antioxidant	Rahman (2001)
Turmeric	Capsaicin	Anti-inflammatory	Vasanthi and Rarameswari (2010)
Onion	Polyphenols, Flavonoids, Flavonols	Antioxidant	Kleijnen et al. (1989)
Red pepper	Curcumin, Capsaicin	Antioxidant, anti-inflammatory	
Fenugreek	Rhaponticin, Isovitexin	Antioxidant, Anticarcinogenic	Bordia et al. (1997)

taste and flavor. Well-being aware purchasers in developed nations lean towards natural products of plant origin instead of cheap synthetic ones. Subsequently, spices are the indispensable structure squares of flavor in food preparations. India actually keeps on being the largest producer, consumer, and exporter of spices in the world. India exports its spices not less than 120 countries of the world; however, a couple of nations lead the importers list for Indian spices in terms of the imported quantity of the spices. It is clear from the region-wise export data that the Asian zone is promptly budding as the foremost destination for Indian spices with 66% in quantity and 48% in worth followed by American zone and European Union countries and their inherent values make them noticeably better in terms of taste, color, and fragrance. USA, Canada, Germany, Japan, Saudi Arabia, Kuwait, Bahrain, and Israel are the central markets for Indian spices. Mexico still being the foremost importer of cinnamon and cassia; while as Saudi Arabia, Bahrain, Kuwait, and Israel are the leading markets for green cardamom, black pepper, ginger, and turmeric. Among the spices cultivated in India, seed spices account for 37% and 17% of the total area and production, respectively. Approximately 24% of export and 10% of the national income from spices is contributed only by seed spices. Out of the 17 seed spices grown, coriander, cumin, fennel, fenugreek, dill seed, celery, and poppy seeds are most important. India possesses many innate advantages like its ample genetic diversity and variability, wide-ranging soil and climatic conditions, and skillful man power over other spice producing countries. Still, productivity of many spice crops is low in India. In spite of the incredible significance of spices, it is quite untoward that the sector has not attained the requisite height of growth and expansion, probably due to the tribulations in the marketing, supply chain, exports, pre- and post-harvesting activities (Bhardwaj et al. 2011).

The altered economic scenario array in the milieu of globalization and liberalization of global trade in agriculture has opened up new-fangled panorama of growth and development. The spices sector is one of the important areas in which India has an intrinsic potency to lead the global markets. Our growth in spices export is noteworthy though not extravagant taking into account the historical significance

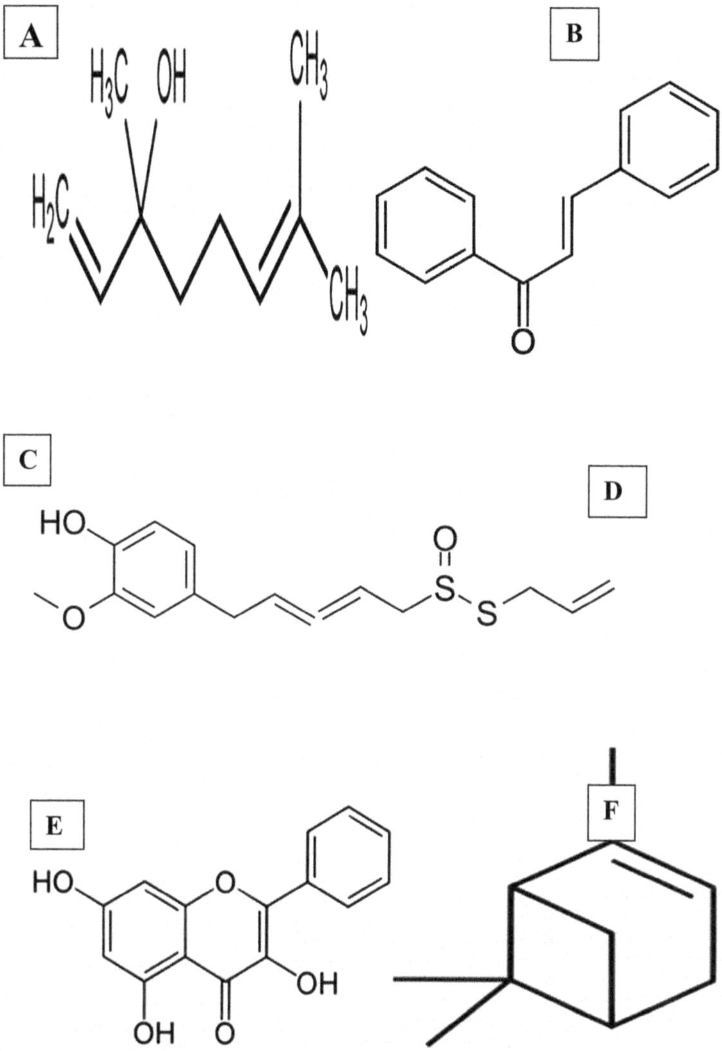

Fig. 11.1 (**a**) Linalool, (**b**) Chalocone, (**c**) Eugenol, (**d**) Allicin, (**e**) Galangin, (**f**) Pinene, (**g**) Borneol, (**h**) Ocimene, (**i**) Quercetin, (**j**) Rutin, (**k**) Limonene, (**l**) Apiol, (**m**) Isothiocynate, (**n**) Thujone, (**o**) Thymol, (**p**) Kaempferol, (**q**) Apigenin, (**r**) Luteolin

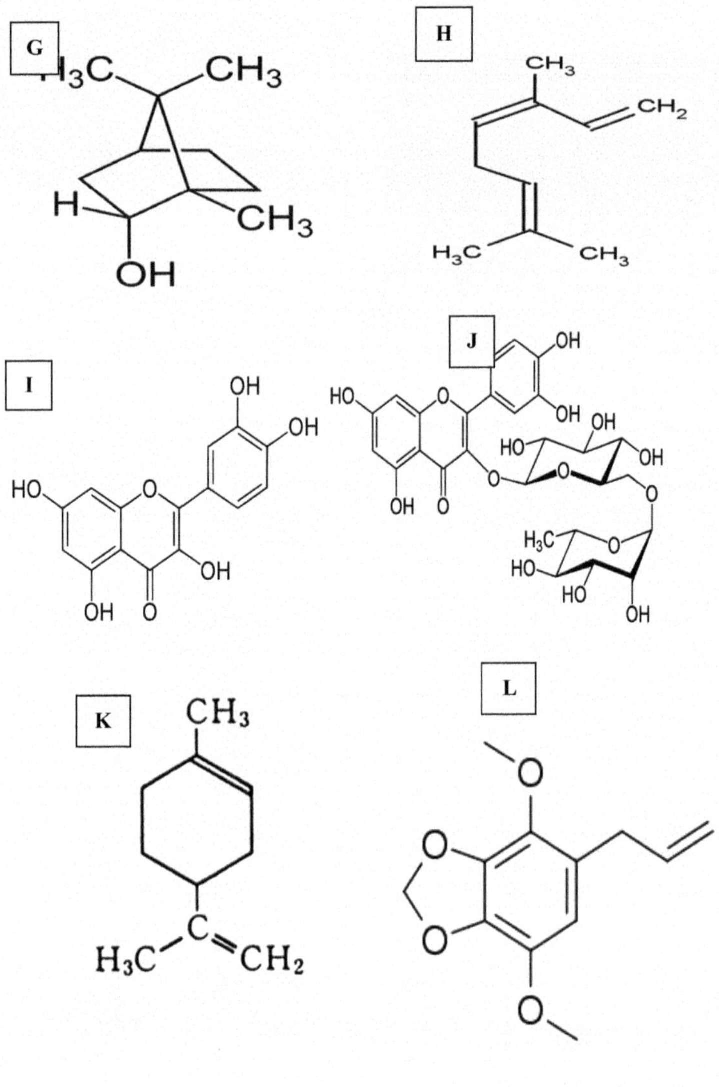

Fig. 11.1 (continued)

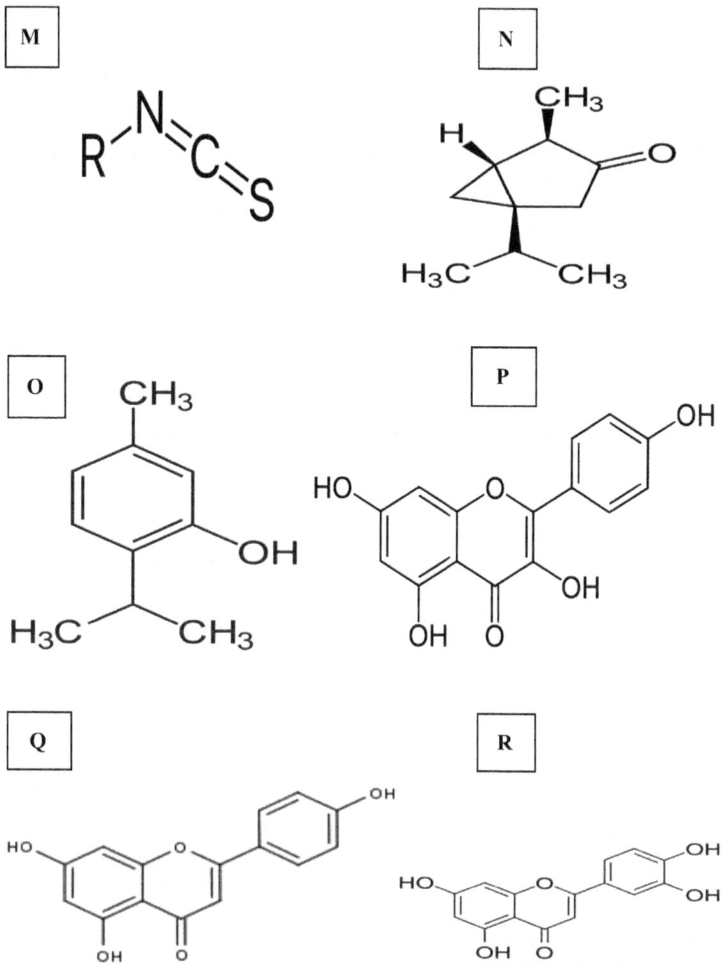

Fig. 11.1 (continued)

of India as a land of spices. Over the years, India's share in world spices market has not cherished a lot and its domination as a purveyor of spices is threatened by nations like China, Brazil, Vietnam, Pakistan, Egypt, Turkey, and other African and Caribbean countries. India also faces dearth of exportable superfluous, due to rising domestic demand. Sharp changes in the quantum and value of exports and in the unit value realization have characterized the spices trade in recent years.

11.1.2.1 Constraints Faced in Production of Spices in India (IISR 2013)

1. Lack of accessibility of worth planting materials and high yielding varieties.
2. Low productivity is due to the presence of varieties of meager inherent potential in terms of quality and quantity.
3. Lack of acceptance and implementation of suggested cultural practices, along with soil and water conservation measures.
4. Non-adoption of integrated pest and disease management practices.
5. Scarce conservatory set-up for effectual conveying of technology.
6. Soaring alterations in prices of the merchandise and the dearth of a prop up price.
7. The prime setback that Indian spices face in the international market are the soaring cost of the product coupled with microbial contaminants including mycotoxins in the finished products.

India can resist the rivalry only by increasing yield and reducing cost of cultivation leading to low cost per unit of production; improving quality of the reducing pesticidal residues and mycotoxins. In order to achieve these goals, sufficient research support is desirable. In India, research and development work in spices cultivation and improvement is carried out by Indian Council of Agricultural Research (ICAR), State Agricultural Universities (SAUs), Spices Board, Directorate of Arecanut and Spices Development, Council of Scientific and Industrial Research (CSIR), and few other organizations.

11.1.3 Importance of Spices

Spices are an integral part of daily foodstuffs as well as are also known for their therapeutic potential. Spices add exceptional natural flavor to cuisines and make food more striking in color. Some of the spices are very fine source of food preservative particularly for industrially processed foods. The importance of spices cannot be overemphasized in terms of food and medicine. Secondary metabolites of *Piper nigrum* have properties, such as anti-apoptotic, anti-bacterial, anti-depressant, anti-diarrheal, anti-inflammatory, antifungal, anti-oxidative, anti-pyretic, antispasmodic, anti-tumor, anti-thyroid, hepato-protective, insecticidal, gastric ailments, and intermittent fever, and others (Ahmad et al. 2013).

Celery (*Apium* graveolens) a biennial herb rich in secondary metabolites and is cultivated worldwide (Sowbhagya 2014). One of the secondary metabolite named polyacetylenes, which is found in celery leaves have positively affects on human cancer cells. It helps in diminution of tumor formation (Christensen and Brandt 2006). Cloves (*Syzygium aromaticum*) are the immature dried buds of the clove tree. Clove oil contains constituents, such as eugenol (70–80%), β-caryophyllene (5–12%), and eugenyl acetate (15%). Secondary metabolites found in cloves are flavonoids, galloyltannins, phenolic acids, and triterpenes. Clove buds are used in traditional medicine particularly as diuretic, odontalgic, stomachic, tonicardiac, aromatic, carminative, and stimulant. It also has antioxidant properties, which can be used for curing of diseases like ulcers, coronary, hyper and hypotension (Mishra and Sharma 2014).

Coriander (*Coriandrum sativum*) seeds possess phytochemicals like linalool, carvone, geraniol, limonene, borneol, camphor, etc. having considerable curative properties. Traditionally it has been used as a remedy for inflammation. It contains antioxidants and inhibits aflatoxin, bacteria (Baratta et al. 1998). In Cumin (*Cuminum cyminum*), using Gas chromatography/Mass spectrometry identified secondary metabolites including α-terpinene-7-al, cumin aldehyde, α-terpinene-7-al, *p*-cymene, and β-pinene, respectively. The unique flavor of cumin is owed to the presence of monoterpenes, such as α-pinene and cis-β-farnesene. Cumin is also used to cure diseases, such as indigestion, jaundice, and diarrhea and possess anti-inflammatory, diuretic, carminative, and antispasmodic properties (Mohammadpour et al. 2012).

Fenugreek (*Trigonella foenum-graecum*) is rich in alkaloids, mainly trigonelline, choline, gentianine, and carpaine, flavonoids, such as apigenin, luteolin, orientin, quercetin, vitexin, isovitexin, saponins, and volatile oils (*n*-alkanes and sesquiterpenes). It contains antioxidant and is used traditionally to cure diabetes, cancer, inflammation, obesity, hypertension, and other disorders (Rathore et al. 2013).

Garlic (*Allium sativum*) has been used for culinary and remedial purposes (Karuppiah and Rajaram 2012). The medicinal properties include hypolipidemic, antithrombotic, anti-hypertensive, ant hyperglycemic, antihypercholesterolemic, and immunomodulatory. Consumption of garlic or its oil helps in reducing bad cholesterol, triglyceride levels, and ultimately cardiovascular risk (Arekemase et al. 2013). Ginger (*Zingiber officinale*) contains constituents, such as gingerols, shagaols, β-carotene, caffeic acid, curcumin, salicylate, and capsaicin. Steam-oil components, such as sesquiterpene hydrocarbons, monoterpene hydrocarbons, and oxygenated monoterpenes are liable to aroma of ginger and non-steam-volatile components such as gingerols is responsible for its pungency (Purseglove et al. 1981). Spices mixtures incur as much to the principle of Ayurveda as much as they do to flavor adjustments (Kumar et al. 2017). Ajwain (*Trachyspermum ammi*) have digestive properties and also used to reduce gaseous and bloating properties of beans and legumes. Fenugreek seed tea helps in treating diabetes and lowering cholesterol. Nutmeg is helpful in treatment of heart related problems and asthma. Pepper helps during cold, cough, infections (Suresh et al. 2013). Ecological disturbances have devastated ample amount of these beneficial materials from its natural habitats. Efforts should be made for the conservation of these valuable germplasm. In vitro gene bank is a safe alternative to preservation and protection of the germplasm of these from the unpredictable change of nature (Ravindran et al. 2004). Crop improvement programs aspire to enhance the yield and eminence of a specific crop to accomplish the demands of growing human population. Major production constraints in spices are deficiency of high yielding, pest and disease resistant varieties, and an inadequate genetic diversity in some crops.

11.2 Biotechnology Research (A Great Potential to Explore Spices)

Biotechnological techniques provide new and exciting opportunities to explore and solve the problems related to spice crop production. Biotechnological tools offer an alternative to conventional crop improvement in spices by identification and isolation of genes valued for crop upgrading, improvement and of industrial values. From past few years, implementation of biotechnological tools helped to combat the problem through commercial propagation, development of new varieties and Marker-Assisted Breeding (Babu et al. 2012). Principle achievements of biotechnology in improvement of spice crops include micropropagation and clonal multiplication in cardamom, black pepper, ginger, turmeric, cinnamon, and vanilla generating excellent quality plantlets, utilizing somaclonal variation and recombinant DNA techniques (Babu et al. 1997; Doyle & Doyle 1990). For developing resistance against rhizome rot in ginger, in vitro selection for resistance to environmental stresses is in progress (Babu et al. 1992; Bhat et al. 1994; Valsala et al. 1997). Synthetic seed technologies for cardamom, ginger, turmeric, and pepper are also available (Bhojwani 1980; Bhat et al. 1992, 1994, 1995; Agarwal 1988; Sim et al. 1998; Rao and Ravishankar 2000; Salvi et al. 2002; Babu et al. 2003, 2005; Chan and Thong 2004; Rahman et al. 2004). With the use of biotechnological techniques, such as commercial propagation, production of new varieties, and breeding lines through somaclonal variation, anther culture, protoplast fusion, bioreactor, and recombinant DNA technologies for advancing, preserving, and exploiting the diversity and escalating the efficacy of spices can be achieved. Biotechnological tools have provided a range of conventional methods for the conservation, characterization, improvement, and utilization for ever-increasing production and productivity of spices.

The techniques of micropropagation are accessible in many spices for generating pathogen free planting material. In many spices somaclonal variation is important, where innate variability is least. The somaclonal variants have been reported in ginger, vanilla, and turmeric. In garlic, saffron, capsicum, black pepper, fennel, fenugreek, and peppermint proplast technology also have been developed in recent years. So many studies are also in progress by using immobilized and transformed cell cultures technique for in vitro production of flavor and also coloring compounds like capsaicin, anethole, crocin, picrocrocin, vanillin, etc. Techniques like molecular taxonomy, identification of duplicate hybrids, estimation of genetic fidelity, crop profiling, fingerprinting, and tagging of genes for marker aided selection (MAS), isolation of useful; genes and development of transgenic plants by using molecular markers are attaining importance (Roychowdhury et al. 2013; Das et al. 2011).

The propagation via somatic embryogenesis and in vitro flowering has been reported (Stephen and Jayabalan 2000). Micropropagation techniques have been adopted for producing virus-free garlic (*Allium sativum*), for abolition of mite-borne virus (Bertaccini et al. 2004). For plant regeneration, anther and microspore cultures have been employed in fennel and celery. Shoot and seed derived callus was also accounted in vanilla for their plant regeneration purpose (Babu et al. 1997; Minoo

2002). According to different studies about 16 microsatellite markers from black pepper have been developed, characterized and been used to study the genetic diversity of 20 varieties from Brazilian collections and also 40 popular genotypes and 4 different species of black pepper from India, respectively (Menezes et al. 2009; Joy et al. 2007, 2011). It has been reported that isolation and characterization of 11 and 9 polymorphic microsatellites loci from a *Piper polysyphonum* and *Piper solmsianum* provide a positive medium to understand structure of population and interrelationships between them (Liao et al. 2009).

11.3 Important Biotechnological Tools for Spices Cultivation: Detailed Analysis

11.3.1 Micropropagation: An Important Approach to Produce Disease-Free Spice Crops

Micropropagation is an in vitro multiplication and/or rejuvenation of explants under aseptic and controlled environmental conditions. A tiny section of a tissue can be used to generate thousands or millions of plantlets for transplantation. Micropropagation and plant regeneration are based on the principle of totipotency. Plant regeneration can be achieved through cultured explants or totipotent cells, which leads to the formation of organs (through explants) or undergo embryonic pathway to form somatic embryos (through totipotent cells), which is further grown to generate whole plants. Micropropagation involves the selection of plants and their growth for about 3 months under aseptic conditions. The culture of these plants is carried out in a suitable culture medium. Explants, such as organs, shoot tips, and axillary buds can be selected and surface sterilized. Then explants are cultured in a culture medium where multiplication of the shoots or rapid embryo formation occurs. Then shoots formed can be transferred to another medium for further growth or can be directly planted to soil to develop roots and further growth (Chang and Criley 1993; Ayabe et al. 1995; Adaniya and Shirai 2001; Aasim et al. 2009). This technique can solve the problems related to spice crop production.

Micropropagation technique using explants, such as shoots whether from mature or from juvenile plants administer positive substitute for such plant species that exhibit opposition to conventional bulk propagation method. It can be applied for extensive production of disease resistant crops and also helps in conservation of germplasm. Immense multiplication percentage coalesced with the increased benefit of attaining pathogen free crops compel micropropagation a reasonable substitute to conventional propagation (Babu et al. 2003).

Black pepper (*Piper nigrum* L.) is an endemic spice crop in India and mentioned as the "King of spices." Micropropagation of black pepper using distinct explants has been reported. Using Benzyl adenine (BA) in the culture medium (Murashige and Skoog, Medium, 1962) with desired concentration of auxins, multiple shoots can be induced (Babu 1997; Babu et al. 2012). Multiple shooting were also stimulated from shoot tips and nodal regions of *Cinnamomum camphora* on

Woody Plant Medium (WPM), from a cotyledonary node on MS medium (Murashige and Skoog medium) (Huang et al. 1998; Azad et al. 2005). The plantlets grown on WPM were found to be considerably superior in average survival rate, increase in height and mean number of leaf per shoot when compared to MS and NN media (Kaji et al. 2013). Various studies have reported somatic embryogenesis by using zygotic embryos, from maternal tissues, mature leaf tissues (Babu et al. 2005; Nair and Dutta Gupta 2003, 2006). Such protocols are useful in transgenic experiments to produce disease resistant varieties. Studies have been reported in Ginger rhizomes that micropropagation results in production of pathogen free planting material infected rhizomes spread diseases.

Micropropagation protocols are also available for coriander, fennel, peppermint, spearmint, celery, thyme, lavender, ocimum, oregano, basil, fennel, garlic (Bhojwani 1980; Ahuja et al. 1982; Cellarova 1992; Patnaik and Chand 1996; Sajina et al. 1997; Iyer and Pai 1998). *Vanilla planifolia* is a source of natural vanillin. It is native to Central America and Mexico. But now it is cultivated in other parts of the tropics. Micropropagation of *Vanilla* has been studied using apical meristem. Apical meristem derived plants found to be disease-free and genetically stable (Minoo 2002; Minoo et al. 2006; Minoo and Nirmal Babu 2009). Micropropagation was also reported in cinamon, Chinese cassia and camphor through seedlings and mature tree explants (Mini et al. 1997; Babu et al. 1997; Huang et al. 1998).

The micropropagation in clove using seedling as explants have been studied. MS medium added with an extra amount of IBA or activated charcoal leads to root formation. No information has been reported towards successful micropropagation of clove from mature shoot explants (Suparman and Blake 1990; Mathew and Hariharan 1990). Studies on micropropagation of curry leaf, pomegranate, and tamarind are also accessible (Rao et al. 1982; Bhuyan et al. 1997; Mathew et al. 1999; Babu et al. 2000; Mehta et al. 2000). Plant regeneration protocols are standardized and the plants have been micropropagated successfully from the callus cultures borrowed from leaf as well as stem explants. Callus cultures have successfully induced the plant regeneration of peppermint, coriander, celery, cumin, fennel, lavender, poppy, oregano, etc. (Chand and Roy 1981; Jha et al. 1982; Hunault and Maatar 1995; Kim et al. 1996; Sajina et al. 1997).

11.3.2 Meristem and Shoot Tip Cultures

Apical meristem is a roof of tissue found at the shoot tip. Apical meristem and young leaf primordia make up the shoot apex. Meristem culture leads to the formation of disease-free plants. For meristem culture, tip of the shoots are isolated from the stem and surface sterilized with 0.1% mercuric chloride. A suitable culture medium (most widely used is MS Medium and White's medium) is prepared and sterilized shoot tips are cultured under aseptic conditions. Larger shoots tips have better survival rate and should be taken from actively region. Growth regulators (Cytokinin, Kinetin, Auxin) concentration should be carefully maintained which will support the growth and development. High level of cytokinin and low level of auxin will form shoot and low cytokinin and high auxin concentration leads to root formation. So, medium used should be different in composition. Optimum temperature for culture should be in the range of 25–26 °C. For good micropropagation, light intensity should be lower and more appropriate (Umadevi et al. 2015). Thus, the technique can solve the problems related to spice crop production.

11.3.3 Protoplast Culture

Protoplast technology is appropriate for genetic conversion by introduction of transgene DNA and somatic hybridization by protoplast fusion of species or subspecies resistant to traditional cross-breeding or isolation of sub-cellular organelles, etc. The protoplast is convenient for an array of manipulations that are impossible with intact cells due to lack of rigid cell wall. Therefore, protoplast is a valuable appliance for parasexual modification of genetic content of cells. For protoplast culture, MS and B5 media can be used with minor modifications. Leaves are used for protoplast culture, because through leaves it is possible to isolate large number of uniform cells. Protoplast can be isolated by using mechanical or enzymatic method (cell wall degrading enzymes). Formation of cell wall around the protoplast is the initial step in protoplast culture, followed by colony formation, due repeated cell divisions. Nutritional and physiological conditions need to be maintained. The cell colonies may further grown into complete plants. Protoplasts can be cultured in semisolid agar or liquid medium. Sometimes, protoplasts are firstly permitted to build up cell wall in liquid medium, and then relocated to agar medium. In Agar culture, Agarose is used to solidify the culture media (0.8% Agarose). Plating of protoplasts can be performed by using Bergmann's cell plating technique. The protoplasts linger at a fixed position in agar culture, multiply and form cell clones. The plus point with agar culture is that clumping of protoplasts can be evaded (Geetha and Shetty 2000; Divakaran et al. 2008). This technique can solve the issues related to spice crop production. Isolation and culture of protoplasts has been reported in *P. nigrum* and *P. colubrinum* (Shaji et al. 1998). Protoplasts could be lucratively isolated from in vitro grown leaf mesophyll tissues up to microcauli stage of cardamom, ginger, and turmeric (Babu 1997; Geetha and Shetty 2000). Protoplast isolation and clonal multiplication of many spice crops can be achieved through in vitro raised plants and leaves collected from field but the best source for explants found to be the taken from the in vitro grown plants. Many workers reported clonal multiplication in ginger (Rout et al. 2001). Isolation and fusion of protoplasts has been reported in many spice crops including from leaves of nutmeg, in fennel, fenugreek, peppermint, garlic, and saffron (Miura and Tabata 1986; Isa et al. 1990; Ayabe et al. 1995; Iyer et al. 2000).

11.3.4 Somaclonal Variations

Somaclonal variations are the genetic variation observed in plant progeny which are regenerated in vitro through culture of somatic cells. It can be caused by chromosomal rearrangements or epigenetic changes in the course of tissue culture. It helps to select herbicide resistant traits for the improvement of crops (Krishna et al. 2016). This technique helps in solving the problems related to spice crop production.

In order to cope with biotic and a biotic stresses, induction and utilization of somaclonal variations have been tried in many spices. Large amount of genotype dependent somaclonal variations was reported in callus cultures of Black pepper and screening procedures have been developed for *Phytophthora* resistant lines. Somaclonal variants providing resistance against *Septoria apiicola, Fusarium* yellow-resistant celery line and resistant to *Alternaria* blight in cumin in callus culture have been accounted by different workers (Evenor et al. 1994; Lacy et al. 1996; Shukla et al. 1997). Somaclonal variations also been attempted in cardamom and this lead to the identification of a few somaclones, which are Katte virus tolerant (Babu et al. 1997). In ginger, few somaclones have been evaluated to develop high yielding lines that are resistant to rhizome rot. Characterization of these somaclones using RAPD as molecular marker manifested profile variations depicting genetic variability (Babu 1997). Morphological characters combined with markers like RAPD and AFLP have been utilized to evaluate the variability and hybrid nature of genotypes and of successful interspecific hybridization and production of hybrids between *Vanilla planifolia* and *V. aphylla* (Minoo et al. 2006). To estimate the degree of genetic diversity and interrelationships between different clones of *V. planifolia* and related species, RAPD marker was used and data revealed limited variation within accessions of *V. planifolia.* This indicates that it has meager genetic base and reflected its affiliation with *V. tahitensis* (Minoo et al. 2008, 2010; Besse et al. 2004; Schluter et al. 2007; Chaveerach et al. 2002; Minoo and Nirmal Babu 2009).

11.3.5 In Vitro Selection for Resistance to Environmental Stresses

Environmental stresses like drought, heat, salinity, and mineral toxicities affect the productivity of crops. For example, high salinity causes osmotic/ionic stress (Hasegawa et al. 2000). Application of biotechnological tools to abiotic constraints faced by these spice crops will entail acquaintance of good biological information concerning the target species along with the mechanisms underlying resistance/tolerance to these stresses (Dita et al. 2006). The protection of crops against salinity-induced damage has become a global challenge. The development of salt-tolerant crops that can withstand increased levels of salts in the soils would be a practical solution of such a problem (Yamaguchi and Blumwald 2006). This is being done using a number of systems in vitro (callus, suspension cultures, somatic embryos, shoot cultures, etc.) which are screened for variation in their ability to bear reasonably high levels of salts in media. In vitro culture of plant cells, tissues or organs on a medium containing selective agents offer the opportunity to select and regenerate plants with desirable characteristics. The technique has also been effectively utilized to induce tolerance, which includes the use of some selective agents that permit the preferential survival and growth of desired phenotypes. The explants are exposed to a broad range of these selective agents like NaCl (for salt-tolerance), PEG or mannitol (for drought-tolerance), specific fungal culture filtrate (FCF) or phytotoxin, such as fusaric acid or the pathogen itself (for disease-resistance), and only the explants capable of sustaining such environments survive in the long run and are selected (Hossain et al. 2007).

11.3.6 In Vitro Production of Secondary Metabolites from Spices

These organic compounds are referred to as secondary metabolites. Phenolic acids and flavonoids function in reduction reactions as reducing agents, scavengers of free radical and helps in quenching singlet oxygen formation. Plant metabolites in spices can be altered by means of tissue culture technology. An alternative to conventional production of plant secondary metabolites, use of biotechnology techniques in plants having medicinal importance can be useful (Babu et al. 2016). Plant tissue culture techniques give consistently well in quality and performance method for isolation of secondary metabolites. Production of secondary metabolites can be achieved using callus cultures and differentiated tissues. Undifferentiated calls are also preferred in culture system in most cases. Parent plant used to initiate culture contains higher amount of secondary product. Selection is based on analyzing cell/organ growth and desired product is quantified by chromatographic and spectroscopic techniques. Production of flavor components and secondary metabolites using bioreactors and culture flasks is ideal for spice crops. Spices manufacture an extensive diversity of organic compounds, e.g. phenolic acids and flavonoids, which do not contribute directly in development and growth but have great significance.

11.4 Conclusion

There is an urgent need to explore the underutilized spices of the Indian Himalayan Region and promote the cultivation and marketing awareness among the masses. Current book chapter may aid in tracing the distribution of spice crops in Indian Himalayas and their phytochemistry, biological activities, marketing strategies, and constraints in production. Furthermore, the role of biotechnological research in promoting and enhancing the cultivation of spices has been discussed. Crop improvement programs aspire to boost the yield and eminence of a target crop to fulfill the demands of growing population. Major production constraints in spices are deficiency of high yielding, pathogen resistant varieties, and an inadequate genetic variability in some crops. With the use of biotechnological techniques, such as commercial propagation, development of new varieties and novel breeding lines via somaclonal variation, anther culture, protoplast fusion, bioreactor, and recombinant DNA technologies for improving, and preserving the diversity and enhancing the utilization of spices can be accomplished. Biotechnological approaches like protoplast fusion, micropropagation, somaclonal variations, production of flavor and coloring components, and development of new transgenics have immense prospective in conservation, utilization, and amplifying the yield of spices. Competent micropropagation systems are accessible for many spices which are being used for propagation, conservation, safe movement and transfer of germplasm, and crop improvement through somaclonal variations. Production of spice varieties that are resistant to environmental stresses and pathogens through recombinant DNA technology has still to go far in order to be efficiently used in crop improvement. Despite the fact that programs have been instigated in many laboratories for in vitro secondary metabolite production, these approaches are to be polished and scaled up for doable industrial production of the products. Thus we conclude that there is an urgent need to explore the underutilized spices of the Indian Himalayan Region.

Acknowledgment Reetika Mahajan is grateful to DST KIRAN WOSB, New Delhi, India for the financial support (project file no. DST/WOS-B/2018/832).

References

Aasim M, Khawar KM, Sancak C, Ozcan S (2009) In vitro shoot regeneration of fenugreek *Trigonellafoenum-graecum* L. Am Eurasian J Sustain Agric 3:135–138

Adaniya S, Shirai D (2001) In vitro induction of tetraploid ginger (*Zingiber officinale*) and its pollen fertilityand germinability. J Hortic 88:277–287

Agarwal S (1988) Shoot tip culture of pepper and its micropropagation. Curr Sci 57:1347–1348

Ahmad N, Abbasi BH, Fazal H (2013) Effect of different in vitro culture extracts of black pepper (*Piper nigrum* L.) on toxic metabolites-producing strains. Toxicol Ind Health 32(3):500–506

Ahuja A, Verma M, Grewal S (1982) Clonal propagation of Ocimum species through tissue culture. Ind J Exp Biol 20:455–458

Anupam KR, Sachan, Kumar S, Kumari K, Singh D (2018) Medicinal uses of spices used in our traditional culture: worldwide. J Med Plants Stud 6:116–122

Arekemase MO, Adetitum DO, Oyeyiola GP (2013) In-vitro sensitivity of selected enteric bacteria to extracts of *Allium sativum* L. Not Sci Biol 5:183–188

Ayabe M, Taniguchi K, Sumi SI (1995) Regeneration of whole plants from protoplasts isolated from tissue cultured shoot primordia of garlic (*Allium sativum* L.). Plant Cell Rep 15:17–21

Azad MAK, Yokota S, Ishiguri F, Yahara S, Yoshizawa N (2005) Large-scale clonal propagation of Cinnamomum camphora(L.) Nees and Eberm. Bull Utsunomiya Univ For 41:101–109

Babu KN (1997) In vitro studies in Zingiber officinale rose. Ph.D Thesis. Calicut University, Kerala, India

Babu KN, Samsudeen K, Ratnambal MJ (1992) In vitroplant regeneration from leaf derived callus in ginger, *Zingiber officinale* rose. Plant Cell Tissue Organ Cult 29:71–74

Babu KN, Ravindran PN, Peter KV (1997) Protocols for micropropagation of spices and aromatic crops. Indian Institute of Spices Research, Calicut

Babu KN, Anu A, Remasree AB, Praveen K (2000) Micropropagation of curry leaf tree Murraya Koenig(L.) Spreng. Plant Cell Tissue Organ Cult 61:199–203

Babu KN, Sajina A, Minoo D, John CZ, Mini PM, Rema J, Ravindran PN (2003) Micropropagation of camphor tree Cinnamomum camphora (Presl.). Plant Cell Tissue Organ Cult 74:179–183

Babu K, George JK, Anandaraj M, Venugopal MN, Nair RR (2005) Improvement of selected spices through Biotechnology tools—black pepper, cardamom, ginger, vanilla. Final Report, Department of Biotechnology, Government of India 111
Babu KN, Nair RR, Saji KV, Parthasarathy VA (2012) Biotechnology. In: Singh HP, Parthasarathy VA, Srinivasan V, Saji KV (eds) Piperaceae. Westville Publishing House, New Delhi, pp 57–81
Babu KN, Divakaran M, Raj RP, Anupama K, Peter KV, Sarma YR (2016) Biotechnological approaches in improvement of spices: a review. Plant biology and biotechnology. In: Plant genomics and biotechnology, vol II. Springer
Baratta MT, Dorman HSJ, Deans SG, Biondi DM, Ruberto G (1998) Chemical composition, antimicrobial and antioxidative activity of laurel, sage, rosemary, oregano and coriander essential oils. J Essent Oil Res 10:618–627
Batta YA (2004) Effect of treatment with Trichoderma harzianum Rifai formulated in invert emulsion on postharvest decay of apple blue mold. Int J Food Microbiol 96:281–288
Bertaccini A, Botti S, Tabanelli D, Dradi G, Fogher C, Previati A (2004) Micropropagation and establishment of mite-borne virus-free garlic (Allium sativum). Acta Cult 631:201–206
Besse P, Bory S, Grisoni M, Duval MF, da Silva D, le Bellec F (2004) RAPD genetic diversity in cultivated vanilla: Vanilla planifolia, and relationships with V. tahitensis and V. pompona. Plant Sci 167:379–385
Bhardwaj RK, Sikka BK, Singh A, Sharma ML, Singh NK (2011) Challenges and constraints of marketing and export of Indian spices in India. In: International conference on technology and business management. pp 28–30
Bhat SR, Kackar A, Chandel KRS (1992) Plant regeneration from callus cultures of Piper longum L. by organogenesis. Plant Cell Rep 11:525–528
Bhat SR, Chandel KSP, Kacker A (1994) In vitro induction of rhizome in ginger. Ind J Exp Biol 32:340–344
Bhat SR, Chandel KRS, Malik SK (1995) Plant regeneration from various explants of cultivated piper species. Plant Cell Rep 14:398–402
Bhojwani SS (1980) In vitro propagation of garlic by shoot proliferation. Sci Hort 13:47–52
Bhuyan AK, Pattnaik S, Chand PK (1997) Micropropagation of curry leaf tree (Murrayakoenigii (L.) Spreng.) by axillary proliferation using intact seedlings. Plant Cell Rep 16:779–782
Bordia A, Verma SK, Srivastava KC (1997) Effect of ginger (Zingiber officinale roscoe) and fenugreek (*Trigonella foenumgraecum* L.) on blood lipids, blood sugar and platelet aggregation in patients with coronary artery disease. Prostaglandins Leukot Essent Fatty Acids 56:379–384
Cellarova E (1992) Micropropagation of Mentha. In: Bajaj YPS (ed) Biotechnology in agriculture and forestry, high tech and micropropagation III, pp 262–275
Chadha KL, Retrinam P (2010) Plantation and spices crops. Advances in horticulture. Malhotra Publishing House, New Delhi, pp 293–345
Chan LK, Thong WH (2004) In vitro propagation of Zingiberaceae species with medicinal properties. Plant Biotech 6:181–188
Chand S, Roy SC (1981) Induction of organogenesisin callus cultures of Nigella sativa L. Ann. Bot 48:1–4
Chang BKW, Criley A (1993) Clonal propagation of pink ginger in vitro. Hort Sci 28:1203
Chaveerach R, Kunitake H, Nuchadomrong S, Sattayasai N, Komatsu H (2002) RAPD patterns as a useful tool to differentiate Thai Piper from morphologically alike Japanese Piper. Sci Asia 28:221–225
Chet I (1987) Trichoderma—application, mode of action and potential as biocontrol agent of soil borne plant pathogenic fungi. In: Chet I (ed) Innovative approaches to plant disease control. Wiley, New York, pp 49–73
Christensen LP, Brandt K (2006) Bioactive polyacetylenes in food plants of the Apiaceae family: occurrence, bioactivity and analysis. J Pharm Biomed Anal 41:683–693
Cook RJ (1985) Biological control of plant pathogens: theory to applications. Phytopathology 75:25–29

Das A, Kesari V, Satyanarayana VM, Parida A, Rangan L (2011) Genetic relationship of Curcuma species from Northeast India using PCR based markers. Mol Biotechnol 49:65–76
Dita MA, Rispail, Prats E, Rubiales D, Singh KB (2006) Biotechnology approaches to overcome biotic and abiotic stress constraints in legumes. Euphytica 147:1–24
Divakaran M, Pillai GS, Babu N, Peter KV (2008) Isolation and fusion of protoplasts in Vanilla species. Curr Sci 94:1
Doyle JJ, Doyle JL (1990) Isolation of plant DNA from fresh tissue. Focus 12:13–15
Dubey S (2017) Indian spices and their medicinal value. Indian J Pharm Educ Res 51:s330–s332
Duke JA, Bogenschutz-Godwin MJ, de Cellier J, Duke PK (2003) Elettaria cardamomum Maton (Zingiberaceae) Cardamon, Malabar or Mysore cardamon. In: CRC handbook of medicinal spices. CRC Press, Washington, pp 120–138
Edison S, Johny AK, Nirmal Babu K, Ramadasan A (1991) Spices varieties. National Research Centre for Spices, Calicut
Evenor D, Pressman E, Ben Yephet Y, Rappaport L (1994) Somaclonal variation in celery and selection by coculturing toward resistance to Septoriaapiicola. Plant Cell Tissue Organ 39:203–210
Fridlender M, Inbar J, Chet I (1999) Biological control of soil borne plant pathogens by a beta 1, 3 glucanase producing Pseudomonas cepacia. Soil Biol Biochem 25:1211–1221
Geetha S, Shetty SA (2000) In vitro propagation of Vanilla planifolia, a tropical orchid. Curr Sci 79:886–889
Gruenwald J, Freder J, Armbruester N (2010) Cinnamon and health. Crit Rev Food Sci Nutr 50:822–834
Haggag WM, Abo Sedera SA (2005) Characteristics of three Trichoderma species in peanut haulms compost involved in biocontrol of cumin wilt disease. Int J Agri Biol 7:222–229
Hasegawa PM, Bressan R, Zhu JK, Bohnert HJ (2000) Plant cellular and molecular responses to high salinity. Annu Rev Plant Biol 51:463–499
Hasler CM, Blumberg JB (1999) Symposium on phytochemicals: biochemistry and physiology. J Nutr 129:756–757
Hossain Z, Mandal AKA, Datta SK, Biswas AK (2007) Development of NaCl tolerant line in Chrysanthemum morifolium Ramat, through shoot organogenesis of selected callus line. J Biotechnol 129:658–667
Huang LC, Huang BL, Murashige T (1998) A micropropagation protocol for *Cinnamomum camphora*. In Vitro Cell Dev Biol Plant 34:141–146
Hunault G, Maatar A (1995) Enhancement of somatic embryogenesis frequency by gibberellic acid in fennel. Plant Cell Tissue Organ Cult 41:171–176
IISR (2013) Vision 5050. www.spices.res.in
Isa T, Ogasawara T, Kaneko H (1990) Regeneration of saffron protoplasts immobilised in Ca–Alginatebeads. Jpn J Breed 40:153–159
Iyer PV, Pai JS (1998) Micropropagation of sweet marjoram. J Spices Aromatic Crops 7:47–49
Iyer IR, Jayaraman C, Gopinath PM (2000) Isolation of protoplasts from leaves of nutmeg *Myristicafragrans* Houtt. Biomedicine 20:200–202
Jha TB, Roy C, Mitra GC (1982) In vitro culture of Cuminum cyminum: regeneration of flowering shoots from calli of hypocotyl and leaf explants. Plant Cell Tissue Organ Cult 2:11–14
Joy N, Abraham Z, Soniya EV (2007) A preliminary assessment of genetic relationships among agronomically important cultivars of black pepper. BMC Genet 8(1):42
Joy N, Prasanth VP, Soniya EV (2011) Microsatellite based analysis of genetic diversity of popular blackpepper genotypes in South India. Genetica 139(8):1033–1043
Kaji BV, Ershadi A, Tohidfar M (2013) In vitro propagation of pomegranate (*Punica granatum* L.). Albanian J Agric Sci 12(1):43–48
Kakani RK, Anwer MM, Meena SS, Saxena SN (2009) Advance production technology of fenugreek. NRCSSTech Release:1–24

Karuppiah P, Rajaram S (2012) Antibacterial effect of Allium sativum cloves and *Zingiber officinale* rhizomes against multiple-drug resistant clinical pathogens. Asian Pac J Trop Biomed 2(8):597–601

Kikuzaki H, Kobayashi H, Nakatani N (1991) Constituents of Zingiberaceae, Diarylheptanoids from rhizomes of *Zingiber officinale*. Phytochemistry 30:3947–3952

Kim SW, Park MK, Liu JR (1996) High frequency plant regeneration via somatic embryogenesis in cell suspension cultures of coriander. Plant Cell Rep 15:751–754

Kleijnen J, Knipschild P, Terriet G (1989) Garlic onions and cardiovascular risk factors. A review of the evidence from human experiments with emphasis on commercially available preparations. Br J Clin Pharmacol 28:535–544

Krishna H, Alizadeh M, Singh D, Singh U, Chauhan N, Eftekhari M, Sadh RK (2016) Somaclonal variations and their applications in horticultural crops improvement. 3 Biotech 6:54

Kumar K, Abdulkhadar, Rangaswamy P, Irrwlappom I (2010) Introduction to spices, plantation crops, medicinal and aromatic plants. Oxford and IBH Publishing Co., New Delhi

Kumar S, Saxena SN, Mistry JG, Fougat RS, Solanki RK, Sharma R (2015) Understanding Cuminum cyminum: an important seed spice crop of arid and semi-arid regions. Int J Seed Spices 5(2):1–19

Kumar S, Dobos GJ, Rampp T (2017) The significance of Ayurvedic medicinal plants. J Evid Based Complement Altern Med 22(3):494–501

Lacy ML, Grumet R, Toh DF, Krebs SL, Cortright BD, Hudgins E (1996) MSU-SHK5: a somaclonally derived *Fusarium* yellows resistant celery line. Hort Sci 31(2):289–290

Liao PC, Gong X, Shih HC, Chiang YC (2009) Isolation and characterization of eleven polymorphic microsatellite loci from an endemic species, Piper polysyphonum (Piperaceae). Conserv Genet 10:1911–1914

Mathew MK, Hariharan M (1990) In vitro multiple shoot formation in Syzygium aromaticum. Ann Bot 65:277–279

Mathew KM, Rao YS, Kumar KP, Sallykutty J, Lakshmanan R, Madhusoodanan KJ (1999) Micropropagation of curry leaf (*Murrayak oenigii* L.). J Spices Arom Crop 8(1):77–79

Mehta UJ, Krishnamurthy KV, Hazra S (2000) Regeneration of plants via adventitious bud formation from mature zygotic embryo axis of tamarind (Tamarindus indica L.). Curr Sci 78:1231–1234

Menezes IC, Cidade FW, Souza AP, Sampaio IC (2009) Isolation and characterization of microsatellite loci in the black pepper, *Piper nigrum* L. (piperaceae). Conserv Genet Resour 81:209–212

Mini PM, John CZ, Samsudeen K, Rema J, Nirmal Babu K, Ravindran PN (1997) Micropropagation of Cinnamomum verum (Bercht & Presl). In: Edison S, Ramana KV, Sasikumar B, Nirmal Babu K, Santhosh JE (eds) Biotechnology of spices, medicinal and aromatic crops. Indian Society for Spices, pp 35–38

Minoo D (2002) Seedling and somaclonal variation and their characterization in Vanilla. PhD thesis, Calicut University, Kerala

Minoo D, Nirmal Babu K (2009) Micropropagation and in vitro conservation of vanilla (Vanilla planifolia Andrews). In: Jain SM, Saxena PK (eds) Springer protocols, methods in molecular biology 547. Protocols for in vitro cultures and secondary metabolite analysis of aromatic and medicinal plants. The Humana Press/Springer, New York, pp 129–138

Minoo D, Babu KN, Ravindran PN, Peter KV (2006) Inter specific hybridization in vanilla and molecular characterization of hybrids and selfed progenies using RAPD and AFLP markers. Sci Hortic 108:414–422

Minoo D, Kumar JVN, Veena SS, Vimala J, Basha A, Saji KV, Babu KN, Peter KV (2008) Genetic variation and interrelationships in *Vanilla planifolia* and few related species as expressed by RAPD polymorphism. Genet Resour Crop Evol 3:459–470

Minoo D, Nirmal Babu K, Grisoni M (2010) Biotechnological applications. In: Eric O, Michel G (eds) Vanilla. CRC Press, Boca Raton, pp 51–73

Mishra RP, Sharma K (2014) Antimicrobial activity of Syzygiumaromaticum L. (clove). Int Res J Biol Sci 3:22–25

Miura Y, Tabata M (1986) Direct somatic embryogenesis from protoplasts of Foeniculum vulgare. Plant Cell Rep 5:310–313

Mohammadpour H, Moghimipour E, Rasooli I, Fakoor MH, Astaneh SA, Moosaie SS, Jalili Z (2012) Chemical composition and antifungal activity of Cuminum cyminum L. Essential oil from Alborz Mountain against Aspergillus species. Jundishapur J Nat Pharm Prod 7:50–55

Nair RR, Dutta Gupta S (2003) Somatic embryogenesis in black pepper (Piper nigrum L.): 1. Direct somatic embryogenesis from the tissues of germinating seeds and ontogeny of somatic embryos. J Hortic Sci Biotechnol 78:416–421

Nair RR, Dutta Gupta S (2006) High frequency plant regeneration through cyclic secondary somatic embryogenesis in black pepper (Piper nigrum L.). Plant Cell Rep 24:699–707

Papavizas GC (1985) Trichoderma and Gliocladium: biology, ecology and potential for biocontrol. Annu Rev Phytopathol 23:23–54

Patnaik S, Chand PK (1996) In vitro propagation of medicinal herbs *Ocimum americanum* L. syn. O. canum Sims (hoary basil) and *Ocimum santum* L.(holy basil). Plant Cell Rep 15:846–851

Phillips S, Hutchinson S, Ruggier R (1993) *Zingiber officinale* does not affect gastric emptying rate. A randomised, placebo controlled, crossover trial. Anaesthesia 48:393–395

Purseglove JW, Brown EG, Green CL, Robbins SRJ (1981) Spices, vol II. Longman, London and New York, pp 736–788

Rahman K (2001) Historical perspective on garlic and cardiovascular disease. J Nutr 131:977–979

Rahman MM, Amin MN, Jahan HS, Ahmed R (2004) In vitro regeneration of plantlets of Curcumalonga Linn: a valuable spice plant in Bangladesh. Asian J Plant Sci 3:306–309

Rao KS, Pant R (2001) Land use dynamics and landscape change pattern in a typical micro watershed in the mid elevation zone of central Himalaya, India. Agric Ecosyst Environ 86:113–123

Rao SR, Ravishankar GA (2000) Vanilla flavour: production by conventional and biotechnological routes. J Sci Food Agric 80:289–304

Rao NKS, Narayanaswamy S, Chacko EK, Swamy RD (1982) Regeneration of plants from callus of Elettaria cardamom Maton. Proc Indian Acad Sci (Plant Sci) 91:37–41

Rathore MS, Shekhawat NS (2008) Incredible spices of India: from traditions to cuisine. Am Eurasian J Bot 1:85–89

Rathore SS, Saxena SN, Singh B (2013) Potential health benefits of major seed spices. Int J Seed spices 3:1–12

Ravindran PN, Rema J, Nirmalbabu K, Peter KV (1996) Tissue culture and in vitro conservation of spices—an overview. In: Proceedings of the national seminar on biotechnology of spices and aromatic plants 24–25 April 1996

Ravindran PN, Babu KN, Saji KV, Geetha SP, Praveen K, Yamuna G (2004) Conservation of spices genetic resources in in vitro gene banks. ICAR project report. Indian Institute of Spices Research, Calicut, p 81

Rout GR, Palai SK, Samantaray S, Das R (2001) Effect of growth regulator and culture conditions on shoot multiplication and rhizome formation in ginger (*Zingiber officinale* rose.) in vitro. In Vitro Cell Dev Biol Plant 37:814–819

Roychowdhury R, Taoutaou A, Abdelgawwad MR, Hakeem KR (2013) Molecular marker assisted technologies for crop improvement. Crop improvement in the era of climate change. International Publication House Pvt. Ltd

Sajina A, Geetha SP, Minoo D, Rema J, Nirmal Babu K, Sadanandan AK, Ravindran PN (1997) Micropropagation of some important herbal spices. In: Edison S, Ramana AV, Sasikumar B (eds) Biotechnology of spices, medocinal and aromatic plants. Indian Society of Spices, Calicut, pp 79–86

Saju KA (2004) Factors affecting the biological control of *Phytophthora capsici* infection in black pepper (*Piper nigrum* L.). University of Calicut, Kerala, p 281

Salvi ND, George L, Eapen S (2002) Micropropagation and field evaluation of micropropagated plants of turmeric. Plant Cell Tissue Organ Cult 68:143–151

Sarma YR (2006) Recent trends in the use of antagonistic organisms for disease management in spice crops. In: Ramanujam B, Rabindra RJ (eds) Current status of biological control of plant diseases using antagonistic organisms in India, proc of the group meeting on antagonistic organisms in plant disease management held at Project Directorate of Biological control, Bangalore, pp 49–73

Sarma YR (2010) Integrated pest and disease management in black pepper (*Piper nigrum* L.). International Pepper Community (IPC) Jakarta, Indonesia and Spices Board, Ministry of Commerce & Industry, Govt of India, Cochin

Sarma YR, Anandaraj M (1998) Biological suppression of diseases of plantation crops and spices. In: Singh SP, Hussaini SS (eds) Biological suppression of diseases of plants, phytoparasitic nematodes and weeds. Project Directorate of Biological Control, Bangalore, pp 24–47

Sarma YR, Saju K (2004) Biological control for the management of foot rot and slow decline diseases of black pepper. Focus on pepper (*Piper nigrum* L.). International Pepper Community, Jakarta

Sarma YR, Manohara D, Premkumar T, Santhosh J (2013) Diseases and insect pests of black pepper (*Pipernigrum*, L). International Pepper Community (IPC), Jakarta

Sarma YR, Devasahayam S, Kumar A, Aravind R (2014) Status of biological control of pest and disease management in spice crops. In: Koul O, Dhaliwal GS, Khokhar S (eds) Biopesticides in sustainable agriculture, progress and potential. Ram Singh Scientific Publishers, Jodhpur, pp 287–333

Schluter P, Soto Arenas M, Harris S (2007) Genetic variation in Vanilla planifolia (Orchidaceous). Econ Bot 61:328–336

Shaji P, Anandaraj M, Sharma YR (1998) Comparative study of protoplast isolation and development in Pipernigrum (black pepper) and P. colubrinum. In: Mathew NM, Jacob CK (eds) Developments in plantation crops research. Allied Publishers, New Delhi, pp 51–53

Sharma ND, Jain AC (1979) Studies on the biological control of *Fusarium oxysporum f.sp. zingiberi*, the causal organism of yellows disease of ginger. Indian Phytopath 31(2):260–261

Shukla MR, Subhash N, Patel DR, Patel SA (1997) In vitro selection for resistance to Alternaria blight in cumin (Cuminum cyminum). In: Edison S, Ramana KV, Sasikumar B, Nirmal Babu K, Eapen SJ (eds) Biotechnology of spices, medicinal and aromatic plants. Indian Society for Spices, Calicut, pp 126–128

Sim SL, Jafar R, Power JB, Davey MR (1998) Development of an Agrobacterium-mediated transformation system for black pepper (*Pipernigrum* L.). Acta Hortic 461:349–354

Singh RD, Choudhary SL, Patel KG (1972) Seed transmission and control of Fusarium wilt of cumin. Phytopathol Mediterr 11:19–24

Sowbhagya HB (2014) Chemistry, technology and nutraceutical functions of celery (Apiumgraveolens L.): an overview. Crit Rev Food Sci Nutr 54:389–398

Stephen R, Jayabalan N (2000) Artificial seed production in coriander (*Coriandrum sativum* L.). Plant Tissue 10:45–49

Suparman U, Blake J (1990) Studies on tissue culture of clove tree plant. Indonesian J Crop Sci 5:67–75

Suresh S, Chung JW, Sung JS, Cho GT, Park JH, Yoon MS, Kim CK, Baek HJ (2013) Analysis of genetic diversity and population structure of 135 dill (Anethumgraveolens L.) accessions using RAPD markers. Genet Resour Crop Evol 60:893–903

Suseela Bhai R (1998) Studies on capsule rot (Azhukal) disease of cardamom. University of Calicut, Calicut

Suseela Bhai R, Thomas J, Naidu R (1993) Biological control of Azhukal disease of cardamom caused by *Phytophthora meadii* Mc Rae. J Plant Crops 21:134–139

Umadevi P, Saji KV, Suraby EJ (2015) Meristem culture for rapid regeneration in black pepper (Piper nigrum Linn.). Ann Plant Sci 4:3

Usman NM, Balakrishnan P, Sarma YR (1996) Biocontrol of rhizome rot of ginger. J Plant Crops 24(Suppl):184–191

Valsala PA, Nair SG, Nazeem PA (1997) In vitro seed set and seed development in ginger. *Zingiber officinale* Rose. In: Edison S (ed) Biotechnology of spices, medicinal and aromatic plants, pp 106–108

Vasanthi HR, Rarameswari RP (2010) Indian spices for healthy heart—an overview. Curr Cardiol Rev 6:274–279

Vyas RK, Mathur K (2002) Introduction of Trichoderma spp. in cumin rhizosphere and their potential in suppressing of wilt. Indian Phytopath 55:451–457

Yamaguchi T, Blumwald E (2006) Developing salt-tolerant crop plants: challenges and opportunities. Trends Plant Sci 10:615–620

The manufacturer's authorised representative in the EU is Springer Nature Customer Service Centre GmbH, Europaplatz 3, 69115 Heidelberg, Germany. If you have any concerns regarding our products, please contact ProductSafety@springernature.com

Printed and bound by CPI Group (UK) Ltd, Croydon, CR0 4YY
30/09/2026
02225241-0002